Conformally Invariant Processes in the Plane

Mathematical
Surveys
and
Monographs

Volume 114

Conformally Invariant Processes in the Plane

Gregory F. Lawler

American Mathematical Society

EDITORIAL COMMITTEE

Jerry L. Bona Peter S. Landweber, Chair
Michael G. Eastwood Michael P. Loss
J. T. Stafford

2000 *Mathematics Subject Classification.* Primary 30C35, 31A15, 60H30, 60J65, 81T40, 82B27.

For additional information and updates on this book, visit
www.ams.org/bookpages/surv-114

Library of Congress Cataloging-in-Publication Data

Lawler, Gregory F., (Gregory Francis), 1955–
  Conformally invariant processes in the plane / Gregory F. Lawler.
    p. cm. — (Mathematical surveys and monographs ; v. 114)
  Includes bibliographical references and index.
  ISBN 0-8218-3677-3 (alk. paper)
  1. Conformal mapping. 2. Potential theory (Mathematics) 3. Stochastic analysis. 4. Markov processes. I. Title. II. Mathematical surveys and monographs ; no. 114.

QA646.L85 2005
515′.9—dc22   2004062341

AMS softcover ISBN 978-0-8218-4624-7

Copying and reprinting. Individual readers of this publication, and nonprofit libraries acting for them, are permitted to make fair use of the material, such as to copy a chapter for use in teaching or research. Permission is granted to quote brief passages from this publication in reviews, provided the customary acknowledgment of the source is given.

Republication, systematic copying, or multiple reproduction of any material in this publication is permitted only under license from the American Mathematical Society. Requests for such permission should be addressed to the Acquisitions Department, American Mathematical Society, 201 Charles Street, Providence, Rhode Island 02904-2294, USA. Requests can also be made by e-mail to reprint-permission@ams.org.

© 2005 by the American Mathematical Society. All rights reserved.
Reprinted by the American Mathematical Society, 2008.
The American Mathematical Society retains all rights
except those granted to the United States Government.
Printed in the United States of America.

∞ The paper used in this book is acid-free and falls within the guidelines
established to ensure permanence and durability.
Visit the AMS home page at http://www.ams.org/

10 9 8 7 6 5 4 3 2 1   13 12 11 10 09 08

Contents

Preface	ix
Some discrete processes	1
0.1. Simple random walk	1
0.2. Loop-erased random walk	3
0.3. Self-avoiding walk	5
0.4. Infinitely growing self-avoiding walk	7
0.5. Percolation exploration process	7
Chapter 1. Stochastic calculus	11
1.1. Definition	11
1.2. Integration with respect to Brownian motion	12
1.3. Itô's formula	17
1.4. Several Brownian motions	18
1.5. Integration with respect to semimartingales	19
1.6. Itô's formula for semimartingales	20
1.7. Time changes of martingales	22
1.8. Examples	22
1.9. Girsanov's transformation	23
1.10. Bessel processes	25
1.11. Diffusions on an interval	30
1.12. A Feynman-Kac formula	39
1.13. Modulus of continuity	39
Chapter 2. Complex Brownian motion	43
2.1. Review of complex analysis	43
2.2. Conformal invariance of Brownian motion	45
2.3. Harmonic functions	46
2.4. Green's function	52
Chapter 3. Conformal mappings	57
3.1. Simply connected domains	57
3.2. Univalent functions	60
3.3. Capacity	66
3.4. Half-plane capacity	69
3.5. Transformations on $\mathbb{D}$	76
3.6. Carathéodory convergence	78
3.7. Extremal distance	80
3.8. Beurling estimate and applications	84
3.9. Conformal annuli	88

Chapter 4. Loewner differential equation — 91
- 4.1. Chordal Loewner equation — 91
- 4.2. Radial Loewner equation — 97
- 4.3. Whole-plane Loewner equation — 100
- 4.4. Chains generated by curves — 104
- 4.5. Distance to the curve — 108
- 4.6. Perturbation by conformal maps — 109
- 4.7. Convergence of Loewner chains — 114

Chapter 5. Brownian measures on paths — 119
- 5.1. Measures on spaces of curves — 119
- 5.2. Brownian measures on $\mathcal{K}$ — 123
- 5.3. $\mathbb{H}$-excursions — 130
- 5.4. One-dimensional excursion measure — 135
- 5.5. Boundary bubbles — 137
- 5.6. Loop measure — 141
- 5.7. Brownian loop soup — 144

Chapter 6. Schramm-Loewner evolution — 147
- 6.1. Chordal SLE — 147
- 6.2. Phases — 150
- 6.3. The locality property for $\kappa = 6$ — 152
- 6.4. The restriction property for $\kappa = 8/3$ — 153
- 6.5. Radial SLE — 156
- 6.6. Whole-plane SLE_κ — 162
- 6.7. Cardy's formula — 163
- 6.8. SLE_6 in an equilateral triangle — 167
- 6.9. Derivative estimates — 169
- 6.10. Crossing exponent for SLE_6 — 171
- 6.11. Derivative estimates, radial case — 174

Chapter 7. More results about SLE — 177
- 7.1. Introduction — 177
- 7.2. The existence of the path — 181
- 7.3. Hölder continuity — 182
- 7.4. Dimension of the path — 183

Chapter 8. Brownian intersection exponent — 187
- 8.1. Dimension of exceptional sets — 187
- 8.2. Subadditivity — 190
- 8.3. Half-plane or rectangle exponent — 191
- 8.4. Whole-plane or annulus exponent — 200

Chapter 9. Restriction measures — 205
- 9.1. Unbounded hulls in $\mathbb{H}$ — 205
- 9.2. Right-restriction measures — 209
- 9.3. The boundary of restriction hulls — 211
- 9.4. Constructing restriction measures — 213

Appendix A. Hausdorff dimension — 217

A.1.	Definition	217
A.2.	Dimension of Brownian paths	220
A.3.	Dimension of random "Cantor sets" in $[0,1]$	221

Appendix B. Hypergeometric functions   229
   B.1.   The case $\alpha = 2/3, \beta = 1/3, \gamma = 4/3$ 230
   B.2.   Confluent hypergeometric functions   230
   B.3.   Another equation   231

Appendix C.   Reflecting Brownian motion   233

Appendix.   Bibliography   237
   Index   240
   Index of symbols   242

Preface

A number of two-dimensional lattice models in statistical physics have continuum limits that are conformally invariant. For example, the limit of simple random walk is Brownian motion. This book will discuss the nature of conformally invariant limits. Most of the processes discussed in this book are derived in one way or another from Brownian motion. The exciting new development in this area is the Schramm-Loewner evolution (SLE), which can be considered as a Brownian motion on the space of conformal maps.

These notes arise from graduate courses at Cornell University given in 2002-2003 on the mathematics behind conformally invariant processes. This may be considered equal doses of probability and conformal mapping. It is assumed that the reader knows the equivalent of first-year graduate courses in real analysis, complex analysis, and probability.

Here is an outline of the book. We start with a quick introduction to some discrete processes which have scaling limits that are conformally invariant. We only present enough here to whet the appetite of the reader, and we will not use this section later in the book. We will not prove any of the important results concerning convergence of discrete processes. A good survey of some of these results is [83].

Chapter 1 gives the necessary facts about one-dimensional Brownian motion and stochastic calculus. We have given an essentially self-contained treatment; in order to do so, we only integrate with respect to continuous semimartingales derived from Brownian motion and we only integrate adapted processes that are continuous (or piecewise continuous). The latter assumption is more restrictive than one generally wants for other applications of stochastic calculus, but it suffices for our needs and avoids having to discuss certain technical aspects of stochastic calculus. More detailed treatments can be found in many books, e.g., [**5, 32, 72, 73**]. Sections 1.10 and 1.11 discuss some particular stochastic differential equations that arise in the analysis of SLE. The reader may wish to skip these sections until Chapter 6 where these equations appear; however, since they discuss properties of one-dimensional equations it logically makes sense to include them in the first chapter.

The next chapter introduces the basics of two-dimensional (i.e., one complex dimension) Brownian motion. It starts with the basic fact (dating back to Lévy [**65**] and implicit in earlier work on harmonic functions) that complex Brownian motion is conformally invariant. Here we collect a number of standard facts about harmonic functions and Green's function for complex Brownian motion. Because much of this material is standard, a number of facts are labeled as exercises.

Conformal mapping is the topic of Chapter 3. The purpose is to present the material about conformal mapping that is needed for SLE, especially material that would not appear in a first course in complex variables. References for much of this chapter are [**2, 23, 30, 71**]. However, our treatment here differs in some ways, most significantly in that it freely uses Brownian motion. We start with simple connectedness and a proof of the Riemann mapping theorem. Although this is really a first-year topic, it is so important to our discussion that it is included here. The next section on univalent functions follows closely the treatment in [**30**]; since the Riemann mapping theorem gives a correspondence between simply connected domains and univalent functions on the disk it is natural to study the latter. We then discuss two kinds of capacity, logarithmic capacity in the plane which is classical and a "half-plane" capacity that is not as well known but similar in spirit. Important uniform estimates about certain conformal transformations are collected here; these are the basis for Loewner differential equations. Extremal distance (extremal length) is an important conformally invariant quantity and is discussed in Section 3.7. The next section discusses the Beurling estimate, which is a corollary of a stronger result, the Beurling projection theorem. This is used to derive a number of estimates about conformal maps of simply connected domains near the boundary; what makes this work is the fact that the boundary of a simply connected domain is connected. The final section discusses annuli, which are important when considering radial or whole-plane processes.

Chapter 4 discusses the Loewner differential equation. We discuss three types, chordal, radial, and whole-plane, although the last two are essentially the same. It is the radial or whole-plane version that Loewner [**66**] developed in trying to study the Bieberbach conjecture and has become a standard technique in conformal mapping theory. The chordal version is less well known; Schramm [**76**] naturally came upon this equation when trying to find a continuous model for loop-erased walks and percolation. The final three sections deal with technical issues concerning the equation. When does the solution of the Loewner equation come from a path? What happens when solutions of the Loewner equation are mapped by a conformal transformation? What does it mean for a sequence of solutions of the Loewner equation to converge? The second of these questions is relevant for understanding the relationship between the chordal and radial Loewner equations.

In Chapter 5 we return to Brownian motion. Some of the most important conformally invariant measures on paths are derived from complex Brownian motion. After discussing a number of well-known measures (with perhaps a slightly different view than usual), we discuss some important measures that have arisen recently: excursion measure, Brownian boundary bubble measure, and the loop measure.

The Schramm-Loewner evolution (SLE_κ), which is the Loewner differential equation driven by Brownian motion, is the topic of Chapter 6. With the Brownian input, the Loewner equation becomes an equation of Bessel type, and much of the analysis of SLE comes from studying such stochastic differential equations. For example, the different "phases" of SLE (simple/non-simple/space-filling) are deduced from properties of the Bessel equation. We discuss two important values of the parameter κ: $\kappa = 6$ which satisfies the locality property and $\kappa = 8/3$ which satisfies the restriction property. One of the main reasons SLE has been so useful is that crossing probabilities (Cardy's formula), crossing exponents, and other derivative exponents can be calculated exactly. These correspond to critical exponents for lattice models. In the case $\kappa = 6$, which corresponds to (among other things) the limit of critical percolation, there is a particularly nice relationship between SLE and Brownian motion that is most easily seen in an equilateral triangle.

More topics about SLE are discussed in Chapter 7, most particularly, the technical problems of showing that SLE generates a random curve and determining the dimension of the path. Many of the results in this chapter first appeared in [**74**]. The techniques are similar to those used in the previous chapter. The main difference is that one considers the Bessel process in the upper half-plane $\mathbb{H}$ rather than just on the real line.

The next chapter gives an application of SLE to Brownian motion. The "intersection exponents" for Brownian motion are examples of critical exponents that give the fractal dimension of certain exceptional sets on the path. They are also nontrivial exponents for a lattice model, simple random walk. The close relationship of SLE_6 and Brownian motion can be used to derive the values of the Brownian exponents from the SLE_6 crossing exponents.

The Schramm-Loewner evolution gives a one-parameter family of conformally invariant measures. There is another important one-parameter family of measures called restriction measures. Roughly speaking, the restriction measure with parameter a corresponds to the union of a Brownian motions (we actually allow a to be any positive real). In Chapter 9 we show the relationship between SLE and restriction measures.

Needless to say, this book would not exist if it were not for Oded Schramm and Wendelin Werner with whom I have had the great opportunity to collaborate. Their ideas permeate this entire book. There are a number of other people who have helped by answering questions or commenting on earlier versions. These include: Christian Beneš, Nathanaël Berestycki, Zhen-Qing Chen, Keith Crank, Rick Durrett, Clifford Earle, Christophe Garban, Lee Gibson, Pavel Gyrya, John Hubbard, Harry Kesten, Evgueni Klebanov, Ming Kou, Michael Kozdron, Robin Pemantle, Melanie Pivarski, José Ramirez, Luke Rogers, Jason Schweinsberg, John Thacker, José Trujillo Ferreras, Brigitta Vermesi. Figures 0.4 and 0.5 were produced by Vincent Beffara and Geoffrey Grimmett, respectively.

During the preparation of this book I have enjoyed extended visits at the Mittag-Leffler Institute, l'Institut Henri Poincaré, the Issac Newton Institute for the Mathematical Sciences, and the Pacific Institute for the Mathematical Sciences at the University of British Columbia, and I have received support from the National Science Foundation.

Finally, and most importantly, I thank Marcia for all her understanding, patience, and support.

Some discrete processes

This book focuses on continuous time conformally invariant processes taking values in $\mathbb{C}$. One of the main motivations for studying such processes is that they arise as limits of simple discrete processes taking values in a planar lattice. While we will not discuss convergence of discrete processes to continuous processes in this book, it will be helpful to the reader to know the discrete processes that motivate some of the processes in this book.

0.1. Simple random walk

Let $\mathbb{Z}^2 = \mathbb{Z} + i\mathbb{Z}$ denote the integer lattice embedded in $\mathbb{C}$. *Simple random walk starting at* $z \in \mathbb{Z}^2$ is the process S_n given by $S_0 = z$ and

$$S_n = z + X_1 + \cdots + X_n,$$

where $X_1, \ldots, X_n$ are independent random variables each uniformly distributed on $\{1, -1, i, -i\}$. We define S_t for positive, noninteger t by linear interpolation. For every $N < \infty$, we can define

$$B_t^{(N)} = N^{-1/2}\, S_{2Nt}.$$

Note that for positive integers k,

$$\mathbf{E}[\mathrm{Re}(B_{k/N}^{(N)})^2] = \mathbf{E}[\mathrm{Im}(B_{k/N}^{(N)})^2] = \frac{k}{N}.$$

As $N \to \infty$, the processes $B_t^{(N)}$ converge to a complex Brownian motion $B_t = B_t^1 + iB_t^2$, where B_t^1, B_t^2 are independent one-dimensional standard Brownian motions.

Let $\mathbb{Z}_+^2 = \mathbb{Z} + i\mathbb{Z}_+ = \{j + ik \in \mathbb{Z}^2 : k > 0\}$ be the discrete upper half plane. If k is an integer, let σ_k denote the stopping time

$$\sigma_k = \min\{n \geq 1 : \mathrm{Im}[S_n] = k\}.$$

A well-known result for one-dimensional walks, sometimes called the *gambler's ruin* estimate, states that if $0 < k < m$,

$$\mathbf{P}\{\sigma_m < \sigma_0 \mid \mathrm{Im}[S_0] = k\} = \frac{k}{m},$$

which implies for $0 < k, k' < m$,

$$\frac{\mathbf{P}\{\sigma_m < \sigma_0 \mid \mathrm{Im}[S_0] = k\}}{\mathbf{P}\{\sigma_m < \sigma_0 \mid \mathrm{Im}[S_0] = k'\}} = \frac{k}{k'}.$$

Let $\tilde{S}_n$ denote the Markov chain that corresponds to simple random walk stopped at time σ_m and conditioned so that $\sigma_m < \sigma_0$. From the last equation we can see the transition probabilities for this chain are given (for $0 < k < m$) by

$$p(j + ik, j - 1 + ik) = p(j + ik, j + 1 + ik) = \frac{1}{4},$$

$$p(j+ik, j+i(k+1)) = \frac{k+1}{4k}, \quad p(j+ik, j+i(k-1)) = \frac{k-1}{4k}.$$

Note that the transition probabilities do not depend on m. Hence, we can let $m \to \infty$ and get a process defined on $\mathbb{Z}_+^2$ called *random walk half-plane excursion*. We can also extend the transition probability to $\mathbb{Z} = \{j+ik : k = 0\}$ by $p(j, j+i) = 1$. If we scale these processes in the same way that we scaled the simple random walk, we get the $\mathbb{H}$-excursion discussed in §5.3.

FIGURE 0.1. A simple random walk and a random walk half-plane excursion.

We call a subset A of $\mathbb{Z}^2$ *simply connected* if A and $\mathbb{Z}^2 \setminus A$ are nonempty connected subsets of $\mathbb{Z}^2$. Let

$$\partial A = \{z \in \mathbb{Z}^2 \setminus A : \text{dist}(z, A) = 1\}.$$

We call a finite path $\omega = [\omega_0, \omega_1, \ldots, \omega_n]$ an *excursion in A (of length n)* if it is a nearest neighbor path (i.e., $|\omega_j - \omega_{j-1}| = 1$ for all j) in $\mathbb{Z}^2$ with $\omega_0, \omega_n \in \partial A$, $\omega_1, \ldots, \omega_{n-1} \in A$. We can consider an excursion of length n as a curve $\omega : [0, n] \to \mathbb{C}$ by linear interpolation. Let $\mathcal{E}_A$ denote the set of excursions in A. The *random walk excursion measure* (see [**43**]) is the measure on $\mathcal{E}_A$ that assigns measure 4^{-n} to each excursion in $\mathcal{E}_A$ of length n. In other words, the excursion measure of $\omega = [\omega_0, \ldots, \omega_n]$ is the probability that the first n steps of a simple random walk starting at ω_0 are the same as ω. Suppose D is a bounded simply connected domain in $\mathbb{C}$ containing the origin. For each $N < \infty$, let D_N denote the connected component containing the origin of the set of $z = x + iy \in \mathbb{Z}^2$ such that $\{x' + iy' : |x - x'| \leq 1/2, |y - y'| \leq 1/2\}$ is contained in ND. For each N, we get a measure on paths by considering the random walk excursion measure on D_N and scaling the excursions by Brownian scaling, $\omega^{(N)}(t) = N^{-1/2}\omega(2tN)$. As $N \to \infty$, these measures approach an infinite measure on paths called *excursion measure* on D, which is discussed in §5.2.

We will see that Brownian motion in $\mathbb{C}$ is invariant under conformal transformation up to a change of time. The excursion measure is also conformally invariant. If z, w are distinct points in ∂D, then conditioning the excursion measure to have endpoints z, w gives a probability distribution on excursions from z to w in D. This is the same (up to a time change) as the conformal image of $\mathbb{H}$-excursions under a conformal transformation of $\mathbb{C}$ onto D sending 0 to z and ∞ to w.

A *random walk loop (of length $2n$)* is a nearest neighbor path $\omega = [\omega_0, \ldots, \omega_{2n}]$ with $\omega_0 = \omega_{2n}$. The *rooted random walk loop measure* (see [**60**]) is the measure that gives each random walk loop of length $2n$ measure $(2n)^{-1} 4^{-2n}$. We can think of this measure as a measure on *unrooted loops* that gives measure 4^{-2n} to every (unrooted) loop of length $2n$ and then chooses a root uniformly among $\{\omega_1, \ldots, \omega_{2n}\}$. By linear interpolation, we can consider loops of length $2n$ as curves $\omega : [0, 2n] \to \mathbb{C}$. For each N we can use Brownian scaling as before, $\omega^{(N)}(t) = N^{-1/2}\omega(2tN)$, to get another measure on loops. As $N \to \infty$, this measure converges to a multiple of the Brownian loop measure discussed in §5.6.

0.2. Loop-erased random walk

Let $\tilde{S}_n$ be a random walk half-plane excursion with $\tilde{S}_0 = 0$. It is not difficult to show that $\tilde{S}_n$ is transient; in fact, $\text{Im}[\tilde{S}_n] \to \infty$. There is a well-defined self-avoiding subpath of $[\tilde{S}_0, \tilde{S}_1, \tilde{S}_2, \ldots]$, which we denote by $[\hat{S}_0, \hat{S}_1, \hat{S}_2, \ldots]$, that is obtained by erasing loops chronologically. To be more precise, let $t_0 = 0$, $\hat{S}_0 = \hat{S}_{t_0} = 0$, and for $n > 0$, let

$$t_n = \max\{m > t_{n-1} : \tilde{S}_m = \tilde{S}_{t_{n-1}+1}\}, \quad \hat{S}_n = \tilde{S}_{t_n} = \tilde{S}_{t_{n-1}+1}.$$

This process is called the *(half-plane) loop-erased random walk*.

The loop-erased random walk can also be defined by giving the (non-Markovian) transition probabilities. If $V \subset \mathbb{Z}_+^2$ is finite, let $q_V(z)$ denote the probability that a random walk half-plane excursion starting at z never visits V; $q_V(z)$ is defined to be 0 if $z \in V$. Then $q_V : \mathbb{Z}_+^2 \cup \mathbb{Z} \to [0,1]$ is the unique solution of the discrete boundary value problem

$$\Delta_e q_V(z) := \sum_{|z-w|=1} p(z,w)\left[q_V(w) - q_V(z)\right] = 0, \quad z \in \mathbb{Z}_+^2 \setminus V,$$

$$q(z) = 0, z \in V \cup \mathbb{Z}; \quad \lim_{k \to \infty} q_V(j + ik) = 1.$$

Here $p(z, w)$ denotes the transition probability for the random walk half-plane excursion. Note that $q_V(z) > 0$ if and only if z is in the unbounded connected component of $\mathbb{Z}_+^2 \setminus V$. It is straightforward to show that

(0.1)
$$\mathbf{P}\{\hat{S}_{n+1} = z \mid \hat{S}_0, \ldots, \hat{S}_n\} = \frac{p(\hat{S}_n, z)\, q_{V_n}(z)}{\sum_{|w-\hat{S}_n|=1} p(\hat{S}_n, w)\, q_{V_n}(w)}, \quad \text{if } |z - \tilde{S}_n| = 1,$$

where $V_n = \{\hat{S}_0, \ldots, \hat{S}_n\}$. For this reason the name *Laplacian random walk* is also given to the process. Given $[\hat{S}_0, \ldots, \hat{S}_n]$ one can obtain the remainder of the loop-erased walk by taking a random walk half-plane excursion starting at $\hat{S}_n$; conditioning on the event that the excursion does not visit V_n after time 0; and erasing loops from this path. Since the excursion measure is conformally invariant and the loop-erasing procedure also seems conformally invariant (since the path obtained from chronological loop-erasure depends only on the order in which one visits points and not the exact parametrization), we might expect that a scaling limit $\gamma(t)$ would have the following properties:

- $\gamma : [0, \infty) \to \mathbb{C}$ is a random simple[1] path with $\gamma(0) = 0$, $\gamma(0, \infty) \subset \mathbb{H} = \{x + iy \in \mathbb{C} : y > 0\}$, $\gamma(t) \to \infty$ as $t \to \infty$.

---

[1] One cannot claim that a limiting process should have simple paths (i.e., paths with no double points) just because the discrete approximations do not intersect. However, for the loop-erased walk and the self-avoiding walk discussed in the next section, there are other reasons to conjecture a limit that is non-self-intersecting. The percolation exploration process discussed in the last section of this chapter is an example of a limit of non-self-intersecting paths whose limit process *does* have self-intersections.

- If $t > 0$ and we have observed $\gamma(s), 0 \leq s \leq t$, then the conditional distribution of $\gamma(s), t \leq s < \infty$, is the same (up to time change) as the original distribution mapped conformally onto $\mathbb{H} \setminus \gamma[0, t]$. In other words, if g is a conformal transformation of $\mathbb{H} \setminus \gamma[0, t]$ onto $\mathbb{H}$ with $g(\gamma(t)) = 0, g(\infty) = \infty$, then the distribution of

$$\eta^{(t)}(s) := g(\gamma(t+s)), \quad 0 \leq s < \infty,$$

is the same as the distribution of $\gamma(s), 0 \leq s \leq \infty$ (modulo time change).

Searching for processes with the above property led Oded Schramm to define the processes that are now called chordal Schramm-Loewner evolutions (chordal SLE_κ). As we will see, this is a one-parameter family of processes parametrized by $\kappa > 0$. For loop-erased random walk in $\mathbb{Z}_+^2$, it has been proved [**57**] that the scaling limit is chordal SLE_2. The determination $\kappa = 2$ comes from particular analysis of the loop-erased walk, i.e., it does not follow only from the assumptions above.

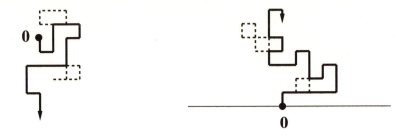

FIGURE 0.2. Loop-erased walks obtained from walks in Figure 0.1.

One can also define the loop-erased random walk in all of $\mathbb{Z}^2$. There is a technical difficult in that the random walk is recurrent and hence one cannot just take an infinite walk and erase the loops. However, any of the reasonable limiting operations one might take wil give the same answer, e.g.:

- Consider simple random walks of n steps and erase the loops. This gives a measure on self-avoiding walks (of a random length $k \leq n$). Take the limit as $n \to \infty$ of this measure.
- Do the same thing, except taking simple random walks until they first visit a point of absolute value at least N.
- Use the transition probabilities as in (0.1) replacing $p(z, w)$ with $\tilde{p} \equiv 1/4$ and q_{V_n} with $\tilde{q}_{V_n}$, which is the unique function from $\mathbb{Z}^2$ into $[0, \infty)$ satisfying

$$\Delta \tilde{q}_{V_n}(z) = 0, \quad z \notin V_n;$$

$$\tilde{q}_{V_n}(z) = 0, \ z \in V_n; \qquad \tilde{q}_{V_n}(z) \sim \log|z|, \ z \to \infty.$$

Here Δ denotes the usual discrete Laplacian,

$$\Delta \tilde{q}_{V_n}(z) = \frac{1}{4} \sum_{|z-w|=1} [\tilde{q}_{V_n}(w) - \tilde{q}_{V_n}(z)].$$

Consider a candidate for the scaling limit of the loop-erased random walk in $\mathbb{Z}^2$. We expect that this is a random, simple $\gamma : [0, \infty) \to \mathbb{C}$ with the following property. Suppose $\gamma[0, t]$ is given. Let $g = g_t$ be a conformal transformation of $\mathbb{C} \setminus \gamma[0, t]$

onto $\{z : |z| > 1\}$ with $g_t(\infty) = \infty$ and $g_t(\gamma(t)) = 1$. Then the distribution of $\eta^{(t)}(s) := g_t(\gamma(s+t))$ should be the same for all t. Also, $\eta^{(t)}(s), 0 \leq s < \infty$, should be independent of $\gamma[0,t]$. The distribution of $\eta^{(t)}$ turns out to be that of radial SLE_2 from 1 to ∞ in $\{z : |z| > 1\}$. The distribution of γ is called whole-plane SLE_2.

0.3. Self-avoiding walk

A *self-avoiding walk (SAW) of length n*, $\omega = [\omega_0, \ldots, \omega_n]$, is a nearest neighbor path such that $\omega_j \neq \omega_k, 0 \leq j < k \leq n$. The term self-avoiding walk is also used for certain measures on SAWs which give each walk of a given length the same measure, e.g.:

- Fix n and consider the uniform probability measure on all SAWs of length n with $\omega_0 = 0$.
- Let $b > 0$ and consider the measure on SAWs that gives each SAW of length n with $\omega_0 = 0$ measure e^{-nb}.

Let C_n denote the number of SAWs of length n starting at the origin. Since $C_{n+m} \leq C_n C_m$, subadditivity of $\log C_n$ shows that $C_n \approx \beta^n$ for some (unknown) $\beta \in (2,3)$ called the connective constant. When $e^b = \beta$, the second measure above is said to be at criticality. (If $\beta \geq e^b$ the measure of the set of SAWs starting at the origin is infinite, and if $\beta < e^b$ the measure is finite.) See [**67**] for more background on SAWs.

For either the uniform measure on SAWs or the critical measure, one can ask what does the "beginning" of a SAW look like. To be more precise, we define the *infinite self-avoiding walk (ISAW)* using the transition probabilities (0.1) where we replace p with $\tilde{p} \equiv 1/4$ and $q_{V_n}(z)$ with

$$q^*_{V_n}(z) := \lim_{m \to \infty} C_m^{-1} \, \#\{\text{SAWs starting at } z \text{ of length } m \text{ that avoid } V_n\}.$$

This definition requires the limit to exist; unfortunately, at this time there is no proof of its existence. However, we will assume the existence and consider possible scaling limits for this process. There is a similar process called the *half-plane infinite self-avoiding walk (IHSAW)* obtained by restricting to SAWs that stay in $\mathbb{Z}^2_+$. Here we use $\tilde{p} \equiv 1/4$ and

$$q^+_{V_n}(z) = \lim_{m \to \infty} \frac{\#\{\text{SAWs starting at } z \text{ of length } m \text{ staying in } \mathbb{Z}^2_+ \text{ that avoid } V_n\}}{\#\{\text{SAWs starting at } 0 \text{ staying in } \mathbb{Z}^2_+\}}.$$

This limit has been established rigorously [**58**], so we know that the process is well defined.

Theoretical physicists have predicted that SAWs have (in some sense) a conformally invariant limit. If this is true, then a scaling limit for the IHSAW should satisfy the same assumptions that we gave for the half-plane loop-erased walk. Similarly the ISAW should satisfy the assumptions given for the whole-plane loop-erased walk. Therefore, we *conjecture* that the IHSAW is chordal SLE_κ and the ISAW is whole-plane SLE_κ for some κ. Unlike the case of the loop-erased random walk, however, there is currently no proof of this fact.

If ISAW and IHSAW do have a limit that is conformally invariant (and hence SLE_κ for some κ), we can actually determine the value of κ. This is because the self-avoiding walk satisfies the *restriction property*. Suppose A is a finite subset of $\mathbb{Z}^2_+$ such that A does not disconnect 0 from infinity in $\mathbb{Z}^2_+$. Consider the IHSAW

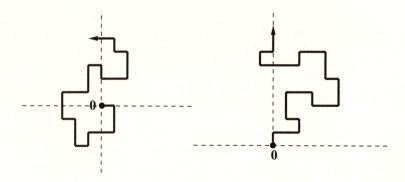

FIGURE 0.3. A self-avoiding walk and a half-space SAW.

conditioned that it stays in $\mathbb{Z}_2^+ \setminus A$. Then the transitions for the conditional process are the same as for the original process, replacing $q_{V_n}^+$ with $q_{V_n \cup A}^+$. In other words, the IHSAW (i.e., the ISAW in $\mathbb{Z}_+^2$) conditioned to avoid A is the same as the ISAW in $\mathbb{Z}_+^2 \setminus A$. A little thought will convince the reader that one would not expect the loop-erased random walk to have this property.

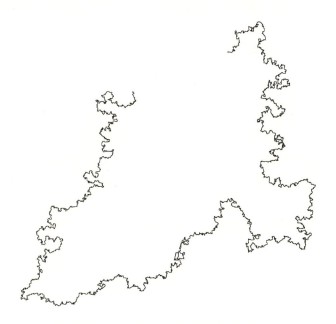

FIGURE 0.4. A long self-avoiding walk.

We are led to look for a chordal SLE_κ that is supported on simple curves and has the following restriction property. Suppose A is a bounded subset of the upper half plane $\mathbb{H}$ such that $0 \notin \overline{A}$ and $\mathbb{H} \setminus A$ is a simply connected domain. Let $\Phi = \Phi_A$ be a conformal transformation of $\mathbb{H} \setminus A$ onto $\mathbb{H}$ fixing 0 and ∞. Then the restriction property asserts that (up to a time change) the distribution of $\Phi \circ \gamma$ given $\gamma[0, \infty) \cap A = \emptyset$ is the same as the distribution of γ. In §6.4 we will show that this implies $\kappa = 8/3$. While it is only a conjecture at this point that IHSAW has a scaling limit of $SLE_{8/3}$, the conjecture is supported by numerical simulation [41].

0.4. Infinitely growing self-avoiding walk

There is another process called the *(half-plane) infinitely growing self-avoiding walk (HIGSAW)*[2] or alternatively the *half-plane Laplacian random walk with exponent 0*. This random walk starts at the origin; moves immediately to i; and from then on chooses uniformly among those nearest neighbors such that the walk will not be trapped. We can write this as in (0.1) using $\tilde{p} \equiv 1/4$ and replacing $q_{V_n}(z)$ with $r_{V_n}(z) := 1\{q_{V_n}(z) > 0\} = \lim_{\epsilon \to 0+} q_{V_n}(z)^\epsilon$. Again, we can conjecture conformal invariance of this process and conjecture that it satisfies most of the assumptions made about the loop-erased walk. The one difference is that it is not expected that the scaling limit of this process will be non-self-intersecting. It is conjectured to be chordal SLE_6, which we will see in Chapter 6 has double points.

Although the HIGSAW in $\mathbb{Z}_+^2$ has not been shown to have a scaling limit of SLE_6, such a limit has been established for the HIGSAW of a different lattice. Let $\mathbb{L}$ denote the *hexagonal or honeycomb* lattice in $\mathbb{C}$. The vertices of the lattice are $\mathbb{L}_e \cup \mathbb{L}_o$ where $\mathbb{L}_e$ is the discrete subgroup generated by $\sqrt{3}$ and $\sqrt{3}\, e^{i\pi/3}$ and $\mathbb{L}_o = e^{i\pi/6} + \mathbb{L}_e$. We make $\mathbb{L}$ into a graph by saying that vertices in $\mathbb{L}$ at distance 1 from each other are adjacent. In other words, each $z \in \mathbb{L}_e$ is adjacent to three points in $\mathbb{L}_o$, $z+e^{i\pi/6}, z+e^{i5\pi/6}, z-i$, and each $w \in \mathbb{L}_o$ is adjacent to three points in $\mathbb{L}_e$, $w+i, w+e^{i7\pi/6}, w+e^{i11\pi/6}$. Note that the edges of $\mathbb{L}$ form regular hexagons of side length 1. The dual graph to the hexagonal lattice is the *triangular lattice* whose vertices are these regular hexagons and where two vertices (hexagons) are called adjacent if they share a boundary edge. Note that each vertex in the triangular lattice has six neighbors.

Simple random walk on $\mathbb{L}$ is the process that at each time jumps to one of the adjacent vertices with equal probability, independently of the steps up to that time. Let $\mathbb{L}_+ = \{x + iy \in \mathbb{L} : y > 0\}$. Then the HIGSAW in $\mathbb{L}_+$ starts at the origin and at each time chooses a vertex uniformly among the adjacent vertices for which there is an infinite path staying in $\mathbb{L}_+$ that avoids the path of the walk up to that point.

0.5. Percolation exploration process

The HIGSAW on $\mathbb{L}$ is also called the *percolation exploration process* because it is closely related to boundaries of site percolation on the triangular lattice. Consider the triangular lattice, i.e., the regular hexagons bounded by edges of $\mathbb{L}$, and suppose we color the hexagons black or white, each hexagon colored black with probability 1/2 independently of the other colors. The boundaries of large clusters of one color look like HIGSAWs. To be more precise, suppose we color all the hexagons whose interior lies entirely in $\mathbb{H}$ black or white, independently, with probability 1/2 for each color. However, for all the hexagons which lie partly in $\mathbb{H}$ and partly in $-\mathbb{H}$ we will color deterministically; those to the left of the origin will be colored black and those to the right will be colored white. The edge $[-i, 0]$ starts a boundary curve between black and white, and we can continue this curve in the upper half plane. A little thought will show that the distribution of the curve is exactly that

---

[2]This is not a good name since this process is much different than the infinite self-avoiding walk of the previous section, but this terminology has been used for this process.

of HIGSAW in $\mathbb{L}$. For example, to find the second step of the boundary curve we need to know the color of the hexagon whose lowest vertex is the origin. If this hexagon is colored black, we move along the edge $[0, e^{i\pi/6}]$; if it is white we move along the edge $[0, e^{i5\pi/6}]$.

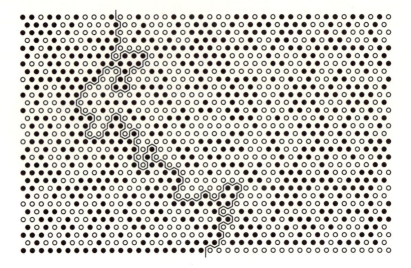

FIGURE 0.5. The percolation exploration process.

As we continue along, every time we traverse an edge the hexagon to the left will be black and the one to the right will be white. After we traverse the edge, we must look at the hexagon that we would enter if we continued. If we do not already know the color of this hexagon we flip a coin to color it black or white and then take a step in the appropriate direction. If the color of the hexagon has been determined, then the edge that separates two hexagons of the same color is a "trapping" edge, i.e., if the random walker on $\mathbb{L}$ went along that edge it would be forced to leave $\mathbb{L}_+$ or to visit a previously visited vertex. In other words, it is a disallowed transition for the HIGSAW.

The HIGSAW or percolation exploration process has a property that we call *locality*. The boundary condition for percolation in the upper half plane was black on the negative real line and white on the positive real line. Imagine that we had a different boundary condition that looked the same near the origin. When we do the exploration process on this different domain, the transition probabilities will not change until we reach a hexagon that is in the boundary of one of the domains but not the other.

It has long been believed (see, e.g., [20, 21, 44]) that percolation at criticality (which in the case of site percolation of the triangular case is when the probability of black is the same as the probability of white) has a conformally invariant scaling limit. This leads us to believe that the percolation exploration process should have a scaling limit satisfying the assumptions for the loop-erased walk except that the paths might not be simple. We also expect that the scaling limit should satisfy the locality property, e.g., if A is a bounded subset of $\mathbb{H}$ with $0 \notin A$ and such that $\mathbb{H} \setminus A$ is a simply connected domain, then we should not be able to distinguish (modulo time change) the process going from 0 to ∞ in $\mathbb{H} \setminus A$ from the one in $\mathbb{H}$ up to the first time it reaches A. In §6.3, we will show that chordal SLE_κ satisfies the locality property only for $\kappa = 6$, so this is the only possibility for the limit.

Smirnov [**79**] proved that percolation on the triangular lattice does have a conformally invariant limit and hence we know that the percolation exploration process converges to SLE_6. While similar results are expected for other models of percolation, e.g., bond percolation on the square lattice, they have not been proved. We will not discuss this proof in this book; however, in §6.8, we will show why SLE_6, i.e., the scaling limit of percolation, likes equilateral triangles. As will be seen in §6.2, the paths of SLE_6 are not simple.

CHAPTER 1

Stochastic calculus

1.1. Definition

A *standard (one-dimensional) Brownian motion* B_t with respect to the filtration $\{\mathcal{F}_t\}$ [1] (i.e., an increasing collection of sub-σ-algebras of $\mathcal{F}$) on the probability space $(\Omega, \mathcal{F}, \mathbf{P})$ is a stochastic process satisfying:

(i) For each $0 < s < t$, the random variable $B_t - B_s$ is $\mathcal{F}_t$-measurable, independent of $\mathcal{F}_s$, and has a normal distribution with mean 0 and variance $t - s$;

(ii) W.p.1[2] $t \mapsto B_t$ is a continuous function.

If B_t is a standard Brownian motion and $b, \sigma \in \mathbb{R}$, then $\tilde{B}_t := B_0 + bt + \sigma(B_t - B_0)$ is called a one-dimensional Brownian motion with *drift b* and *variance (parameter) σ^2*. It satisfies the same conditions as B_t except that $\tilde{B}_t - \tilde{B}_s$ has a normal distribution with mean $b(t-s)$ and variance $\sigma^2(t-s)$.

A *standard d-dimensional Brownian motion* with respect to $\mathcal{F}_t$ is a process $B_t = (B_t^1, \ldots, B_t^d)$ where $B_t^1, \ldots, B_t^d$ are independent (one-dimensional) standard Brownian motions adapted to $\mathcal{F}_t$. If $d = 2$, we can write B_t in complex form, $B_t = B_t^1 + iB_t^2$, and call B_t a *(standard) complex Brownian motion*. (Whenever we write complex Brownian motion the word *standard* will be implicit.) If $z \in \mathbb{R}^d$ (or $z \in \mathbb{C}$ for $d = 2$), we will write $\mathbf{P}^z, \mathbf{E}^z$ for probabilities and expectations for Brownian motions assuming $B_0 = z$.

A *domain* will be a connected, open subset of $\mathbb{R}^d$ or $\mathbb{C}$. If D is a domain in $\mathbb{R}^d$, we let τ_D be the first time that Brownian motion is not in D,
$$\tau_D = \inf\{t > 0 : B_t \notin D\}.$$

EXERCISE 1.1. Suppose B_t is a standard d-dimensional Brownian motion starting at the origin, $A : \mathbb{R}^d \to \mathbb{R}^d$ is an orthogonal transformation, and $r > 0$. Show that $X_t := A B_t$ and $Y_t := r^{-1/2} B_{rt}$ are standard (d-dimensional) Brownian motions.

In this chapter, we will give a short introduction to integration with respect to Brownian motion and continuous semimartingales. Throughout this chapter we assume that $B_t, B_t^1, B_t^2, \ldots$ are independent Brownian motions with respect to the same filtration $\mathcal{F}_t$. If H_t is another process we say that H_t is:

- *adapted* if each H_t is $\mathcal{F}_t$-measurable;
- *bounded* if there is an $N < \infty$ such that w.p.1 $|H_t| \leq N$ for all t;
- *continuous* if w.p.1 the function $t \mapsto H_t$ is continuous;
- a *martingale* (with respect to $\mathcal{F}_t$) if for each t, $\mathbf{E}[|H_t|] < \infty$ and $\mathbf{E}[H_t \mid \mathcal{F}_s] = H_s$ for all $s \leq t$;

---

[1] For technical reasons, we assume that $\{\mathcal{F}_t\}$ is right-continuous ($\mathcal{F}_t = \cap_{s>t} \mathcal{F}_s$) and each $\mathcal{F}_t$ contains all the null sets.

[2] W.p.1 stands for "with probability one".

- a *square integrable martingale* if H_t is a martingale with $\mathbf{E}[H_t^2] < \infty$ for each t;
- a *local martingale* if there exists a sequence of stopping times $\tau_1 < \tau_2 < \cdots$ with respect to $\mathcal{F}_t$ such that w.p.1 $\tau_j \to \infty$ and such that for each j, $H_{t \wedge \tau_j}$ is a martingale.

Let $\mathcal{I}$ denote the set of adapted, continuous processes. Let $\mathcal{M}, \mathcal{M}^2, \mathcal{LM}$ denote the collection of continuous martingales, square integrable martingales, and local martingales, respectively. We write $b\mathcal{I}, b\mathcal{M}$ for the collection of bounded processes in $\mathcal{I}, \mathcal{M}$, respectively.

Throughout this chapter, we will use partitions of intervals $[0, t]$. If t is fixed, we will use Π_n to denote a sequence of partitions, i.e., times

$$0 = t_0^n < t_1^n < t_2^n < \cdots < t_{k_n}^n = t.$$

In order to simplify the notation, we will write just t_j for t_j^n. We write $\|\Pi_n\|$ for the mesh of the partition, i.e., the maximal value of $t_j - t_{j-1}$. We write $\overline{\Pi}_n$ for the sequence of dyadic partitions, $t_j = j2^{-n}$ with an appropriate correction for t_{k_n} if t is not a dyadic rational. Note that $\|\overline{\Pi}_n\| = 2^{-n}$.

1.2. Integration with respect to Brownian motion

We call H a *simple process* if it is of the form

$$H_s = \sum_{j=1}^n C_j \, 1_{[t_{j-1}, t_j)}(s),$$

where $t_0 < t_1 < \cdots < t_n$ and C_j is a bounded $\mathcal{F}_{t_{j-1}}$-measurable random variable. If $n = 1$, we define

$$Z_t = \int_0^t H_s \, dB_s = \begin{cases} 0, & t \leq t_0, \\ C_1 [B_t - B_{t_0}], & t_0 \leq t \leq t_1, \\ C_1 [B_{t_1} - B_{t_0}], & t \geq t_1. \end{cases}$$

For $n > 1$, we define

$$Z_t = \int_0^t H_s \, dB_s$$

by linearity. It is easy to check that this definition does not depend on how the simple process is written. Let $\mathcal{I}_s$ denote the collection of simple processes.

PROPOSITION 1.2.
- If $H, K \in \mathcal{I}_s$, $a, b \in \mathbb{R}$, then $aH + bK \in \mathcal{I}_s$ and

$$\int_0^t (aH_s + bK_s) \, dB_s = a \int_0^t H_s \, dB_s + b \int_0^t K_s \, dB_s.$$

- If $H_s \in \mathcal{I}_s$, and

$$Z_t = \int_0^t H_s \, dB_s$$

then $Z_t \in \mathcal{M}^2$. If we define the quadratic variation $\langle Z \rangle_t$ by

$$\langle Z \rangle_t = \int_0^t H_s^2 \, ds,$$

then $Z_t^2 - \langle Z \rangle_t \in \mathcal{M}$. In particular,

(1.1) $$\mathbf{E}[Z_t^2] = \mathbf{E}[\langle Z \rangle_t] = \int_0^t \mathbf{E}[H_s^2]\, ds.$$

PROOF. This can be proved directly from the definition and is left to the reader. □

The next proposition shows that "quadratic variation" is a good term for $\langle Z \rangle_t$.

PROPOSITION 1.3. *If $H \in \mathcal{I}_s$, $Z_t = \int_0^t H_s\, dB_s$, and Π_n is a sequence of partitions with $\|\Pi_n\| \to 0$, then*

$$\lim_{n \to \infty} \sum_{j=1}^{k_n} (Z_{t_j} - Z_{t_{j-1}})^2 = \langle Z \rangle_t$$

in L^2 (and hence in probability). If $\Pi_n = \overline{\Pi}_n$ is the sequence of dyadic partitions or any other sequence such that $\sum \|\Pi_n\| < \infty$, then the limit can be taken w.p.1.

PROOF. We will prove the result for $H_t = 1_{[0,t)}$; the extension to other $H \in \mathcal{I}_s$ is immediate from the definition of the integral. Let

$$Q_n = [\sum_{j=1}^{k_n}(B_{t_j} - B_{t_{j-1}})^2] - t = \sum_{j=1}^{k_n} \Delta(j, n)$$

where $\Delta(j, n) = (B_{t_j} - B_{t_{j-1}})^2 - (t_j - t_{j-1})$. For fixed n, the random variables $\Delta(1, n), \ldots, \Delta(k_n, n)$ are independent and each $\Delta(j, n)$ has the distribution of $(t_j - t_{j-1})(N^2 - 1)$ where N is a standard normal random variable. Therefore,

$$\mathbf{E}[Q_n^2] = \text{Var}[Q_n] = \sum_{j=1}^{k_n} (t_j - t_{j-1})^2 \, \mathbf{E}[(N^2 - 1)^2] \leq \|\Pi_n\|\, t\, \mathbf{E}[(N^2 - 1)^2],$$

and hence $Q_n \to 0$ in L^2. Also, Chebyshev's inequality gives

$$\mathbf{P}\{Q_n \geq \epsilon\} \leq \epsilon^{-2}\, \mathbf{E}[Q_n^2] \leq \epsilon^{-2}\, \|\Pi_n\|\, t\, \mathbf{E}[(N^2 - 1)^2].$$

If $\sum \|\Pi_n\| < \infty$, the Borel-Cantelli Lemma tells us that with probability one, for all large n, $Q_n \leq \epsilon$. Hence $Q_n \to 0$ w.p.1. □

We now define $\int_0^t H_s\, dB_s$ for $H \in \mathcal{I}$. Our strategy is as follows. For dyadic rational t, we will define

(1.2) $$Z_t := \int_0^t H_s\, dB_s = \lim_{n \to \infty} \sum_{j=1}^{k_n} H_{t_{j-1}} [B_{t_j} - B_{t_{j-1}}],$$

after showing that the limit exists in probability for fixed t. We then show that Z_t, restricted to dyadic rational t, is uniformly continuous on compact intervals w.p.1. We then define Z_t for other t by continuity and then (1.2) holds for all t.

We first consider $H \in b\mathcal{I}$. Fix t, let $\|H\|_\infty = \|H\|_{\infty,t} = \sup_{0 \leq s \leq t} \|H_s\|_\infty$, and

$$\text{osc}(H, \delta) = \text{osc}(H, \delta, t) = \sup\{|H_s - H_r| : 0 \leq r, s \leq t, |r - s| \leq \delta\}.$$

If $H \in b\mathcal{I}$ and Π_n is a sequence of partitions with $\|\Pi_n\| \to 0$, let $H^{(n)} \in \mathcal{I}_s$ be defined by $H_s^{(n)} = H_{t_{j-1}}$, $t_{j-1} \leq s < t_j$, and $H_s^{(n)} = 0$ for $s \geq t$. Note that

$$|H_s - H_s^{(n)}| \leq \text{osc}(H, \|\Pi_n\|), \quad 0 \leq s < t.$$

Let $O_n = O_{n,t} = \text{osc}(H, \|\Pi_n\|)$. Since the H_s are continuous, $O_n \to 0$ w.p.1. Since $O_n \leq 2\|H\|_\infty$, we know from the dominated convergence theorem that $\mathbf{E}[O_n^2] \to 0$. Let

$$Z_t^{(n)} = \int_0^t H_s^{(n)} \, dB_s.$$

Since $|H_t^{(n)} - H_t^{(m)}| \leq O_n + O_m$, by (1.1),

$$\mathbf{E}[(Z_t^{(n)} - Z_t^{(m)})^2] \leq t \, \mathbf{E}[(O_n + O_m)^2].$$

In particular, $\{Z_t^{(n)}\}, n = 1, 2, \ldots$, is a Cauchy sequence in $L^2(\Omega, \mathcal{F}_t, \mathbf{P})$ and we can define

$$Z_t = \int_0^t H_s \, dB_s = \lim_{n \to \infty} Z_t^{(n)},$$

where the limit is taken in the L^2 sense. In particular (1.2) holds where the limit can be taken in L^2 or in probability. Note that

$$\mathbf{E}[Z_t^2] = \lim_{n \to \infty} \mathbf{E}[(Z_t^{(n)})^2] = \lim_{n \to \infty} \int_0^t \mathbf{E}[(H_s^{(n)})^2] \, ds = \int_0^t \mathbf{E}[H_s^2] \, ds.$$

The last equality uses the dominated convergence theorem. As before, we define the quadratic variation

$$\langle Z \rangle_t = \int_0^t H_s^2 \, ds.$$

Note that

$$Z_t^2 - \langle Z \rangle_t = \lim_{n \to \infty} (Z_t^{(n)})^2 - \langle Z^{(n)} \rangle_t,$$

where the convergence is in L^1 (this holds because $(Z_t^{(n)})^2 \to Z_t^2$ in L^1 and $\langle Z^{(n)} \rangle_t$ is uniformly bounded and converges almost surely to $\langle Z \rangle_t$). In particular, if $s < t$,

$$\mathbf{E}[Z_t \mid \mathcal{F}_s] = Z_s, \quad \mathbf{E}[Z_t^2 - \langle Z \rangle_t \mid \mathcal{F}_s] = Z_s^2 - \langle Z \rangle_s.$$

Now, for the moment, restrict our consideration to t in the dyadic rationals D. Using a diagonalization argument, we can find a subsequence Π_{n_i}, that we will denote as just Π_n, such that w.p.1 for all $t \in D$,

$$Z_t = \int_0^t H_s \, dB_s = \lim_{n \to \infty} \sum_{j=1}^{k_n} H_{t_{j-1}} [B_{t_j} - B_{t_{j-1}}].$$

LEMMA 1.4. *Suppose $H \in b\mathcal{I}$ and Z_t is defined as above. Then w.p.1, for each $t \in D$, $s \mapsto Z_s$ is a uniformly continuous function on $D \cap [0, t]$.*

PROOF. If $M_s, s \in D$ is any square integrable martingale, then (see, e.g., [31, Chapter 4, (4.2)])

$$\mathbf{P}\{\sup\{|M_s - M_0| : 0 \leq s \leq t, s \in D\} \geq \epsilon\} \leq \epsilon^{-2} \mathbf{E}[M_t^2].$$

Applying this to $Z_s - Z_s^{(n)}$, we get that

$$\mathbf{P}\{\sup\{|Z_s - Z_s^{(n)}| : 0 \leq s \leq t, s \in D\} \geq \epsilon\} \leq \epsilon^{-2} \mathbf{E}[(Z_t - Z_t^{(n)})^2] \to 0.$$

By taking a subsequence if necessary, we can assume that $\sum \mathbf{E}[(Z_t - Z_t^{(n)})^2] < \infty$ and hence, using the Borel-Cantelli Lemma, that with probability one $Z_s, 0 \leq s \leq t$, is a uniform limit of $Z_s^{(n)}, 0 \leq s \leq t$. Since the uniform limit of continuous functions is continuous, the result is proved. □

We now *define* Z_t for all t by continuity. It is an easy exercise to verify that this is the same as (1.2) (up to an event of probability zero). Moreover, if Π_n is any sequence of partitions, then there is a subsequence (which can depend on H but does not depend on ω) such that w.p.1, for all $s \leq t$,

$$(1.3) \qquad \int_0^s H_r \, dB_r = \lim_{n \to \infty} \sum_{t_j \leq s} H_{t_{j-1}} [B_{t_j} - B_{t_{j-1}}].$$

PROPOSITION 1.5. *If $H \in b\mathcal{I}$, $Z_t = \int_0^t H_s \, dB_s$, and Π_n is a sequence of partitions as above, then*

$$\lim_{n \to \infty} \sum_{j=1}^{k_n} [Z_{t_j} - Z_{t_{j-1}}]^2 = \langle Z \rangle_t = \int_0^t H_s^2 \, ds$$

in probability.

PROOF. Consider a different sequence of partitions Π_l^*, which we write $0 = s_0 < s_1 < \ldots < s_l = t$. Let $Z_s^{(l)}$ be the approximation of Z_s by simple processes using this sequence. For a fixed l,

$$\lim_{n \to \infty} \sum_{j=1}^{k_n} [Z_{t_j}^{(l)} - Z_{t_{j-1}}^{(l)}]^2 = \langle Z^l \rangle_t = \int_0^t (H_s^{(l)})^2 \, ds.$$

But

$$\|\sum_{j=1}^{k_n} [Z_{t_j} - Z_{t_{j-1}}]^2 - [Z_{t_j}^{(l)} - Z_{t_{j-1}}^{(l)}]^2 \|_2 \leq c_l,$$

for some $c_l \to 0$. We note that by taking a subsequence we can make it an almost sure limit for all $t \in D$, and hence (by monotonicity) for all t. □

We will now define the integral for $H \in \mathcal{I}$. Let

$$\phi_N(x) = \begin{cases} -N, & x \leq -N \\ x, & -N \leq x \leq N \\ N, & x \geq N. \end{cases}$$

Let

$$Z_{t,N} = \int_0^t \phi_N(H_s) \, dB_s.$$

Define the stochastic integral Z_t by

$$Z_t = \lim_{N \to \infty} Z_{t,N}.$$

This limit exists trivially (on the event where H is continuous). Also, if Π_n is any sequence of partitions,

$$(1.4) \qquad Z_t = \lim_{n \to \infty} \sum_{j=1}^{k_n} H_{t_{j-1}} [B_{t_j} - B_{t_{j-1}}],$$

where the limit is taken in probability. Moreover, we can find a subsequence (depending on H) such that the limit is an almost sure limit. As before, let
$$\langle Z \rangle_t = \int_0^t H_s^2 \, ds.$$
Then,
$$\lim_{n \to \infty} \sum_{j=1}^{k_n} [Z_{t_j} - Z_{t_{j-1}}]^2 = \langle Z \rangle_t$$
in probability. By taking a subsequence, the limit can be made almost sure for every t. For later reference, we note that it also follows that if G_s is any continuous process (not necessarily adapted), and $s_j \in [t_{j-1}, t_j]$, then

(1.5) $$\lim_{n \to \infty} \sum_{j=1}^{k_n} G_{s_j} [Z_{t_j} - Z_{t_{j-1}}]^2 = \int_0^t G_s \, d\langle Z \rangle_s = \int_0^t G_s H_s^2 \, ds,$$

where the limit is in probability (or w.p.1 along a subsequence).

PROPOSITION 1.6. *If $H \in \mathcal{I}$, then Z_t and $Z_t^2 - \langle Z \rangle_t$ are local martingales. If*
$$\mathbf{E}[\langle Z \rangle_t] = \int_0^t \mathbf{E}[H_s^2] \, ds < \infty,$$
then $Z_t \in \mathcal{M}^2, Z_t^2 - \langle Z \rangle_t \in \mathcal{M}$.

PROOF. If
$$\tau_N = \inf\{s : |H_s| \geq N\},$$
then $Z_{t \wedge \tau_N} = Z_{t \wedge \tau_N, N}$. Also $\tau_1 < \tau_2 < \cdots$ and $\tau_N \to \infty$. □

PROPOSITION 1.7. *If $H, K \in \mathcal{I}$, and $a, b \in \mathbb{R}$, then*
$$\int_0^t (aH_s + bK_s) \, dB_s = a \int_0^t H_s \, dB_s + b \int_0^t K_s \, dB_s.$$

PROOF. Immediate from the definition. □

If $H_t, K_t \in \mathcal{I}$ and $Z_t = \int_0^t H_s \, dB_s$, we define
$$\int_0^t K_s \, dZ_s = \int_0^t K_s H_s \, dB_s.$$
If Π_n is any sequence of partitions,
$$\int_0^t K_s \, dZ_s = \lim_{n \to \infty} \sum_{j=1}^{k_n} K_{t_{j-1}} [Z_{t_j} - Z_{t_{j-1}}],$$
where the limit is taken in probability. This can be seen from (1.4) using the continuity of K_t.

PROPOSITION 1.8. *If $H \in \mathcal{I}, Z_t = \int_0^t H_s \, dB_s$, then*
$$Z_t^2 = 2 \int_0^t Z_s \, dZ_s + \langle Z \rangle_t.$$

PROOF. Let Π_n be a sequence of partitions as above. Then (recalling that $Z_0 = 0$),

$$Z_t^2 = \sum_{j=1}^{k_n}[Z_{t_j}^2 - Z_{t_{j-1}}^2] = 2\sum_{j=1}^{k_n} Z_{t_{j-1}}[Z_{t_j} - Z_{t_{j-1}}] + \sum_{j=1}^{k_n}[Z_{t_j} - Z_{t_{j-1}}]^2.$$

As $n \to \infty$, the right hand side converges in probability to

$$2\int_0^t Z_s\, dZ_s + \langle Z\rangle_t.$$

□

Suppose $H_s, K_s \in \mathcal{I}$ and

$$Z_t = \int_0^t H_s\, dB_s, \quad Y_t = \int_0^t K_s\, dB_s.$$

The covariance process is defined by

$$\langle Z, Y\rangle_t = \int_0^t H_s K_s\, ds.$$

Under this definition, $\langle Z\rangle_t = \langle Z, Z\rangle_t$. Note that

$$4\langle Z, Y\rangle_t = \langle Z+Y\rangle_t - \langle Z-Y\rangle_t.$$

This implies that $Z_t Y_t - \langle Z, Y\rangle_t$ is a local martingale and if Π_n is a sequence of partitions as above,

$$\lim_{n\to\infty} \sum_{j=1}^{k_n}(Z_{t_j} - Z_{t_{j-1}})(Y_{t_j} - Y_{t_{j-1}}) = \langle Z, Y\rangle_t,$$

in probability. In fact,

$$Z_t Y_t - \langle Z, Y\rangle_t = \int_0^t Z_s\, dY_s + \int_0^t Y_s\, dZ_s.$$

1.3. Itô's formula

Suppose $h(t, x)$ is a collection of random variables indexed by $(t, x) \in [0, \infty) \times \mathbb{R}$. We say that h is an *adapted continuous function* if:
- With probability one, $h(t, x)$ is a continuous function on $[0, \infty) \times \mathbb{R}$,
- For each rational x, $h(t, x) \in \mathcal{I}$.

The assumption of continuity implies that $h(t, x)$ is determined by the values of $h(t, x)$ for rational x. Let $\mathcal{A}$ be the set of adapted continuous functions, and let $\mathcal{A}_{1,2}$ be the set of adapted continuous functions $h(t, x)$ such that with probability one, $\dot{h}, h', h''$ exist and are in $\mathcal{A}$ (here we use dots for time derivatives and $'$," for x derivatives). A deterministic function h is in $\mathcal{A}_{1,2}$ if it is C^1 in t and C^2 in x.

PROPOSITION 1.9 (Itô's Formula). *Suppose $H \in \mathcal{I}$, $Z_t = \int_0^t H_s\, dB_s$ and $h \in \mathcal{A}_{1,2}$. Then,*

$$h(t, Z_t) - h(0, Z_0) = \int_0^t h'(s, Z_s) H_s\, dB_s + \int_0^t [\dot{h}(s, Z_s) + \frac{1}{2} h''(s, Z_s) H_s^2]\, ds.$$

PROOF. We may assume that $h(t,x)$ is zero outside a compact interval in $\mathbb{R}$ (If not, we can let $h_N(t,x) = h(t,x)g_N(x)$ where g is a C^∞ function that is 1 on $[-N,N]$, 0 on $[-N-1, N+1]^c$ and $0 \leq g \leq 1$. If we have the result for each h_N we have the general result.) Fix t, and let Π_n be a sequence of partitions as above. We write the telescoping sum

$$h(t, Z_t) - h(0, Z_0) = \sum_{j=1}^{k_n} [h(t_j, Z_{t_j}) - h(t_{j-1}, Z_{t_{j-1}})].$$

By the mean value theorem, we can write

$$h(t_j, Z_{t_j}) - h(t_{j-1}, Z_{t_{j-1}}) = [h(t_j, Z_{t_j}) - h(t_{j-1}, Z_{t_j})]$$

$$+ h'(t_{j-1}, Z_{t_{j-1}})[Z_{t_j} - Z_{t_{j-1}}] + \frac{1}{2} h''(t_{j-1}, Z_{s_j})[Z_{t_j} - Z_{t_{j-1}}]^2,$$

for some $s_j \in [t_{j-1}, t_j]$.

Let

$$O_n = \sup\{|\dot{h}(s,x) - \dot{h}(r,x)| : x \in \mathbb{R}, 0 \leq r, s \leq t, |s-r| \leq \|\Pi_n\|\},$$

$$O_n^1 = \sup\{|Z_s - Z_r| : 0 \leq r, s \leq t, |s-r| \leq \|\Pi_n\|\},$$

$$O_n^2 = \sup\{|h''(r,x) - h''(s,y)| : 0 \leq r, s \leq t, |s-r| \leq \|\Pi_n\|, |x-y| \leq O_n^1\}.$$

Then $O_n + O_n^1 + O_n^2 \to 0$ w.p.1. Since

$$h(t, Z_{t_j}) - h(t_{j-1}, Z_{t_j}) = [\dot{h}(t_{j-1}, Z_{t_{j-1}}) + \epsilon_j](t_j - t_{j-1}),$$

for some $|\epsilon_j| \leq O_n$, it follows that w.p.1,

$$\lim_{n \to \infty} \sum_{j=1}^{k_n} [h(t_j, Z_{t_j}) - h(t_{j-1}, Z_{t_j})] = \int_0^t \dot{h}(s, Z_s)\, ds.$$

By (1.3) we know that

$$\lim_{n \to \infty} \sum_{j=1}^{k_n} h'(t_{j-1}, Z_{t_{j-1}})[Z_{t_j} - Z_{t_{j-1}}] = \int_0^t h'(s, Z_s)\, dZ_s,$$

in probability. Finally, since $|h''(t_{j-1}, Z_{s_j}) - h''(t_{j-1}, Z_{t_{j-1}})| \leq O_n^2$, it follows from (1.5) that w.p.1

$$\lim_{n \to \infty} \sum_{j=1}^{k_n} h''(t_{j-1}, Z_{s_j})[Z_{t_j} - Z_{t_{j-1}}]^2 = \int_0^t h''(s, Z_s)\, d\langle Z \rangle_s = \int_0^t h''(s, Z_s) H_s^2\, ds.$$

□

1.4. Several Brownian motions

Assume that a d-dimensional Brownian motion $\bar{B}_t = (B_t^1, \ldots, B_t^d)$ is defined with respect to the filtration $\mathcal{F}_t$. We have already shown how to define the stochastic integral

$$Z_t = \int_0^t H_s\, dB_s^i,$$

where $H_t \in \mathcal{I}$. We have the following covariance rule:
$$\langle \int_0^t H_s\, dB_s^i, \int_0^t K_s\, dB_s^i \rangle_t = \int_0^t H_s\, K_s\, ds.$$

PROPOSITION 1.10. *If $H, K \in \mathcal{I}$, $i \neq l$, and $Z_t = \int_0^t H_s\, dB_s^i$, $Y_t = \int_0^t K_s\, dB_s^l$ then $Z_t Y_t$ is a local martingale. Also, if Π_n is a sequence of partitions,*
$$\lim_{n\to\infty} \sum_{j=1}^{k_n} [Z_{t_j} - Z_{t_{j-1}}][Y_{t_j} - Y_{t_{j-1}}] = 0,$$

in probability.

PROOF. Do this first for simple processes, then bounded processes, then general processes. The second result uses the fact that
$$\lim_{n\to\infty} \sum_{j=1}^{k_n} [B_{t_j}^i - B_{t_{j-1}}^i][B_{t_j}^l - B_{t_{j-1}}^l] = 0,$$

in L^2. This follows easily from
$$\mathbf{E}[\,(B_{t_j}^i - B_{t_{j-1}}^i)(B_{t_j}^l - B_{t_{j-1}}^l)\,] = 0,$$
$$\mathrm{Var}[\,(B_{t_j}^i - B_{t_{j-1}}^i)(B_{t_j}^l - B_{t_{j-1}}^l)\,] = (t_j - t_{j-1})^2.$$
$\square$

More generally, if
$$Z_t^1 = \sum_{j=1}^d \int_0^t H_s^j\, dB_s^j, \quad Z_t^2 = \sum_{j=1}^d \int_0^t K_s^j\, dB_s^j.$$
are two continuous local martingales, and
$$\langle Z^1, Z^2 \rangle_t = \int_0^t [\sum_{j=1}^d H_s^j\, K_s^j]\, ds,$$
then $Z_t^1 Z_t^2 - \langle Z^1, Z^2 \rangle_t$ is a local martingale and
$$\lim_{n\to\infty} \sum_{j=1}^{k_n} [Z_{t_j}^1 - Z_{t_{j-1}}^1][Z_{t_j}^2 - Z_{t_{j-1}}^2] = \langle Z^1, Z^2 \rangle_t,$$

in probability.

1.5. Integration with respect to semimartingales

Let $\mathcal{SM}$ denote the set of all processes Z_t of the form
$$Z_t = R_t + \sum_{j=1}^d \int_0^t H_s^j\, dB_s^j,$$
where $R_s, H_s^j \in \mathcal{I}$ and such that w.p.1 the function $t \mapsto R_t$ has bounded variation on each bounded interval. These processes are called continuous *semimartingales*. We write this in shorthand by
$$dZ_t = dR_t + \sum_{j=1}^d H_t^j\, dB_t^j = dR_t + \bar{H}_t \cdot d\bar{B}_t,$$

where $\bar{H}_t = (H_t^1, \ldots, H_t^d), \bar{B}_t = (B_t^1, \ldots, B_t^d)$. The quadratic variation of Z_t is defined to be the quadratic variation of the martingale part,

$$\langle Z \rangle_t = \langle \sum_{j=1}^d \int_0^t H_s^j \, dB_s^j \rangle_t = \int_0^t [\sum_{j=1}^d (H_s^j)^2] \, ds$$

If

$$Y_t = S_t + \sum_{j=1}^d \int_0^t K_s^j \, dB_s^j,$$

ia another semimartingale in $\mathcal{SM}$, the covariance process is defined by

$$\langle Z, Y \rangle_t = \langle \sum_{j=1}^d \int_0^t H_s^j \, dB_s^j, \sum_{j=1}^d \int_0^t K_s^j \, dB_s^j \rangle_t = \int_0^t [\sum_{j=1}^d H_s^j \, K_s^j] \, ds.$$

If Π_n is a sequence of partitions as above, then

$$\lim_{n \to \infty} \sum_{j=1}^{k_n} [Z_{t_j} - Z_{t_{j-1}}][Y_{t_j} - Y_{t_{j-1}}] = \langle Z, Y \rangle_t,$$

in probability. If $J_s \in \mathcal{I}$, we *define* the integral with respect to Z:

$$\int_0^t J_s \, dZ_s = \int_0^t J_s \, dR_s + \sum_{j=1}^d \int_0^t J_s \, H_s^j \, dB_s^j.$$

The first integral is a standard Riemann-Stieljes integral. Then,

$$\langle \int_0^t J_s \, dZ_s, \int_0^t F_s \, dZ_s \rangle_t = \int_0^t [\sum_{j=1}^d J_s \, F_s \, (H_s^j)^2] \, ds.$$

Also,

$$\lim_{n \to \infty} \sum_{j=1}^{k_n} J_{t_{j-1}} [Z_{t_j} - Z_{t_{j-1}}] = \int_0^t J_s \, dZ_s,$$

in probability and the following product rule holds,

$$Z_t Y_t = \int_0^t Z_s \, dY_s + \int_0^t Y_s \, dZ_s + \langle Z, Y \rangle_t.$$

This can be written

(1.6) $$d(Z_t Y_t) = Z_t \, dY_t + Y_t \, dZ_t + d\langle Z, Y \rangle_t.$$

1.6. Itô's formula for semimartingales

Let m be a positive integer and suppose that

$$K_s^i, \quad i = 1, \ldots, m; \qquad H_s^{i,j}, \quad i = 1, \ldots, m; \; j = 1, \ldots, d,$$

are in $\mathcal{I}$ and suppose that the paths of K_s^i are of bounded variation (on each bounded interval). Write $\bar{H}_s^i = (H_s^{i,1}, \ldots, H_s^{i,d})$. Let $Z^1, \ldots, Z^m$ be semimartingales of the form

$$Z_t^i = Z_0^i + K_t^i + \int_0^t \bar{H}_s^i \cdot d\bar{B}_s = Z_0^i + K_t^i + \sum_{j=1}^d \int_0^t H_s^{i,j} \, dB_s^j,$$

and let $\vec{Z}_t = (Z_t^1, \ldots, Z_t^m)$. Let $\mathcal{A}^m = \underbrace{\mathcal{A} \times \cdots \times \mathcal{A}}_{m}$ denote the set of adapted continuous functions $h(t, x^1, \ldots, x^m)$, and let $\mathcal{A}_{1,2}^m$ denote the set of $h(t, x^1, \ldots, x^m)$ such that for all $1 \leq i, l \leq m$ the derivatives

$$\dot{h}(t, x^1, \ldots, x^m), \quad h_i(t, x^1, \ldots, x^m), \quad h_{il}(t, x^1, \ldots, x^m)$$

exist and are in $\mathcal{A}^m$. Here we write h_i for differentiation with respect to x^i and h_{il} for the corresponding double partials. We state an extension of Itô's formula which can be proved in the same way as the previous one.

PROPOSITION 1.11. *Suppose $Z^1, \ldots, Z^m$ are as above and $h \in \mathcal{A}_{1,2}^m$. Then*

$$h(t, \vec{Z}_t) - h(0, \vec{Z}_0) =$$

$$\int_0^t \dot{h}(s, \vec{Z}_s) \, ds + \sum_{i=1}^m \int_0^t h_i(s, \vec{Z}_s) \, dZ_s^i + \frac{1}{2} \sum_{i=1}^m \sum_{l=1}^m \int_0^t h_{il}(s, \vec{Z}_s) \, d\langle Z^i, Z^l \rangle_s.$$

One can generalize this proposition. For $1 \leq b \leq m+1$, let $\mathcal{A}_{1,2}^{m,b}$ be the set of adapted continuous functions $h(t, z^1, \ldots, z^m)$ such that for all $1 \leq i \leq m$ and $b \leq i, l \leq m$ the derivatives

$$\dot{h}(t, x^1, \ldots, x^m), \quad h_i(t, x^1, \ldots, x^m), \quad h_{il}(t, x^1, \ldots, x^m)$$

exist and are in $\mathcal{A}^m$. Under this definition, $\mathcal{A}_{1,2}^m = \mathcal{A}_{1,2}^{m,1}$. Then we get the following.

PROPOSITION 1.12. *Suppose $Z^1, \ldots, Z^m$ are as above and $h \in \mathcal{A}_{1,2}^{m,b}$. Suppose $\bar{H}_i \equiv 0$ for $i < b$. Then*

$$h(t, \vec{Z}_t) - h(0, \vec{Z}_0) =$$

$$\int_0^t \dot{h}(s, \vec{Z}_s) \, ds + \sum_{i=1}^m \int_0^t h_i(s, \vec{Z}_s) \, dZ_s^i + \frac{1}{2} \sum_{i=b}^m \sum_{l=b}^m \int_0^t h_{il}(s, \vec{Z}_s) \, d\langle Z^i, Z^l \rangle_s.$$

If the semimartingales are just independent Brownian motions, the form of Itô's formula is easier. If $h(t, x^1, \ldots, x^d)$ is a function on $[0, \infty) \times \mathbb{R}^d$, write Δ for the Laplacian in the space variables:

$$\Delta h(t, \bar{x}) = \sum_{j=1}^d h_{jj}(t, \bar{x}).$$

PROPOSITION 1.13. *Suppose $\bar{B}_t = (B_t^1, \ldots, B_d^t)$ is a standard d-dimensional Brownian motion and $h(t, x^1, \ldots, x^d)$ is a function that is C^1 in t and C^2 in $x^1, \ldots, x^d$. Then,*

$$h(t, \bar{B}_t) - h(0, \bar{B}_0) = \sum_{j=1}^d \int_0^t h_j(s, \bar{B}_s) \, dB_s^j + \int_0^t [\dot{h}(s, \bar{B}_s) + \frac{1}{2} \Delta h(s, \bar{B}_s)] \, ds.$$

1.7. Time changes of martingales

Suppose $H \in \mathcal{I}$ and $Z_t = \sum_{j=1}^d \int_0^t H_s^j \, dB_s^j$ is a continuous local martingale. Suppose that w.p.1,

$$\lim_{t \to \infty} \langle Z \rangle_t = \int_0^\infty \sum_{j=1}^d (H_s^j)^2 \, ds = \infty,$$

and define stopping times τ_r by

$$\tau_r = \inf\{t : \langle Z \rangle_t = r\}.$$

PROPOSITION 1.14. *Let $W_r = Z_{\tau_r}$. Then W_r is a standard Brownian motion with respect to the filtration $\mathcal{F}_{\tau_r}$.*

PROOF. Obviously $W_0 = 0$ and W_r has continuous paths almost surely. It suffices, therefore, to show that for every $r_0 < r$, that the distribution of $W_r - W_{r_0}$ conditioned on $\mathcal{F}_{\tau_{r_0}}$ is normal, mean zero, variance $r - r_0$. From the strong Markov property of Brownian motion, it suffices to prove this when $r_0 = 0$.

If $y \in \mathbb{R}$, let $Y_t = \exp\{iyZ_t + y^2 \langle Z \rangle_t / 2\}$. Itô's formula shows that this is a local martingale (here we need to apply Itô's formula to a complex function, but we just apply it to the real and imaginary parts separately). For $t \leq \tau_r$, Y_t is uniformly bounded. Hence we can use the optional sampling theorem (see, e.g., [**39**, Theorem 1.3.22]) to conclude that $1 = \mathbf{E}[Y_0] = \mathbf{E}[Y_{\tau_r}]$. This implies $\mathbf{E}[e^{iyZ_{\tau_r}}] = e^{-ry^2/2}$ for each $y \in \mathbb{R}$ and hence Z_{τ_r} has a normal distribution with mean zero and variance r. □

1.8. Examples

1.8.1. Martingales from harmonic functions.
A function $f : \mathbb{R}^d \to \mathbb{R}$ is *harmonic* in a domain D if

$$\Delta f(\bar{x}) := \sum_{j=1}^d f_{jj}(\bar{x}) = 0, \quad \bar{x} \in D.$$

Suppose f is a harmonic function on $\mathbb{R}^d$. Let $\bar{B}_t$ be a standard d-dimensional Brownian motion. Then Itô's formula shows that

$$f(\bar{B}_t) - f(\bar{B}_0) = \sum_{j=1}^d \int_0^t f_j(\bar{B}_s) \, dB_s^j.$$

In particular, $Y_t = f(\bar{B}_t)$ is a local martingale. If D is an open set in $\mathbb{R}$,

$$\tau = \tau_D = \inf\{t \geq 0 : \bar{B}_t \notin D\},$$

and $\Delta f(\bar{x}) = 0$ for $\bar{x} \in D$, then we can similarly show that $Y_{t \wedge \tau}$ is a local martingale. If D is a bounded domain and f is continuous on $\overline{D}$ (so that f is bounded on $\overline{D}$), then $Y_{t \wedge \tau}$ is a bounded martingale.

1.8.2. Exponential martingale.
Suppose that $Z_t = \sum_{j=1}^d \int_0^t H_s \, dB_s$ is a continuous local martingale. Then applying Itô's formula with $h(t, x) = e^x$, gives

$$e^{Z_t} - 1 = \int_0^t e^{Z_s} \, dZ_s + \int_0^t \frac{1}{2} e^{Z_s} \, d\langle Z \rangle_s.$$

If $Y_t = Z_t - \langle Z \rangle_t/2$, then
$$e^{Y_t} - 1 = \int_0^t e^{Y_s}\, dZ_s,$$
i.e., $M_t = e^{Y_t} = \exp\{Z_t - \langle Z \rangle_t/2\}$ satisfies the exponential differential equation $dM_t = M_t\, dZ_t$. In particular, M_t is a local martingale called the *exponential martingale* derived from Z_t.

1.8.3. Bessel process. Let $\bar{B}_t$ be a standard d-dimensional Brownian motion $(d > 1)$ with $\bar{B}_0 \neq 0$. Applying Itô's formula to
$$f(x^1, \ldots, x^d) = \sqrt{(x^1)^2 + \cdots + (x^d)^2},$$
gives
$$|\bar{B}_t| = |\bar{B}_0| + \sum_{j=1}^d \int_0^t \frac{B_s^j}{|\bar{B}_s|}\, dB_s^j + \int_0^t \frac{d-1}{2\,|\bar{B}_s|}\, ds.$$

Since w.p.1 $\bar{B}_t \neq 0$ for all t, there is no problem with the integrals. Note that
$$\tilde{B}_t := \sum_{j=1}^d \int_0^t \frac{B_s^j}{|\bar{B}_s|}\, dB_s^j$$
is a continuous local martingale with
$$\langle \tilde{B} \rangle_t = \sum_{j=1}^d \int_0^t \left[\frac{B_s^j}{|\bar{B}_s|}\right]^2 ds = t.$$

Hence by Proposition 1.14, $\tilde{B}_t$ is a standard Brownian motion, and $Y_t := |\bar{B}_t|$ satisfies the stochastic differential equation
$$dY_t = \frac{a}{Y_t}\, dt + d\tilde{B}_t,$$
where $a = (d-1)/2$. This equation is called the Bessel SDE and the solution Y_t is called a Bessel-d process.

1.9. Girsanov's transformation

Suppose $K_s \in b\mathcal{I}$ with $|K_s| \leq N$ for all s, and let
$$M_t = \exp\left\{\int_0^t K_s\, dB_s - \frac{1}{2}\int_0^t K_s^2\, ds\right\}.$$

Itô's formula shows that M_t is a local martingale satisfying the equation $dM_t = K_t M_t\, dB_t$. Since $M_t \leq e^{N B_t}$, M_t is uniformly integrable on every bounded interval and hence M_t is a positive martingale. Let $\mathbf{Q}_t$ denote the measure on $(\Omega, \mathcal{F}_t)$ whose Radon-Nikodym derivative with respect to $\mathbf{P}$ is M_t. If we let $\mathbf{E}_{\mathbf{Q}_t}$ denote expectations with respect to $\mathbf{Q}_t$, then for every $\mathcal{F}_t$ measurable Y, $\mathbf{E}_{\mathbf{Q}_t}[Y] = \mathbf{E}[Y\, M_t]$. If $s < t$ and Y is $\mathcal{F}_s$-measurable, then
$$\mathbf{E}[Y\, M_t] = \mathbf{E}[\,\mathbf{E}[Y\, M_t \mid \mathcal{F}_s]\,] = \mathbf{E}[\,Y\, \mathbf{E}[M_t \mid \mathcal{F}_s]\,] = \mathbf{E}[Y\, M_s],$$
so $\mathbf{Q}_t$ restricted to $\mathcal{F}_s$ is $\mathbf{Q}_s$. Hence there exists a measure that we denote by $\mathbf{Q}$ such that for each t, $\mathbf{Q}$ restricted to $\mathcal{F}_t$ is $\mathbf{Q}_t$.

PROPOSITION 1.15. *If*
$$X_t = B_t - \int_0^t K_r\, dr,$$
then X_t is a **Q**-martingale with respect to $\mathcal{F}_t$, i.e., if $s < t$, then $\mathbf{E_Q}[X_t \mid \mathcal{F}_s] = X_s$.

PROOF. We first recall that the conditional expectation $\mathbf{E_Q}[X_t \mid \mathcal{F}_s]$ is the unique $\mathcal{F}_s$-measurable random variable Y such that for all $A \in \mathcal{F}_s$, $\mathbf{E_Q}[Y\, 1_A] = \mathbf{E_Q}[X_t\, 1_A]$. In other words,
$$\mathbf{E}[Y\, 1_A\, M_s] = \mathbf{E}[X_t\, 1_A\, M_t].$$
Hence to prove the result, it suffices to show that $X_t M_t$ is a **P**-martingale. The product formula gives
$$d(X_t M_t) = X_t\, dM_t + M_t\, dX_t + d\langle X, M\rangle_t.$$
The "dt" terms cancel. □

More generally, let
$$Z_t = \sum_{j=1}^d \int_0^t H_s^j\, dB_s^j,$$
and let $M_t = \exp\{Z_t - \langle Z\rangle_t/2\}$ be the corresponding exponential martingale satisfying $dM_t = M_t\, dZ_t$. Assume sufficient boundedness so that M_t is a martingale (not just a local martingale).

PROPOSITION 1.16. *If*
$$X_t = Z_t - \int_0^t M_r^{-1}\, d\langle Z\rangle_r,$$
then X_t is a **Q**-martingale with respect to $\mathcal{F}_t$, i.e., if $s < t$, then $\mathbf{E_Q}[X_t \mid \mathcal{F}_s] = X_s$.

EXAMPLE 1.17. Suppose $B_0 = x > 0$, let $M_t = B_t$ and let $T = \inf\{t : B_t = 0\}$. Then for $0 \le t < T$, $dM_t = dB_t$ and hence $K_t = 1/B_t$. Let $\mathbf{Q}_t$ be the measure on paths whose Radon-Nikodym derivative with respect to **P** is $x^{-1} M_{t \wedge T} = x^{-1} B_t\, 1_{0 \notin B[0,t]}$. Note that $\mathbf{Q}_t$ is concentrated on paths with $T > t$. With respect to the measure $\mathbf{Q}_t$, the process
$$\tilde{B}_s = B_s - \int_0^s \frac{1}{B_r}\, dr, \quad 0 \le s \le t$$
is a standard Brownian motion. In other words, with respect to $\mathbf{Q}_t$,
$$dB_s = \frac{1}{B_s}\, ds + d\tilde{B}_s.$$
This is the Bessel equation for $d = 3$. Fix some $r > x$, and consider this process stopped at $T_r = \inf\{t : B_t = r\}$. Again, for each t, $\mathbf{Q}_t$ concentrates on paths with $t \wedge T_r < T$. As $t \to \infty$, the $\mathbf{Q}_t$ probability that $T_r < t$ goes to one. We therefore get a limiting measure $\mathbf{Q} = \mathbf{Q}_\infty$ on paths starting at x and ending at the first time they reach r. From the construction we can see this is the same as the conditional measure on usual Brownian motion given that $T_r < T$. Hence, the Bessel-3 process can be considered as "Brownian motion conditioned to stay positive always." We can start a Bessel-3 process with $B_0 = 0$ — the easiest way is to let $B_t = |\bar{B}_t|$ where $\bar{B}_t$ is a three-dimensional Brownian motion with respect to **Q**.

EXAMPLE 1.18. More generally, let $q > 0$ and let
$$M_t = B_t^q \exp\left\{-\frac{1}{2}q(q-1)\int_0^t \frac{1}{B_s^2}\,ds\right\}.$$
Itô's formula shows that M_t is a martingale at least up to time $T(\epsilon) = \inf\{t : B_t \leq \epsilon\}$ with
$$dM_t = \frac{q}{B_t} M_t\, dB_t.$$
If we let $\mathbf{Q}$ be the measure with $d\mathbf{Q}_t = x^{-q} M_t\, d\mathbf{P}$, then with respect to $\mathbf{Q}$,
$$\tilde{B}_t = B_t - \int_0^t \frac{q}{B_s}\,ds,$$
is a standard Brownian motion, i.e., with respect to $\mathbf{Q}$, B_t satisfies the Bessel equation
$$dB_t = \frac{q}{B_t}\,dt + d\tilde{B}_t.$$

EXAMPLE 1.19. Suppose $D \subset \mathbb{R}^d$ is an open set and h is a positive, harmonic function on D. For the moment, let us assume that D is bounded and that h has an extension as a positive harmonic function in an open set containing $\bar{D}$. Let $\bar{B}_t = (B_t^1, \ldots, B_t^d)$ be a standard d-dimensional Brownian motion starting at $z \in D$ and let $M_t = h(\bar{B}_{t \wedge \tau_D})$ where $\tau_D = \inf\{t : \bar{B}_t \notin D\}$. Itô's formula shows that M_t satisfies
$$dM_t = \nabla h(\bar{B}_t) \cdot d\bar{B}_t = K_t\, M_t \cdot d\bar{B}_t, \quad t < \tau_D,$$
where $K_t = \nabla h(\bar{B}_t)/h(\bar{B}_t)$. If we let $\mathbf{Q}$ be as above, then $\bar{B}_{t \wedge \tau_D} - \int_0^{t \wedge \tau_D} K_s\,ds$ is a $\mathbf{Q}$-martingale, i.e., $\bar{B}_t$ satisfies the stochastic differential equation

(1.7) $$d\bar{B}_t = \frac{\nabla h(\bar{B}_t)}{h(\bar{B}_t)}\,dt + d\bar{W}_t,$$

where $\bar{W}$ is a standard d-dimensional Brownian motion with respect to $\mathbf{Q}$. The assumption that h is defined on a neighborhood of D can be eased by taking a sequence of bounded domains D_n increasing to D. This process is often called an *h-process* or *h-transform*.

1.10. Bessel processes

Bessel processes arise frequently in analysis of the Schramm-Loewner evolution, so it will be convenient to collect some facts about them. Let B_t be a standard Brownian motion with $B_0 = 0$, and for each $x > 0$ let X_t^x denote a Bessel-d process starting at x driven by B_t, i.e., the solution of
$$dX_t^x = \frac{a}{X_t^x}\,dt + dB_t, \quad X_0^x = x,$$
where $a = (d-1)/2$. We write just X_t for X_t^1. This solution is well defined at least up to $T_x := \inf\{t : X_t^x = 0\}$. Note that we have chosen the same Brownian motion for each x. If $t \leq T_x$, then
$$X_t^x = x + B_t + a\int_0^t \frac{ds}{X_s^x}.$$

26   1. STOCHASTIC CALCULUS

If $x < y$, then $X_t^x < X_t^y$ for all $t < T_x$; in particular $T_x \leq T_y$ (we will see below that $T_x = T_y < \infty$ is possible for some values of a). As was noted in §1.8.3, the absolute value of a d-dimensional Brownian motion is a Bessel-d process.

LEMMA 1.20 (Scaling). *If $x > 0$ and $Y_t = x^{-1} X_{x^2 t}^x$, then Y_t has the same distribution as X_t.*

PROOF. Clearly $Y_0 = 1$, and
$$dY_t = \frac{a}{Y_t}\, dt + d\tilde{B}_t,$$
where $\tilde{B}_t = x^{-1} B_{x^2 t}$. □

PROPOSITION 1.21.
- *If $a \geq 1/2$, then w.p.1 $T_x = \infty$ for all $x > 0$.*
- *If $a = 1/2$, then w.p.1 $\inf_t X_t^x = 0$ for all $x > 0$.*
- *If $a > 1/2$, then w.p.1 $X_t^x \to \infty$ for all $x > 0$.*
- *If $a < 1/2$, then w.p.1 $T_x < \infty$ for all $x > 0$.*
- *If $1/4 < a < 1/2$ and $x < y$, then $\mathbf{P}\{T_x = T_y\} > 0$.*
- *If $a \leq 1/4$, then w.p.1 $T_x < T_y$ for all x, y.*

PROOF. Suppose $0 < x_1 < x < x_2 < \infty$, and let $\sigma = \inf\{t : X_t^x \in \{x_1, x_2\}\}$, $\phi(x) = \phi(x; x_1, x_2) = \mathbf{P}\{X_\sigma^x = x_2\}$. Note that $\phi(X_{t \wedge \sigma}^x) = \mathbf{E}[\phi(X_\sigma^x) \mid \mathcal{F}_t]$ is a martingale. Assuming for the moment that ϕ is C^2, we can use Itô's formula to show that ϕ satisfies
$$\frac{1}{2}\phi''(x) + \frac{a}{x}\phi'(x) = 0, \quad x_1 < x < x_2.$$
Moreover, ϕ is continuous on $[x_1, x_2]$ with $\phi(x_1) = 0, \phi(x_2) = 1$. The unique solution to this boundary value problem is
$$\phi_0(x) = \phi_0(x; x_1, x_2) = \frac{x^{1-2a} - x_1^{1-2a}}{x_2^{1-2a} - x_1^{1-2a}}, \quad a \neq \frac{1}{2},$$
$$\phi_0(x) = \phi_0(x; x_1, x_2) = \frac{\log x - \log x_1}{\log x_2 - \log x_1}, \quad a = \frac{1}{2}.$$
Itô's formula shows that $M_t := \phi_0(X_{t \wedge \sigma}^x)$ is a bounded martingale and hence by the optional sampling theorem,
$$\mathbf{P}\{X_\sigma^x = x_2\} = \mathbf{E}[M_\infty \mid \mathcal{F}_0] = \phi_0(x; x_1, x_2).$$
The first three assertions are then obtained by considering
$$\lim_{x_1 \to 0+} \phi_0(x; x_1, x_2), \quad \lim_{x_2 \to \infty} \phi_0(x; x_1, x_2).$$

For the remainder of this proof, assume $a < 1/2$ and $0 < x < y$ so that with probability one $T_x \leq T_y < \infty$. Let $q(x, y) = q(x, y; a) = \mathbf{P}\{T_x = T_y\}$. By scaling, $q(x, y) = q(1, y/x)$. Also $\lim_{r \to \infty} q(1, r) = 0$ since for all fixed t, $\lim_{r \to \infty} \mathbf{P}\{T_r < t\} = 0$. Assume for ease that $x = 1$, and let $Y_t = X_t^y$. We claim that the event $\{T_1 = T_y\}$ is the same up to an event of probability zero as the event

(1.8) $$\sup_{t < T_1} \frac{Y_t - X_t}{X_t} < \infty.$$

To see this, note that (1.8) implies $T_y = T_1$, but the strong Markov property implies
$$\mathbf{P}\left\{T_y = T_1; \sup_{t<T_1} \frac{Y_t - X_t}{X_t} \geq r\right\} \leq q(1, 1+r),$$
and $q(1, 1+r) \to 0$ as $r \to \infty$. Let $Z_t = \log[(Y_t - X_t)/X_t]$. By Itô's formula,

(1.9) $$dZ_t = \left[(\frac{1}{2} - 2a)\frac{1}{X_t^2} + a\frac{Y_t - X_t}{X_t^2 Y_t}\right] dt - \frac{1}{X_t} dB_t.$$

Define a time change $r(t)$ by
$$\int_0^{r(t)} \frac{ds}{X_s^2} = t,$$
and let $\tilde{Z}_t = Z_{r(t)}$. Under this time change it takes X_t an infinite amount of time to reach the origin, i.e., $r(T_x) = \infty$ (see Lemma 1.23 below). Note that $\tilde{Z}_t$ satisfies
$$d\tilde{Z}_t = \left[(\frac{1}{2} - 2a) + a\frac{Y_{r(t)} - X_{r(t)}}{Y_{r(t)}}\right] dt + d\tilde{B}_t,$$
where $\tilde{B}_t = -\int_0^{r(t)} X_s^{-1} dB_s$ is a standard Brownian motion. By integrating, we can write
$$\tilde{Z}_t = \tilde{Z}_0 + \tilde{B}_t + (\frac{1}{2} - 2a)t + a \int_0^t \frac{Y_{r(s)} - X_{r(s)}}{Y_{r(s)}} ds.$$

If $a \leq 1/4$, then $\tilde{Z}_t$ takes on arbitrarily large values. In particular, $\sup_{t<T_r} e^{Z_t} = \infty$. If $1/4 < a < 1/2$, choose $b \in (1/4, a)$ and let $\epsilon = 2(a-b)/a$. Suppose $x = 1, y = 1 + (\epsilon/2)$ and let σ be the first time that $Y_{r(t)} - X_{r(t)} = \epsilon X_{r(t)}$. For $0 \leq t < T_1 \wedge \sigma$, $\tilde{Z}_t < \tilde{Z}_t^*$ where $\tilde{Z}_t^*$ satisfies
$$d\tilde{Z}_t^* = (\frac{1}{2} - 2b) dt + d\tilde{B}_t.$$

Since $(1/2) - 2b < 0$, there is a positive probability that $\tilde{Z}_t^*$ never reaches $\log \epsilon$ given that it starts at $\log(\epsilon/2)$. On this event, $\tilde{Z}_t$ also stays below $\log \epsilon$ and hence (1.8) holds. This shows that $q(1, 1 + (\epsilon/2)) > 0$. It is not difficult from this to show that $q(x, y) > 0$ for all $x, y > 0$ since $Y_t^y - Y_t^x$ decreases with t for $t < T_1$.

We have established the result for fixed $x < y$, but if we consider rational x, y, we get the proposition. □

REMARK 1.22. If $1/4 < a < 1/2$, the function $q(1, 1+r)$ can be computed. See Proposition 6.34.

For the remainder of this section we will study functionals of the Bessel process that arise in the study of derivatives of the conformal maps in chordal SLE.

LEMMA 1.23. If $a < 1/2$ and $x > 0$, then w.p.1,
$$I := \int_0^{T_x} \frac{1}{(X_t^x)^2} dt = \infty.$$
If $a \geq 1/2$ and $x > 0$, then w.p.1,
$$\int_0^\infty \frac{1}{(X_t^x)^2} dt = \infty.$$

PROOF. We may assume $x = 1$, and we write $X_t = X_t^1$. First, assume $a < 1/2$. Let t_j be the first time t that $X_t = 2^{-j}$ and let

$$Y_j = \int_{t_{j-1}}^{t_j} \frac{1}{X_t^2}\, dt.$$

Then $I = Y_1 + Y_2 + \cdots$. Using Lemma 1.20 and the strong Markov property, we see that $Y_1, Y_2, \ldots$ are i.i.d. with positive expectation. Hence $I = \infty$ w.p.1. If $a \geq 1/2$, a similar argument is used with t_j being the first t such that $X_t = 2^j$. □

Let $X_t = X_t^x$ be a Bessel-$(2a+1)$ process as above starting at $x \in (0,1)$ and let $\sigma = \inf\{t : X_t = 0 \text{ or } 1\}$. For every $\beta > 0$, let

$$F(x) = \int_0^\sigma \frac{1}{X_t^2}\, dt, \quad \phi_a(x;\beta) = \mathbf{E}[e^{-\beta F(x)}] = \mathbf{E}[e^{-\beta F(x)}; X_\sigma = 1].$$

The last equality follows from the first part of Lemma 1.23. If $\beta = 0$, we define $\phi_a(x;\beta)$ using the rule $0^0 = 0$, i.e.,

$$\phi_a(x;0) = \mathbf{P}\{F(x) < \infty\} = \mathbf{P}\{T_x > \sigma\}.$$

We will now compute $\phi(x) = \phi_a(x;\beta)$. Let X_t be a Bessel-$(2a+1)$ process starting at $x \in (0,1)$, i.e.,

$$dX_t = \frac{a}{X_t}\, dt + dB_t, \quad X_0 = x.$$

Let

$$Z_t = \exp\left\{-\beta \int_0^{t \wedge T_x} \frac{ds}{X_s^2}\right\},$$

$$Y_t = \phi(X_{t \wedge \sigma})\, Z_{t \wedge \sigma} = \mathbf{E}[Z_\sigma \mid \mathcal{F}_t].$$

Here $\mathcal{F}_t$ is the filtration of B_t. Clearly Y_t is a martingale. Itô's formula gives for $t < \sigma$,[3]

$$dY_t = Z_t \left[-\frac{\beta}{X_t^2} \phi(X_t) + \frac{a}{X_t} \phi'(X_t) + \frac{1}{2} \phi''(X_t)\right] dt + Z_t\, \phi'(X_t)\, dB_t.$$

Since Y_t is a martingale, we see that $\phi(x)$ must satisfy

$$\phi''(x) + \frac{2a}{x} \phi'(x) - \frac{2\beta}{x^2} \phi(x) = 0.$$

There exist linearly independent solutions of this equation of the form $\phi(x) = x^q$. Plugging in, we see that q satisfies

(1.10) $$q(q-1) + 2aq - 2\beta = 0.$$

The only solution with boundary conditions $\phi(0) = 0, \phi(1) = 1$ is $\phi(x) = x^q$ where

(1.11) $$q = q(a,\beta) = \frac{1 - 2a + \sqrt{(1-2a)^2 + 8\beta}}{2}.$$

This is also valid for $\beta = 0$ if $a < 1/2$, giving $q(a,0) = 1 - 2a$.

---

[3]This use of Itô's formula assumes that $\phi(x)$ is C^2 in $(0,1)$. However, once a solution to the differential equation is found, one can justify this step in a similar way as in the first paragraph of the proof of Proposition 1.21.

REMARK 1.24. Assume $a \geq 1/2$ so that $\mathbf{P}\{T_x < \infty\} = 0$. For every $q \geq 0$, let
$$M_t = M_{t,q,a} = X_t^q \exp\left\{-\beta \int_0^t \frac{1}{X_s^2} \, ds\right\},$$
where $\beta = \beta_{q,a} = aq + (1/2)q(q-1)$. Itô's formula gives
$$dM_t = \frac{M_t}{X_t^2}\left[aq + \frac{1}{2}q(q-1) - \beta\right] dt + M_t \frac{q}{X_t} dB_t = M_t \frac{q}{X_t} dB_t,$$
so we see that β has been chosen so that M_t is a martingale. Note that
$$x^q = \mathbf{E}[M_0] = \mathbf{E}[M_\sigma] = \mathbf{E}\left[\exp\left\{-\beta \int_0^\sigma \frac{1}{X_s^2} \, ds\right\}\right].$$
This gives another derivation of (1.11). If $\mathbf{Q}$ denotes the measure with $d\mathbf{Q}_t = x^{-q} M_t \, d\mathbf{P}$, then Girsanov's transformation shows that with respect to $\mathbf{Q}$,
$$\tilde{B}_t = B_t - \int_0^t \frac{q}{X_s} \, ds,$$
is a standard Brownian motion, i.e., X_t satisfies
$$dX_t = \frac{a+q}{X_t} dt + d\tilde{B}_t.$$
In other words, if we take a Bessel-$(2a+1)$ process and weight the paths by $e^{-\beta F}$, normalized to be a probability measure, then the paths we get have the distribution of a Bessel-$(2(a+q)+1)$ process. Suppose we start with a Bessel-$(2a+1)$ process. If we wish to weight paths by $\exp\{-(\beta+\beta')\int_0^\sigma X_s^{-2} \, ds\}$ we can first weight paths by $\exp\{-\beta \int_0^\sigma X_s^{-2} \, ds\}$ and then weight those paths by $\exp\{-\beta' \int_0^\sigma X_s^{-2} \, ds\}$. This gives a consistency condition
$$q(a, \beta+\beta') = q(a,\beta) + q(a + q(a,\beta), \beta').$$
One can check that (1.11) satisfies this.

We now make a similar calculation for fixed times. Fix $a > 0, \beta \geq 0$; let
$$S_t = \exp\left\{-\beta \int_0^{t \wedge T_x} \frac{ds}{X_s^2}\right\},$$
and let $\psi(t,x) = \mathbf{E}[S_t; T_x > t]$. Scaling tells us that $\psi(t,x) = \psi(x/\sqrt{t})$ where $\psi(x) = \psi(1,x)$. The Markov property tells us that for fixed t_0, $M_t := S_t \psi(t_0-t, X_t)$ is a martingale for $t \leq t_0$. Using Itô's formula (again assuming sufficient smoothness of ψ) we see that this implies that $\psi(t,x)$ satisfies
$$-\dot{\psi}(t,x) + \frac{1}{2}\psi''(t,x) + \psi'(t,x)\frac{a}{x} - \psi(t,x)\frac{\beta}{x^2} = 0,$$
and hence $\psi(x)$ satisfies the ODE
$$\psi''(x) + \left[\frac{2a}{x} + x\right]\psi'(x) - \frac{2\beta}{x^2}\psi(x) = 0.$$
The solution satisfying the conditions $\psi(0) = 0, \psi(\infty) = 1$ is

(1.12) $\quad \psi(x) = x^q \dfrac{\Gamma(\frac{1}{2} + \frac{q}{2} + a)}{2^{q/2} \, \Gamma(\frac{1}{2} + q + a)} e^{-x^2/2} \, \Phi\left(\dfrac{1}{2} + \dfrac{q}{2} + a, \dfrac{1}{2} + q + a, \dfrac{1}{2}x^2\right),$

where Φ denotes the confluent hypergeometric function of the first kind, see §B.2. The exact form of ψ is not important; we only need that it is increasing in x and $\psi(x) \sim c\, x^q$ as $x \to 0+$.

1.11. Diffusions on an interval

Suppose that X_t satisfies

$$(1.13) \qquad dX_t = \frac{1}{2} v(X_t)\, dt + dB_t,$$

with $X_0 \in (0, 2\pi)$. Here B_t is a standard Brownian motion, $v : (0, 2\pi) \to \mathbb{R}$ is a C^1 function with $v' \leq 0$, and the process is "killed" (stopped) at time $T = \inf\{t : X_t \in \{0, 2\pi\}\}$. We do not assume that v is bounded, and it is possible that $T = \infty$ w.p.1. Let

$$\psi(x) = \int_\pi^x e^{-V(u)}\, du,$$

where $V(x)$ is the antiderivative of v with $V(\pi) = 0$. Then ψ is strictly increasing and satisfies the differential equation

$$\psi''(x) + v(x)\, \psi'(x) = 0, \qquad 0 < x < 2\pi.$$

LEMMA 1.25. $\mathbf{P}^x\{T < \infty; X_T = 0\} > 0$ if and only if $\psi(0+) > -\infty$. $\mathbf{P}^x\{T < \infty; X_T = 2\pi\} > 0$ if and only if $\psi(2\pi-) < \infty$.

PROOF. We will prove the first assertion; the second is proved similarly. Suppose $0 < y_1 < x < y_2 < 2\pi$ and let $\sigma = \sigma(y_1, y_2)$ be the smallest t with $X_t \in \{y_1, y_2\}$. It is easy to see that $\sigma < \infty$ w.p.1 since v is bounded on $[y_1, y_2]$. Itô's formula shows that $M_t := \psi(X_{t \wedge \sigma})$ is a bounded martingale, and hence by the optional sampling theorem,

$$\psi(x) = \mathbf{E}^x[M_0] = \mathbf{E}^x[M_\sigma] = \mathbf{P}^x\{X_\sigma = y_1\}\, \psi(y_1) + \mathbf{P}^x\{X_\sigma = y_2\}\, \psi(y_2).$$

Hence,

$$\mathbf{P}^x\{X_\sigma = y_1\} = \frac{\psi(y_2) - \psi(x)}{\psi(y_2) - \psi(y_1)}.$$

The condition $v' \leq 0$ implies that $\mathbf{P}^x\{\sigma(0, y_2) < \infty\} > 0$ for every $y_2 < 2\pi$. (To see this, if v is always positive, $X_t \geq B_t$ for $t \leq \sigma(0, y_2)$; if v can be zero, then a similar comparison shows that the process either leaves $(0, 2\pi)$ or keeps returning to points x where $v(x) = 0$. If it keeps returning to such x, then it also keeps returning to y_2 since one can clearly reach y_2 from x in finite time with positive probability.) Using this we see that

$$\begin{aligned}
\mathbf{P}^x\{T < \infty; X_T = 0\} &= \lim_{y_2 \to 2\pi-} \mathbf{P}^x\{X_{\sigma(0, y_2)} = 0\} \\
&= \lim_{y_2 \to 2\pi-} \lim_{y_1 \to 0+} \mathbf{P}\{X_\sigma = y_1\} \\
&= \lim_{y_2 \to 2\pi-} \frac{\psi(y_2) - \psi(x)}{\psi(y_2) - \psi(0+)}.
\end{aligned}$$

If $\psi(0+) = -\infty$, then the limit equals zero since all the terms are zero. If $\psi(0+) > -\infty$, the terms are strictly positive for all $y_2 \in (x, 2\pi)$, and since the terms increase with y_2, the limit is strictly positive. $\square$

1.11. DIFFUSIONS ON AN INTERVAL

Let $p(t, x, y), t > 0, 0 < x, y < 2\pi$ denote the transition probability density, i.e., the density (in y) of the random variable $X_t \mathbf{1}\{T > t\}$ assuming $X_0 = x$. If f, g are C^2 functions on $[0, 2\pi]$, define

$$P_t f(x) = \mathbf{E}^x[f(X_t)] = \int_0^{2\pi} p(t, x, y) f(y)\, dy,$$

$$P_t^* g(y) = \int_0^{2\pi} p(t, x, y)\, g(x)\, dx.$$

If g is a probability density, then $P_t^* g$ is the density of $X_t \mathbf{1}\{T > t\}$ assuming X_0 has density g. Note that if $t < t_0$, $M_t := \mathbf{E}[f(X_{t_0}) \mid \mathcal{F}_t] = P_{t_0-t} f(X_t)$ is a martingale for $0 \le t \le t_0$. Hence by Itô's formula[4], we can see that p must satisfy

$$2\dot{p}(t, x, y) = \partial_{xx} p(t, x, y) + v(x)\, \partial_x p(t, x, y).$$

LEMMA 1.26. *Suppose g is a continuous function on $[0, 2\pi]$ and $q(t, y) = P_t^* g(y)$. Then q satisfies the adjoint equation*

(1.14) $$2\dot{q}(t, y) = \partial_{yy} q(t, y) - \partial_y [v(y)\, q(t, y)].$$

PROOF. Suppose f is a C^2 function whose (closed) support is contained in the open interval $(0, 2\pi)$. Using the Chapman-Kolmogorov equation

$$p(t + s, x, y) = \int_0^{2\pi} p(t, x, z)\, p(s, z, y)\, dz,$$

one can easily check that

$$\int_0^{2\pi} P_{t+s}^* g(y)\, f(y)\, dy = \int_0^{2\pi} [P_t^* g(y)]\, [P_s f(y)]\, dy.$$

Hence,

$$\frac{d}{dt} \int_0^{2\pi} P_t^* g(y)\, f(y)\, dy = \int_0^{2\pi} P_t^* g(y)\, [\frac{d}{ds} P_s f(y)\mid_{s=0}]\, dy$$

$$= \frac{1}{2} \int_0^{2\pi} P_t^* g(y)\, [f''(y) + v(y)\, f'(y)]\, dy$$

$$= \frac{1}{2} \int_0^{2\pi} f(y)\, [\, \partial_{yy} P_t^* g(y) - \partial_y [P_t^* g(y)\, v(y)]\,]\, dy.$$

The last inequality uses integration by parts and the fact that f and f' vanish at the endpoints. Since this holds for every f, we get the lemma. □

It follows from the lemma that p satisfies the adjoint equation

$$2\dot{p}(t, x, y) = \partial_{yy} p(t, x, y) - v(y)\, \partial_y p(t, x, y) - v'(y)\, p(t, x, y).$$

Suppose that ϕ_0 is a positive eigenfunction of the adjoint equation with Dirichlet boundary conditions, i.e., $\phi_0 : (0, 2\pi) \to (0, \infty)$ is a C^2 function with $\phi_0(0+) = \phi_0(2\pi-) = 0$, and

(1.15) $$\phi_0''(y) - v(y)\, \phi_0'(y) - v'(y)\, \phi_0(y) = -2\lambda\, \phi_0(y),$$

---

[4]Here we are assuming that $p(t, x, y)$ is C^1 in t and C^2 in x which can be derived directly using the conditions on v.

for some $\lambda \geq 0$. Assume that ϕ_0 has been normalized so that $\int_0^{2\pi} \phi_0(y)\, dy = 1$. Then $q_0(t,y) := e^{-\lambda t}\phi_0(y)$ is the solution of (1.14) with $q_0(0,y) = \phi_0(y)$, i.e., $P_t^*\phi_0(y) = e^{-\lambda t}\phi_0(y)$. Also, if f is a C^2 function with support in $(0, 2\pi)$,

$$\int_0^{2\pi} [P_t f(x)]\, \phi_0(x)\, dx = \int_0^{2\pi} f(x)\, P_t^*\phi_0(x)\, dx$$
$$= e^{-\lambda t} \int_0^{2\pi} f(x)\, \phi_0(x)\, dx.$$

Hence ϕ_0 is an invariant probability density in the sense,

$$\int_0^{2\pi} p(t,x,y)\, \phi_0(x)\, dx = e^{-\lambda t} \phi_0(y).$$

1.11.1. A Bessel-like process on an interval. We will encounter a particular Bessel-like process in the study of the radial Schramm-Loewner evolution. Suppose X_t satisfies (1.13) with $v(x) = a\cot(x/2)$, i.e.,

(1.16) $\qquad dX_t = (a/2)\cot(X_t/2)\, dt + dB_t, \qquad X_0 = x \in (0, 2\pi),$

where $a > 0$, and the process is run until $T = T_x := \inf\{t : X_t \in \{0, 2\pi\}\}$. Note that as $x \to 0+$, $(a/2)\cot(x/2) \sim a/x$, so the process should behave like a Bessel-$(2a+1)$ process.

LEMMA 1.27. *Suppose $x \in (0, 2\pi)$ and X_t satisfies (1.16). If $a \geq 1/2$, w.p.1, $T = \infty$. If $a < 1/2$, w.p.1, $T < \infty$.*

PROOF. In the notation of Lemma 1.25,

$$v(x) = a\cot(x/2), \qquad V(x) = 2a\log(\sin(x/2)),$$

$$\psi(x) = \int_\pi^x [\sin(y/2)]^{-2a}\, dy.$$

Since the integral converges at 0 and 2π if and only if $a < 1/2$, the lemma follows from Lemma 1.25. $\square$

LEMMA 1.28. *Let $p(t,x,y)$, $0 < x, y < 2\pi$, denote the density (in y) of X_t, where X_t satisfies (1.16) and is stopped at $\{0, 2\pi\}$. Then,*

$$\int_0^{2\pi} p(t,x,y) \sin^{2a}(x/2)\, dx = \sin^{2a}(y/2), \qquad a \geq 1/2,$$

$$\int_0^{2\pi} p(t,x,y) \sin(x/2)\, dx = e^{-\lambda_a t}\sin(y/2), \qquad a \leq 1/2,$$

where $\lambda_a = (1/2) - a$.

PROOF. This can be verified using (1.15). $\square$

Let

$$F_t = -\frac{1}{2}\int_0^{t\wedge T} v'(X_s)\, ds = \frac{a}{2}\int_0^{t\wedge T} \frac{ds}{2\sin^2(X_s/2)}.$$

Since $\sin x \sim x$ as $x \to 0+$, comparison to a Bessel process and Proposition 1.23 show that w.p.1 $F_T = \infty$ for $a < 1/2$.

For the remainder of this section we will fix an $a > 0$ and a nonnegative number b. We assume $b > 0$ if $a \geq 1/2$ or $b \geq 0$ if $a < 1/2$. In the latter case, we will abuse notation and write $e^{-0 \cdot F_t} = 1\{F_t < \infty\} = 1\{T > t\}$. We set $x_n = 2^{-n}\pi$, $\sigma_y = \inf\{t : X_t \in \{0, y\}\}$, $\sigma^n = \sigma_{x_n}$, and $\sigma = \sigma^0 = \sigma_\pi$. Let

$$\psi_0(t, x) = \mathbf{E}^x[e^{-bF_t}].$$

Clearly, ψ_0 is decreasing in t and $\psi_0(t, \pi - x) = \psi_0(t, \pi + x)$. Also, we claim that $\psi(t, x)$ is strictly increasing in x for $0 < x \le \pi, t > 0$. (One can see this by a coupling argument — if X_t, Y_t are processes satisfying (1.16) starting at $0 < x < y \le \pi$, respectively, let them run independently until the first time that $|\pi - X_t| = |\pi - Y_t|$ and from then on couple the processes so that $|\pi - X_t| = |\pi - Y_t|$.) In the other direction, we have

(1.17) $$\psi_0(t, x) \ge \mathbf{E}^x[e^{-bF_\sigma}]\psi_0(t, \pi), \quad 0 < x \le \pi.$$

LEMMA 1.29. *There exists a continuous ψ_0 on $[0, 2\pi)$ such that*

$$\psi_0''(x) + a \cot(x/2)\,\psi_0'(x) - \frac{ab}{2\sin^2(x/2)}\psi_0(x) = 0, \quad 0 < x < 2\pi$$

and

$$\psi_0(x) = x^q + O(x^{q+1}), \quad x \to 0+,$$

where

(1.18) $$q = q(a, b) = \frac{1 - 2a + \sqrt{(1 - 2a)^2 + 8ab}}{2} > 0.$$

PROOF. The function ψ_0 is

$$\sin^q(x/2)\,[1 + \cos(x/2)]^{(1/2) - a - q}\,F(\frac{1}{2} - a, \frac{1}{2} + a, q + \frac{1}{2} - a; [1 - \cos(x/2)]/2),$$

where F denotes the hypergeometric function (see §B.3). Note that the assertion that $q > 0$ follows from the assumption that either $b > 0$ or $a < 1/2$. □

PROPOSITION 1.30. *If $0 < x < y < 2\pi$, then*

$$\mathbf{E}^x[e^{-bF_{\sigma_y}}] = \psi_0(x)/\psi_0(y).$$

PROOF. Itô's formula shows that $M_t := \exp\{-bF_{t \wedge \sigma_y}\}\psi_0(X_{t \wedge \sigma_y})$ is a bounded martingale. Hence, by the optional sampling theorem,

$$\psi_0(x) = \mathbf{E}^x[M_0] = \mathbf{E}^x[M_{\sigma_y}] = \psi_0(y)\,\mathbf{E}^x[e^{-bF_{\sigma_y}}].$$

□

It follows that there is a c such that

$$\mathbf{E}^x[e^{-bF_\sigma}] \sim c\,x^q, \quad x \to 0+,$$

where q is as in (1.18). Using (1.17) we see that

$$\psi_0(t, x) \ge c\,x^q\,\psi_0(t, \pi).$$

Let $\phi_0(x) = c_0 \sin^q(x/2)$, $\phi^*(x) = c_1 \sin^{2(a+q)}(x/2)$, where c_0, c_1 is chosen so that $\int_0^{2\pi} \phi_0(x)\,dx = \int_0^{2\pi} \phi^*(x)\,dx = 1$.

LEMMA 1.31. *Let $q = q(a,b)$ be as in (1.18) and ϕ_0 as above. Then*

$$\phi_0''(x) + a \cot(x/2)\, \phi_0'(x) - \frac{ab}{2\sin^2(x/2)}\, \phi_0(x) = -2\lambda\, \phi_0(x),$$

where

(1.19) $$\lambda = \lambda(a,b) = \frac{q + 2ab}{8} = \frac{1 - 2a + 4ab + \sqrt{(1-2a)^2 + 8ab}}{16}.$$

If $\phi(t,x) = e^{-\lambda t}\, \phi_0(x)$, then

$$2\dot{\phi}(t,x) = \phi''(t,x) + a \cot(x/2)\, \phi'(t,x) - \frac{ab}{2\sin^2(x/2)}\, \phi(t,x).$$

If $t \geq 0$,

(1.20) $$\int_0^{2\pi} \phi^*(x)\, \psi_0(t,x)\, dx = e^{-\lambda t}.$$

In particular, there exist c_1, c_2 such that

(1.21) $$c_1\, e^{-\lambda t} \leq \int_0^\pi x^{2(q+a)}\, \psi_0(x,t)\, dx \leq c_2\, e^{-\lambda t}.$$

PROOF. An easy computation establishes the differential equations. Itô's formula shows that $M_t := e^{\lambda t}\, e^{-bF_t}\, \phi_0(X_t)$ is a martingale; in fact,

$$dM_t = \frac{q}{2} \cot(X_t/2)\, M_t\, dB_t.$$

By Girsanov's transformation, under the measure $d\mathbf{Q}_t := M_t\, d\mathbf{P}_t$, X_t satisfies the stochastic differential equation

$$dX_t = \frac{a+q}{2} \cot(X_t/2)\, dt + d\tilde{B}_t,$$

for a **Q**-Brownian motion $\tilde{B}_t$. Note that $(a+q)/2 > 1/2$. Lemma 1.28 tells us that ϕ^* is the invariant probability density for this process. This implies (1.20). The final assertion comes from $\sin(x/2) \sim x/2$. □

PROPOSITION 1.32. *There exist c_1, c_2 such that if $0 < x \leq \pi$,*

$$\mathbf{E}^x[e^{-bF_\sigma}; \sigma \leq 1] \geq c_1\, \mathbf{E}^x[e^{-bF_\sigma}].$$

$$c_1\, \psi_0(x) \leq \psi_0(1,x) \leq c_2\, \psi_0(x).$$

PROOF. We first note that it is easy to show that for every $y \in (0,\pi]$ and every $t > 0$, there is a $c(y,t) > 0$ such that

$$\mathbf{E}^x[e^{-bF_\sigma}; \sigma \leq t] \geq c(y,t), \quad y \leq x \leq \pi.$$

For example, on the event

$$\{B_s \geq -x/2,\ 0 \leq s \leq t;\ B_t \geq 1\},$$

we have $\sigma \leq t$; $X_s \geq B_s, 0 \leq s \leq \sigma \wedge t$; and $e^{-bF_\sigma} \geq \exp\{-abt\sin^{-2}(x/4)\}$. If $x \leq x_{n+1}$, the strong Markov property gives

$$\mathbf{E}^x[e^{-bF_\sigma}; \sigma^n - \sigma^{n+1} \geq n2^{-2n}]$$
$$= \frac{\psi_0(x)}{\psi_0(x_{n+1})} \mathbf{E}^{x_{n+1}}[e^{-bF_\sigma}; \sigma^n \geq n2^{-2n}]$$
$$\leq \frac{\psi_0(x)}{\psi_0(x_{n+1})} \exp\{-\frac{ba}{2} \int_0^{n2^{-2n}} \frac{ds}{2\sin^2(x_n/2)}\} \psi_0(x_n)$$
$$\leq c\psi_0(x) e^{-\beta n},$$

for some constants c, β. Therefore, there exists an m such that for all $x \leq x_{m+1}$,

$$\mathbf{E}^x[e^{-bF_\sigma}; \sigma^n - \sigma^{n+1} \leq n2^{-2n} \text{ for all } n \geq m] \geq (1/2)\,\psi_0(x).$$

By choosing a larger m if necessary we may assume $\sum_{n \geq m} n\, 2^{-2n} \leq 1/2$. Therefore, for all $0 < x < x_m$,

$$\mathbf{E}^x[e^{-bF_{\sigma^m}}; \sigma^m \leq 1/2] \geq \mathbf{E}^x[e^{-bF_\sigma}; \sigma^m \leq 1/2] \geq (1/2)\,\psi_0(x),$$

and hence

$$\mathbf{E}^x[e^{-bF_\sigma}; \sigma \leq 1] \geq (1/2)\,\psi_0(x)\, c(x_m, 1/2).$$

This proves the first assertion.

The inequality $\psi_0(1, x) \geq \psi_0(x)\psi_0(1, \pi)$ gives the lower bound for the second assertion. Let $\alpha = [\sum_{n \geq 1} n^2\, 2^{-2n}]^{-1}$, and let $t_0 = 0, t_{n+1} = t_n + \alpha\, n^2\, 2^{-2n}$. To show the upper bound it suffices to show that $\psi_0(x_n, t_n)/\psi_0(x_n)$ is uniformly bounded. But,

$$\psi(x_{n+1}, t_{n+1}) \leq \mathbf{E}^{x_{n+1}}[e^{-bF_{\sigma^n}}]\,\psi_0(x_n, t_n) + \mathbf{E}^{x_{n+1}}[e^{-bF_{\sigma^n}}; \sigma^n \geq \alpha\, n^2\, 2^{-2n}]$$
$$\leq \frac{\psi_0(x_{n+1})}{\psi_0(x_n)} \psi_0(x_n, t_n) + c e^{-\beta n^2},$$

for some β. But we know that $\psi_0(x_n, t_n) \geq c2^{-nq}$ and $\psi_0(x_{n+1})/\psi_0(x_n) \geq c$. Hence we can write

$$\frac{\psi(x_{n+1}, t_{n+1})}{\psi_0(x_{n+1})} \leq \frac{\psi(x_n, t_n)}{\psi_0(x_n)}[1 + \delta_n],$$

where $\sum \delta_n < \infty$. Since $\psi_0(x_0, t_0) = \psi_0(x_0) = 1$, this implies

$$\frac{\psi_0(x_n, t_n)}{\psi_0(x_n)} \leq \prod_{j=0}^{\infty}(1 + \delta_j) < \infty.$$

$\square$

PROPOSITION 1.33. *There exist c_1, c_2 such that if $t \geq 1$, $0 < x \leq \pi$,*

(1.22) $$c_1\, x^q\, e^{-\lambda t} \leq \psi_0(t, x) \leq c_2\, x^q\, e^{-\lambda t}.$$

PROOF. We have shown that $c_1 x^q \psi(t+1, \pi) \leq \psi(t+1, x) \leq c_2\, x^q\, \psi(t, \pi)$. Since $\psi_0(t, x) \geq c\psi_0(t, \pi)$ for $x \in [\pi/2, \pi]$, (1.21) gives $\pi(t, \pi) \asymp e^{-\lambda t}$. $\square$

Note also that this implies that

(1.23) $$\mathbf{E}^x[e^{-bF_{\sigma \wedge 1}}] \leq \mathbf{E}^x[e^{-bF_\sigma}] + \mathbf{E}^x[e^{-bF_1}] \leq c x^q.$$

PROPOSITION 1.34. *For every x, the limit*
$$\lim_{t\to\infty} e^{\lambda t}\,\psi_0(t,x) = \psi_*(x)$$
exists. In fact, there is an $\alpha > 0$ such that
$$\psi_0(t,x) = e^{-\lambda t}\,\psi_*(x)\,[1 + O(e^{-\alpha t})], \qquad t \geq 1.$$

PROOF. Let
$$\bar{\psi}_0(t,x,y) = \frac{\psi_0(t,x,y)}{\int_0^{2\pi} \psi_0(t,x,y')\,dy'}.$$
This is the density of X_t assuming $X_0 = x$ and that the paths are weighted by $e^{-bF_t}/\mathbf{E}[e^{-bF_t}]$. It suffices to show that

$$(1.24) \qquad \int_0^{2\pi} |\bar{\psi}_0(t,x,y) - \phi^*(y)|\,dy \leq c\,e^{-\alpha t}, \qquad t \geq 1,$$

where c, α are independent of t, x. To see this, note that $0 \leq s \leq 1$,
$$\frac{\psi_0(t+s,x)}{\psi_0(t,x)} = \int_0^{2\pi} \bar{\psi}_0(t,x,y)\,\psi_0(s,y)\,dy,$$
and
$$\int_0^{2\pi} \phi^*(y)\,\psi_0(s,y)\,dy = e^{-\lambda s}.$$

Hence, it suffices to prove (1.24). But recall from the proof of Lemma 1.31, that the conditioned process moves like a solution of

$$(1.25) \qquad dX_t = \frac{a+q}{2}\cot(X_t/2)\,dt + dB_t.$$

Hence (1.24) can be established by a standard coupling argument for processes satisfying (1.25), one with initial condition $X_0 = x$ and the other with (invariant) initial density ϕ^*. $\square$

1.11.2. A related estimate. In this subsection we will consider a family of processes X_t^x all satisfying (1.16) with the same Brownian motion B_t. Let $T^x = \inf\{t : X_t^x \in \{0, 2\pi\}\}$ and $T = \sup_x T^x$. We will write just X_t for X_t^π. We fix a, b as in the previous subsection and allow constants to depend on a, b. Let
$$F_t^x = \frac{a}{2}\int_0^{t\wedge T_x} \frac{ds}{2\sin^2(X_s^x/2)}, \qquad R_t(x) = e^{-F_t^x}, \qquad R_t = \int_0^{2\pi} R_t(x)\,dx.$$
Note that $\mathbf{E}[R_t(x)^b] = \psi_0(t,x)$. If $t < T$, define $Y_{t,-} = \inf\{X_t^x : T_x > t\}$, $Y_{t,+} = \sup\{X_t^x : T_x > t\}$. Note that $Y_{t,-}$ follows the SDE (1.16) when it is in $(0, 2\pi)$ and "reflects" off of 0; similarly, $Y_{t,+}$ follows (1.16) with reflection at 2π. (The exact nature of the reflection will not be important to us.) Note that T is the first time that $Y_{t,-} = Y_{t,+}$.

LEMMA 1.35. *If $t < T$, $R_t = Y_{t,+} - Y_{t,-}$.*

PROOF. Choose x, y with $T^x, T^y > t$. Then (1.16) implies that
$$\frac{d}{dt}[X_t^y - X_t^x] = (a/2)\,[\cot(X_t^y/2) - \cot(X_t^x/2)].$$
Hence if $v_t(x) = \log[(d/dx)X_t^x]$,
$$\dot{v}_t(x) = -(a/2)\,[2\sin^2(X_t^x/2)]^{-1}.$$

1.11. DIFFUSIONS ON AN INTERVAL

Hence, $v_t(x) = -F_t^x = \log R_t(x)$. But

$$Y_{t,+} - Y_{t,-} = \int_0^{2\pi} e^{v_t(x)}\, dx.$$

□

PROPOSITION 1.36. *There exist c_1, c_2 such that*

$$c_1 e^{-\lambda t} \leq \mathbf{E}[R_t^b] \leq c_2 e^{-\lambda t},$$

where $\lambda = \lambda(a,b)$ is as in (1.19).

PROOF. If $b \geq 1$, then $R_t^b \leq (2\pi)^{b-1} \int_0^{2\pi} R_t(x)^b\, dx$, and if $b \leq 1$, then $R_t^b \geq (2\pi)^{b-1} \int_0^{2\pi} R_t(x)^b\, dx$. Hence the upper bound for $b \geq 1$ and the lower bound for $b \leq 1$ follow from (1.22).

To find a lower bound for $b \geq 1$, it suffices to find a c such that

(1.26) $\mathbf{E}[\, R_t(\pi)^b\,;\, R_t(x) \geq c\, R_t(\pi) \text{ for } \pi - c \leq x \leq \pi\,] \geq c e^{-\lambda t}.$

We first claim that there exist c_1, c_2 such that for all t

(1.27) $\mathbf{E}[R_t(\pi)^b; V_t] \geq c_2\, e^{-\lambda t},$

where $V_t = V_t(c_1)$ is the event

$$V_t = \{c_1\, s^{-2/q} \leq X_s \leq 2\pi - c_1 s^{-2/q} \text{ for } 0 \leq s \leq t\}.$$

It suffices to show this for integer t. Let

$$U_j = \{|X_s - \pi| \geq \pi - c_1\, (j-1)^{-2/q} \text{ for some } j-1 \leq s \leq j\},$$

and let c_3 be such that $\mathbf{E}[R_{t+1}(\pi)^b] \geq c_3\, e^{-\lambda t}$. By the strong Markov property and (1.22),

$$\begin{aligned}\mathbf{E}[R_{t+1}(\pi)^b; U_j] &\leq \psi(j-1,\pi)\, \psi(t+1-j, c_1\,(j-1)^{-2/q}) \\ &\leq c_4\, e^{-\lambda t}\, c_1^q\, (j-1)^{-2}.\end{aligned}$$

Hence if we choose c_1 sufficiently small, we get $\mathbf{E}[\, R_{t+1}(\pi)^b\,;\, V_t^c\,] \leq (c_3/2)\, e^{-\lambda t}$, which implies (1.27) with $c_2 = c_3/2$. Note that if $0 < x < \pi$ and $t < T^x \wedge T^\pi$,

(1.28) $\dfrac{d}{dt}[X_t - X_t^x] = \dfrac{a}{2}[\cot(\dfrac{X_t}{2}) - \cot(\dfrac{X_t^x}{2})] \leq -\dfrac{a}{4}[X_t - X_t^x],$

and hence

$$X_t - X_t^x \leq e^{-at/4}\,(\pi - x), \quad t \leq T^x \wedge T^\pi.$$

Choose c_4 sufficiently small so that $c_4\, e^{-at/4} \leq (c_1/2)\, t^{-2/q}$ for all t. Then if $\pi - c_4 \leq c \leq \pi$, on the event V_t, we have $T^x > t$ and

$$\sin^2(X_t^x/2) \geq \sin^2(X_t/2) - \dfrac{1}{2}(X_t - X_t^x) \geq \sin^2(X_t/2) - c_4\, e^{-at/4},$$

$$\sin^{-2}(X_t^x/2) \leq \sin^{-2}(X_t/2) + c_5\, e^{-at/4}\, t^{8/q},$$

$$R_t(x) \geq R_t(\pi) \exp\{-c \int_0^\infty e^{-at/4}\, t^{8/q}\, dt\}.$$

This gives (1.26).

We will now prove the upper bound for $b < 1$. We have already noted that $R_t = Y_{t,+} - Y_{t,-} \leq 2\pi\, e^{-at/4}$. If $0 \leq x \leq y \leq 2\pi$, let $R_t(x,y) = \int_x^y R_t(u)\, du$. and $e_t(x,y) = \mathbf{E}[R_t(x,y)^b]$. Note that if $s < t$, $\mathbf{E}[R_t^b \mid \mathcal{F}_s] = 1\{T > s\}\, e_{t-s}(Y_{s,-}, Y_{s,+})$.

Also, (1.23) and monotonicity imply that if $0 \leq x < y \leq \pi$, then $e_1(x,y) \leq c\,(y-x)^b\,y^q$. .

For positive integer k, let V_k be the event that $T > k$ and $[Y_{t,-}, Y_{t,+}] \subset [0, e^{-ak/16}] \cup [2\pi - e^{-ak/16}, 2\pi]$ for some $t \in [k, k+1]$. We will show that there exist constants c, β such that for any t and any collection of integers $k_1 < k_2 < \cdots < k_m \leq t$,

(1.29) $\quad F(t; k_1, \ldots, k_m) := \mathbf{E}[R_t^b; V_{k_1} \cap V_{k_2} \cap \cdots \cap V_{k_m}] \leq c^m\, e^{-\beta(k_1+\cdots+k_m)}\, e^{-\lambda t}.$

Since

$$\sum_{m=0}^{\infty} c^m \sum_{k_1 < k_2 < \cdots < k_m} e^{-\beta(k_1+\cdots+k_m)} < \infty,$$

this implies the upper bound. It suffices to find an integer K such that (1.29) holds for all collections of integers satisfying $k_{j+1} \geq k_j + 2$, $j = 1, \ldots, m-1$; $k_1 \geq K$; and $t \geq k_m + 2$. We will delay our choice of K, but we willl require that K is sufficiently large so that $e^{-at/16} \geq 2e^{-at/8} \geq 4\pi e^{-at/4}$ for $t \geq K$. If $k \geq K$, $Y_{k,+} - Y_{k,-} \leq 2\pi\, e^{-ak/4} \leq e^{-ak/8} \leq (1/2)\,e^{-ak/16}$; in particular, if $k_j \leq t \leq k_j + 1$ and $[Y_{t,-}, Y_{t,+}] \not\subset [0, e^{-ak_j/16}] \cup [2\pi - e^{-ak_j/16}, 2\pi]$, then $[Y_{t,-}, Y_{t,+}] \subset [e^{-ak_j/8}, 2\pi - e^{-ak_j/8}]$. Note that to prove (1.29) it suffices to prove two estimates:

- there exist c, β such that

(1.30) $\quad F(k_m + 2; k_1, \ldots, k_m) \leq c\, e^{-\beta k_m}\, F(k_m; k_1, \ldots, k_{m-1});$

- there is a K such that if $k_m \geq K$, $t \geq k_m + 2$,

(1.31) $\quad F(t; k_1, \ldots, k_m) \leq c\, e^{-\lambda(t - k_m)}\, F(k_m + 2; k_1, \ldots, k_m).$

To get (1.30), choose $k_1, \ldots, k_m$ and let $k = k_m$. We will assume that there is a $k \leq t \leq k+1$ with $Y_{t,+} \leq e^{-ak/16}$; a similar argument holds if $Y_{t,-} \geq 2\pi - e^{-ak/16}$ for some $t \in [k, k+1]$. Let τ be the first $t \geq k$ with $Y_{t,+} \leq e^{-ak/16}$ and σ the first time $t' > t$ with $Y_{t,+} = \pi$. By monotonicity, for every $y \in [Y_{t,-}, Y_{t,+}]$, and $\tau \leq s \leq \sigma$

$$\frac{R_s(y)}{R_\tau(y)} \leq \frac{R_s(Y_{t,+})}{R_\tau(Y_{t,+})}.$$

But we know that $\mathbf{E}[R(x)^b] \leq c\,x^q$. Hence, $\mathbf{E}[R_{k+2}^b\, 1_{V_k} \mid \mathcal{F}_k] \leq c\,R_k^b\,[e^{-ak/16}]^q$. This establishes (1.30) with $\beta = aq/16$.

For every $Y_{s,-} \leq y_1 < y_2 \leq Y_{s,+}$,

$$|\sin^2(y_1/2) - \sin^2(y_2/2)| \leq \frac{1}{2}(y_2 - y_1) \leq \pi\,e^{-as/4}.$$

Hence if

(1.32) $\quad [Y_{s,-}, Y_{s,+}] \subset [e^{-a(s-1)/16}, 2\pi - e^{-a(s-1)/16}],$

then,

$$|\sin^{-2}(y_1/2) - \sin^{-2}(y_2/2)| \leq c\,e^{-as/8}.$$

In particular, if (1.32) holds for all $k_M + 2 \leq s \leq t$,

$$\frac{R_t}{R_{k_m+2}} \leq c\,\frac{R(Y_{t,+})}{R(Y_{k_m+2,+})}.$$

Hence, if $k_m + 2 \leq t$,
$$\mathbf{E}[R_t^b \, \mathbf{1}_{V_{k_m+2}^c \cap V_{k_m+3}^c \cap \cdots \cap V_{k_m+[t]}^c} \mid \mathcal{F}_{k_m+2}] \leq c \, R_{k_m+2}^b \, e^{-\lambda(t-(k_m+2))}.$$
This gives (1.31) and finishes the proof. □

1.12. A Feynman-Kac formula

We have already been using a number of results that go under the name the Feynman-Kac formula. In this section we will give another example. Suppose Y_t satisfies the SDE
$$dY_t = v(Y_t)\, dt + dB_t.$$
and $V : \mathbb{R} \to [0, \infty)$ is a continuous function such that w.p.1
$$S := \int_0^\infty V(Y_t)\, dt < \infty.$$
Suppose that $\psi : \mathbb{R} \to \mathbb{C}$ is a bounded C^2 function satisfying

(1.33) $$\frac{1}{2}\psi''(y) + v(y)\,\psi'(y) + i\,V(y)\,\psi(y) = 0.$$

Then Itô's formula shows that
$$M_t := \psi(Y_t) \, \exp\left\{ i \int_0^t V(Y_s)\, ds \right\},$$
is a bounded martingale. If furthermore, we know that $Y_t \to \infty$ w.p.1 and $\psi(y) \to 1$ as $y \to \infty$, then the martingale convergence theorem shows that

(1.34) $$\psi(y) = \mathbf{E}^y[M_0] = \mathbf{E}^y[M_\infty] = \mathbf{E}^y\left[\exp\left\{i \int_0^\infty V(Y_t)\, dt\right\}\right] = \mathbf{E}^y[e^{iS}].$$

In other words, the characteristic function of S can be computed by finding solutions to the equation (1.33).

1.13. Modulus of continuity

If $f : [0, \infty) \to \mathbb{R}^d$ is a function, then the *modulus of continuity* or *oscillation* is defined by
$$\mathrm{osc}(f, \delta) = \mathrm{osc}(f, \delta, t_0) = \sup\{|f(t) - f(s)| : 0 \leq s, t \leq t_0, |t - s| \leq \delta\}.$$
In this section we will consider $\mathrm{osc}(B_\delta) = \mathrm{osc}(B, \delta, 1)$, the modulus of continuity of Brownian motion. We will restrict ourselves to the case $d = 1$; results for other d follow from considering the different components separately. It will be easier to consider the random variable
$$M_n = \max_{j=0,\ldots,2^n-1} \sup_{0 \leq t \leq 2^{-n}} |B_{j\,2^{-n}+t} - B_{j\,2^{-n}}|.$$
Note that

(1.35) $$M_{n+1} \leq \mathrm{osc}(B, \delta) \leq 3 M_n \quad \text{for } 2^{-(n+1)} \leq \delta \leq 2^{-n}.$$

PROPOSITION 1.37. *If B_t is a one-dimensional Brownian motion and M_n is defined as above, then w.p.1,*
$$\lim_{n \to \infty} \frac{M_n}{\sqrt{2^{-n} \log(2^n)}} = \sqrt{2}.$$

PROOF. If X is a $N(0,1)$ random variable and $a \geq 1$, then it is easy to verify that
$$e^{-(a+1)^2/2} \leq \sqrt{2\pi}\, \mathbf{P}\{X \geq a\} \leq e^{-a^2/2}.$$
If $c > 0$, then
$$\begin{aligned}
\mathbf{P}\{M_n \leq c\, 2^{-n/2}\, \sqrt{n}\} &\leq \prod_{j=0}^{2^n-1} \mathbf{P}\{|B_{(j+1)\,2^{-n}} - B_{j\,2^{-n}}| \leq c\, 2^{-n/2}\, \sqrt{n}\} \\
&= [\mathbf{P}\{|B_1| \leq c\, \sqrt{n}\}]^{2^n} \\
&\leq [\, 1 - (\sqrt{2/\pi})e^{-(c\sqrt{n}+1)^2/2}\,]^{2^n} \\
&\leq \exp\{-2^n\, (\sqrt{2/\pi})\, e^{-(c\sqrt{n}+1)^2/2}\}
\end{aligned}$$
If we set $C_n = \sqrt{2\log 2} - 2\, n^{-1/2}$, then
$$\sum_{n=1}^{\infty} \mathbf{P}\{M_n \leq C_n\, 2^{-n/2}\, \sqrt{n}\} < \infty,$$
and hence by the Borel-Cantelli Lemma, w.p.1
$$\liminf_{n \to \infty} \frac{M_n}{\sqrt{2^{-n}\, n}} \geq \sqrt{2\, \log 2}.$$

For the other direction, we will use the "reflection principle" for one-dimensional Brownian motion (see, e.g., [**5**, Theorem I.3.8]): if $a > 0$, then
$$\mathbf{P}\{\sup_{0 \leq s \leq t} B_s \geq a\} = 2\, \mathbf{P}\{B_t \geq a\}.$$
Then if $c > \sqrt{2\, \log 2}$,
$$\begin{aligned}
\mathbf{P}\{M_n \geq c\, 2^{-n/2}\, \sqrt{n}\} &\leq 2^n\, \mathbf{P}\{\sup_{0 \leq t \leq 2^{-n}} |B_t| \geq c\, 2^{-n/2}\, \sqrt{n}\} \\
&\leq 2^n \cdot 4\, \mathbf{P}\{B_1 \geq c\, \sqrt{n}\} \\
&\leq (4/\sqrt{2\pi})\, 2^n\, e^{-c^2 n/2} \\
(1.36) \qquad &\leq (4/\sqrt{2\pi})\, \exp\{-n\, (c - \sqrt{2\, \log 2})^2/2\}
\end{aligned}$$
If we choose $C_n = \sqrt{2\log 2} + n^{-1/4}$, we get
$$\sum_{n=1}^{\infty} \mathbf{P}\{M_n \geq C_n\, 2^{-n/2}\, \sqrt{n}\} < \infty,$$
and hence by the Borel-Cantelli Lemma, w.p.1
$$\limsup_{n \to \infty} \frac{M_n}{\sqrt{2^{-n}\, n}} \leq \sqrt{2\, \log 2}.$$
$\square$

COROLLARY 1.38. *If B_t is a one-dimensional Brownian motion, then w.p.1,*
$$1 \leq \liminf_{\delta \to 0+} \frac{\mathrm{osc}(B, \delta)}{\sqrt{\delta\, \log(1/\delta)}} \leq \limsup_{\delta \to 0+} \frac{\mathrm{osc}(B, \delta)}{\sqrt{\delta\, \log(1/\delta)}} \leq 6.$$
Moreover, if $a \geq 12\, \sqrt{\log 2}$ and $\delta \in (0, 1)$,
$$\mathbf{P}\{\mathrm{osc}(B, \delta) \geq a\, \sqrt{\delta\, \log(1/\delta)}\} \leq (4/\sqrt{2\pi})\, (2\delta)^{a^2/144}.$$

1.13. MODULUS OF CONTINUITY

PROOF. If $2^{-(n+1)} \leq \delta \leq 2^{-n}$, then (1.35) gives
$$\frac{M_{n+1}}{\sqrt{2^{-n}\log 2^{n+1}}} \leq \frac{\operatorname{osc}(B,\delta,1)}{\sqrt{\delta\ \log(1/\delta)}} \leq \frac{3M_n}{\sqrt{2^{-(n+1)}\log 2^n}}.$$
So the first statement follows from Proposition 1.37. Note that (1.36) shows that for all $a \geq 2\sqrt{2\log 2}$ and all n,
$$\mathbf{P}\{M_n \geq a\, 2^{-n/2}\sqrt{n}\} \leq (4/\sqrt{2\pi})\exp\{-a^2 n/8\}.$$
Hence, if $a \geq 12\sqrt{\log 2}$
$$\begin{aligned}
\mathbf{P}\{\operatorname{osc}(B,\delta) \geq a\sqrt{\delta\ \log(1/\delta)}\} &\leq \mathbf{P}\{M_n \geq (3\sqrt{2})^{-1} a\, 2^{-n/2}\sqrt{\log 2^n}\} \\
&\leq (4/\sqrt{2\pi})\exp\{-a^2(\log 2)n/144\} \\
&= (4/\sqrt{2\pi})(2^{-n})^{a^2/144} \leq (4/\sqrt{2\pi})(2\delta)^{a^2/144}
\end{aligned}$$
$\square$

REMARK 1.39. The last corollary can be improved to get Lévy's theorem on the modulus of continuity of Brownian motion, with probability one,
$$\lim_{\delta \to 0+} \frac{\operatorname{osc}(B,\delta)}{\sqrt{\delta\ \log(1/\delta)}} = \sqrt{2}.$$
See, for example, [**39**, Theorem 2.9.25]. Corollary 1.38 will suffice for applications in this book.

A function $f : [0,T] \to \mathbb{C}$ is *Hölder α-continuous* if there is a c such that for all $0 \leq s, t \leq T$, $|f(t) - f(s)| \leq c\,|t-s|^\alpha$, i.e., if $\operatorname{osc}(f,\delta,T) \leq c\,\delta^\alpha$ for all $\delta > 0$. The smallest such c is sometimes called the *Hölder-α norm* of f. The following corollary follows immediately from Corollary 1.38.

COROLLARY 1.40. *With probability one, $B_t, 0 \leq t \leq 1$, is Hölder α-continuous for all $\alpha < 1/2$, but not Hölder $1/2$-continuous.*

CHAPTER 2

Complex Brownian motion

2.1. Review of complex analysis

A *domain* in $\mathbb{C}$ is an open, connected subset of $\mathbb{C}$. If D is a domain, a function $f : D \to \mathbb{C}$ is *analytic* or *holomorphic* in D if the complex derivative

$$f'(z) = \lim_{w \to z} \frac{f(w) - f(z)}{w - z}$$

exists at every $z \in D$. If $f = u + iv$, then f is analytic at z if and only if f is C^1 and satisfies the Cauchy-Riemann equations $\partial_x u = \partial_y v, \partial_y u = -\partial_x v$. If f is analytic at 0, $f(0) = 0$ and $f'(0) = re^{i\theta}$ with $r > 0$, then locally at 0, f looks like a dilation by r and a rotation by θ.

A *curve* in $\mathbb{C}$ will be a continuous function $\gamma : [a, b] \to \mathbb{C}$; here $[a, b]$ is a closed interval which can be infinite in either direction. The term curve will always mean continuous curve but will imply no further smoothness. The curve is C^k or *smooth* if γ is C^k or infinitely differentiable, respectively, for all $a < t < b$. The curve is *closed* if $-\infty < a < b < \infty$ and $\gamma(a) = \gamma(b)$. The curve is *simple* if $\gamma(s) \neq \gamma(t)$ for $a \leq s < t < b$. A simple closed curve is also called a *Jordan curve*.

If $\gamma : [a, b] \to \mathbb{C}$ is a piecewise C^1 curve, then the integral of f along γ is defined:

$$\int_\gamma f(z)\, dz = \int_a^b f(\gamma(t))\, \gamma'(t)\, dt.$$

This is independent of the choice of parametrization, in the sense that if $h : [c, d] \to [a, b]$ is a smooth, increasing homeomorphism and $\tilde\gamma(t) = \gamma(h(t))$, then

$$\int_{\tilde\gamma} f(z)\, dz = \int_\gamma f(z)\, dz.$$

If f is analytic in D, $D' \subset D$, and $\partial D'$ is a closed C^1 curve γ, then $\int_\gamma f(z)\, dz = 0$. Let $\mathcal{B}(z, \epsilon)$ denote the open disk and $\overline{\mathcal{B}}(z, \epsilon)$ the closed disk of radius ϵ centered at z. If f is analytic in D and $\overline{\mathcal{B}}(z, \epsilon) \subset D$, then the power series representation

$$f(w) = \sum_{n=0}^\infty a_n (w - z)^n, \quad a_n = \frac{f^{(n)}(z)}{n!},$$

is valid for $w \in \overline{\mathcal{B}}(z, \epsilon)$. It follows that f is locally one-to-one at z if and only if $f'(z) \neq 0$. By dividing by $(w - z)$ and integrating we get the Cauchy integral formula

$$f(z) = \frac{1}{2\pi i} \int_{\partial \mathcal{B}(z,\epsilon)} \frac{f(w)}{w - z}\, dw.$$

Here we use the convention that the curve $\partial \mathcal{B}(z,\epsilon)$ is traversed exactly one time in the counterclockwise direction. By differentiating, we get the more general form

$$f^{(n)}(z) = \frac{n!}{2\pi i} \int_{\partial \mathcal{B}(z,\epsilon)} \frac{f(w)}{(w-z)^{n+1}} \, dw,$$

which, in particular, implies that all derivatives of analytic functions are analytic. The Cauchy-Riemann equations imply that $\Delta u = \Delta v = 0$, i.e., u, v are harmonic on D. The Cauchy integral formula can be used to prove the *maximum principle* which states that $|f(z)|$ has no local maxima in D unless f is constant. We say that $f: D \to D'$ is a *conformal transformation* if f is analytic, one-to-one, and onto. It follows that $f'(z) \neq 0$ for $z \in D$, and $f^{-1}: D' \to D$ is also a conformal transformation.

If $\Phi = (\Phi_1, \Phi_2)$ is a vector field in $\mathbb{R}^2 \, (= \mathbb{C})$ we write

$$\int_\gamma \Phi \cdot dz = \int_a^b [\, \Phi_1(\gamma(t)) \, x'(t) + \Phi_2(\gamma(t)) \, y'(t) \,] \, dt,$$

where $\gamma(t) = x(t) + iy(t)$. If $f = u + iv$, then

$$(2.1) \qquad \int_\gamma f(z) \, dz = \int_\gamma (u, -v) \cdot dz + i \int_\gamma (v, u) \cdot dz.$$

If γ is a simple closed curve traversed counterclockwise bounding the bounded domain D', Green's Theorem states

$$\int_\gamma \Phi \cdot dz = \int_{D'} [\partial_x \Phi_2 - \partial_y \Phi_1] \, dA,$$

where A denotes area.

The *Riemann sphere* is the set $\hat{\mathbb{C}} = \mathbb{C} \cup \{\infty\}$ with the usual topology where a set $\mathcal{N}$ is a neighborhood of ∞ if $\mathbb{C} \setminus \mathcal{N}$ is bounded. A *linear fractional transformation* is a function of the form

$$f(z) = \frac{az+b}{cz+d}, \quad a,b,c,d \in \mathbb{C}, \; ad - bc \neq 0.$$

If we choose the representative of f such that $ad - bc = 1$, we get a one-to-one correspondence between the set of linear functionals and $SL_2(\mathbb{C})$, the set of 2×2 complex-valued matrices with determinant one. This correspondence is, in fact, a group isomorphism, i.e., $f_{A_1} \circ f_{A_2} = f_{A_1 A_2}$. Linear fractional transformations map $\hat{\mathbb{C}}$ one-to-one onto $\hat{\mathbb{C}}$ and map circles to circles (lines in $\mathbb{C}$ are circles in $\hat{\mathbb{C}}$). We will use the term *Möbius transformation* for linear fractional transformations from $\mathbb{D}$ onto $\mathbb{D}$ or from $\mathbb{H}$ onto $\mathbb{H}$ where

$$\mathbb{D} = \{z \in \mathbb{C} : |z| < 1\}, \quad \mathbb{H} = \{z \in \mathbb{C} : \mathrm{Im}(z) > 0\}.$$

The Möbius transformations from $\mathbb{D}$ onto $\mathbb{D}$ are of the form

$$(2.2) \qquad f(z) = e^{ia} \frac{z - w}{1 - \overline{w} z}, \quad w \in \mathbb{D}, a \in \mathbb{R}.$$

The Möbius transformations from $\mathbb{H}$ onto $\mathbb{H}$ are of the form

$$f(z) = \frac{az+b}{cz+d}, \quad a,b,c,d \in \mathbb{R}, \; ad - bc > 0.$$

In other words the Möbius transformations from $\mathbb{H}$ onto $\mathbb{H}$ correspond to $SL_2(\mathbb{R})$.

LEMMA 2.1 (Schwarz lemma). *If $f : \mathbb{D} \to \mathbb{D}$ is analytic with $f(0) = 0$, then $|f(z)| \leq |z|$ for all z. Moreover, if $|f(z)| = |z|$ for some $z \neq 0$, then $f(z) = e^{i\theta} z$ for some $\theta \in \mathbb{R}$.*

PROOF. Apply the maximum principle to $f(z)/z$. □

As a corollary, the only conformal transformations of $\mathbb{D}$ onto $\mathbb{D}$ fixing the origin are rotations, and, more generally, the only conformal transformations of $\mathbb{D}$ onto $\mathbb{D}$ are the Möbius transformations.

2.2. Conformal invariance of Brownian motion

Let $B_t = B_t^1 + iB_t^2$ be a complex Brownian motion. Suppose $D \subset \mathbb{C}$ is a domain and $f : D \to \mathbb{C}$ is a nonconstant holomorphic function. Let
$$\tau_D = \inf\{t \geq 0 : B_t \notin D\}.$$
If $z \in D$ with $f'(z) \neq 0$, then locally f looks like a dilation by $|f'(z)|$ composed with a rotation. Brownian motion is invariant under rotations, and dilations only change the time parametrization (Exercise 1.1). This is the basic idea in the proof of the following fundamental fact about complex Brownian motion.

THEOREM 2.2. *Suppose D is a domain and $f : D \to \mathbb{C}$ is a nonconstant analytic function. Suppose B_t is a (complex) Brownian motion starting at $z \in D$, and define*
$$S_t = \int_0^t |f'(B_r)|^2 \, dr, \quad 0 \leq t < \tau_D.$$
Let $\sigma_s = S_s^{-1}$, i.e.,
$$\int_0^{\sigma_s} |f'(B_r)|^2 \, dr = s.$$
Then
$$Y_s := f(B_{\sigma_s}), \quad 0 \leq s < S_{\tau_D},$$
has the same distribution as that of a Brownian motion starting at $f(z)$ stopped at S_{τ_D}.

PROOF. Write $f = u + iv$. The Cauchy-Riemann equations imply that $u_x = v_y, u_y = -v_x$; in particular, u and v are harmonic. Itô's formula and the Cauchy-Riemann equations give
$$d[u(B_t)] = u_1(B_t) \, dB_t^1 + u_2(B_t) \, dB_t^2,$$
$$d[v(B_t)] = -u_2(B_t) \, dB_t^1 + u_1(B_t) \, dB_t^2.$$
Then $u(B_t)$ and $v(B_t)$ are local martingales with
$$\langle u(B) \rangle_t = \langle v(B) \rangle_t = \int_0^t ([u_1(B_s)]^2 + [u_2(B_s)]^2) \, ds = \int_0^t |f'(B_s)|^2 \, ds.$$
Note also that $\langle u(B), v(B) \rangle_t = 0$. The remainder of the proof follows that in Proposition 1.14. □

REMARK 2.3. If f is nonconstant on D, then the zeroes of f' are isolated, and hence S_t is strictly increasing for $0 \leq t < \tau_D$. Therefore S_s^{-1} is well defined. Also, if f is one-to-one, then $S_{\tau_D} = \tau_{f(D)}$.

2.3. Harmonic functions

Suppose D is a domain with boundary ∂D. A point $z \in \partial D$ is called *regular (for D)* if
$$\mathbf{P}^z\{\tilde{\tau}_D = 0\} = 1,$$
where $\tilde{\tau}_D = \inf\{t > 0 : B_t \notin D\}$. An example of a point that is not regular is the origin if $D = \{0 < |z| < 1\}$. A domain D is called regular if ∂D is nonempty and every point in ∂D is regular. We say that ∂D *contains a curve at z* if there exists a curve $\gamma : [0,1] \to \mathbb{C}$ with $\gamma(0) = z, \gamma(1) \neq z$, $\gamma[0,1] \subset \partial D$.

EXERCISE 2.4. Show that $z \in \partial D$ is regular if and only if for every $\delta > 0$,
$$\lim_{\epsilon \to 0+} \sup_{|w-z| < \epsilon} \mathbf{P}^w\{|B_{\tau_D} - z| \geq \delta\} = 0.$$

EXERCISE 2.5. Show that there exist $0 < a, c < \infty$ such that the probability that a complex Brownian motion starting at ϵ makes a closed loop about the origin before reaching the circle of radius 1 is at least $1 - c\epsilon^a$. (Hint: consider the probability that a Brownian motion starting at 2^{-n} makes a closed loop about 0 before reaching the circle of radius 2^{-n+1}.) [Note: it is a harder problem to determine the best a for this result. See Chapter 8.]

EXERCISE 2.6. Suppose ∂D contains a curve at z. Show that z is a regular point.

EXERCISE 2.7. There exists a $c > 0$, such that if K is a connected subset of $\mathbb{D}$, then
$$\mathbf{P}^0\{B[0, \tau_{\mathbb{D}}] \cap K \neq \emptyset\} \geq c \operatorname{diam}(K).$$
(Hint: Consider two cases depending on whether or not $K \subset \{|z| \geq 1 - \operatorname{diam}(K)\}$.)

Recall that a function $u : D \to \mathbb{R}$ is *harmonic* if it is C^2 and
$$\Delta u(z) := \partial_{xx} u(z) + \partial_{yy} u(z) = 0$$
for all $z \in D$. If $\overline{\mathcal{B}}(z, \epsilon) \subset D$, define the spherical mean value $MV(u, z, \epsilon)$ by
$$MV(u, z, \epsilon) = \mathbf{E}^z[u(B_\eta)] = \frac{1}{2\pi} \int_0^{2\pi} u(z + \epsilon e^{i\theta})\, d\theta,$$
where $\eta = \eta_{z,\epsilon} = \inf\{t \geq 0 : |B_t - z| = \epsilon\}$. Recall (see §1.8.1), that if u is harmonic in D, then $u(B_{t \wedge \tau_D})$ is a local martingale. By letting $B_0 = z$ and stopping the martingale at time η, we see that u satisfies the *(spherical) mean value property*, i.e., $MV(u, z, \epsilon) = u(z)$ for all ϵ such that $\overline{\mathcal{B}}(z, \epsilon) \subset D$. The next two exercises give the converse to this statement.

EXERCISE 2.8. Show that if $u : D \to \mathbb{R}$ is C^2 and $z \in D$, then
$$\Delta u(z) = \lim_{\epsilon \to 0+} \epsilon^{-2}[MV(u, z, \epsilon) - u(z)].$$

EXERCISE 2.9. (i) Suppose $u : D \to \mathbb{R}$ is a bounded measurable function that satisfies the spherical mean value property. Show that u is C^∞. Hint: fix $z \in D$ and assume $\mathcal{B}(z, 2\epsilon) \subset D$. Let ψ be a nonnegative radially symmetric C^∞ function

on $\mathbb{C}$ that vanishes on $\{w : |w| \geq \epsilon\}$ and whose integral (with respect to area) is 1. Show that in a neighborhood of z,
$$u(z) = \int_{\mathbb{C}} \psi(w-z)\, u(w)\, dA(w),$$
where A denotes area. Show that the right-hand side is C^∞.

(ii) Let $u : D \to \mathbb{R}$ such that for every $z \in D$ there is a neighborhood around z such that u is bounded and satisfies the mean value property. Show that u is harmonic in D. (Hint: use (i) and Exercise 2.8.)

PROPOSITION 2.10. *Suppose D is a domain such that ∂D has at least one regular point, and $F : \partial D \to \mathbb{R}$ is a bounded, measurable function. Define $u : \overline{D} \to \mathbb{R}$ by $u(z) = F(z)$ for $z \in \partial D$ and*

(2.3) $$u(z) = \mathbf{E}^z[F(B_{\tau_D})], \quad z \in D.$$

Then u is a bounded, harmonic function in D and is continuous at all regular points $z \in \partial D$ at which F is continuous.

PROOF. It is obvious that $\|u\|_\infty = \|F\|_\infty < \infty$. The rotational symmetry of Brownian motion and the strong Markov property show that u satisfies the mean value property in D and hence is harmonic in D. Finally, the continuity at regular points follows from Exercise 2.4. □

PROPOSITION 2.11. *Suppose D is a regular domain, and $F : \partial D \to \mathbb{R}$ is a bounded, continuous function. There is a unique bounded, continuous function $u : \overline{D} \to \mathbb{R}$ that is harmonic in D and agrees with F on ∂D. For $z \in D$, u is given by (2.3).*

PROOF. We have already seen that u as defined by (2.3) is continuous in $\overline{D}$ and harmonic in D. Suppose g were another such function. Then (see §1.8.1), $M_t = g(B_{t \wedge \tau_D})$ is a bounded, continuous martingale. By the optional sampling theorem, if $z \in D$,
$$g(z) = \mathbf{E}^z[M_0] = \mathbf{E}^z[M_{\tau_D}] = \mathbf{E}^z[F(B_{\tau_D})].$$
□

REMARK 2.12. It is important in the last proposition that we assumed that u is bounded. For example, if $D = \{|z| > 1\}$ and $F \equiv 0$, the bounded solution is $u \equiv 0$, but the function $u(x) = \log|x|$ is also harmonic in D, continuous on $\overline{D}$, and equal to F on ∂D. If $u : \overline{D} \to [0, \infty)$ is continuous and harmonic in D, it is true that

(2.4) $$u(z) \geq \mathbf{E}^z[u(B_{\tau_D})].$$

To see this, let $T_n = \inf\{t : |B_t| = n\}$. Since u is bounded on $\overline{D} \cap \{|z| \leq n\}$, if $z \in D$ and $|z| < n$, then
$$u(z) = \mathbf{E}^z[u(B_{\tau_D \wedge T_n})] \geq \mathbf{E}^z[u(B_{\tau_D}); \tau_D \leq T_n].$$
Letting $n \to \infty$ and using monotone convergence, we get (2.4).

EXERCISE 2.13. Let $D = \{z : r < |z| < R\}$ and let $z \in D$. Show that
$$\mathbf{P}^z\{|B_{\tau_D}| = R\} = \frac{\log|z| - \log r}{\log R - \log r}.$$

DEFINITION 2.14. If D is a domain such that ∂D has at least one regular point and $z \in D$, then *harmonic measure in D from z* is the probability measure on ∂D, $\operatorname{hm}(z, D; \cdot)$ given by
$$\operatorname{hm}(z, D; V) = \mathbf{P}^z\{B_{\tau_D} \in V\}.$$

REMARK 2.15. A standard notation for hm in complex analysis is ω, but this has other meanings in probability. One can also say harmonic measure "on ∂D" instead of "in D". Equation (2.3) can be written
$$u(z) = \int_{\partial D} F(w) \operatorname{hm}(z, D; dw).$$

We say that ∂D is *locally analytic* at $z \in \partial D$ if there exists a one-to-one analytic function $f : \mathbb{D} \to \mathbb{C}$ with $f(0) = z$ and such that
$$f(\mathbb{D}) \cap D = f(\{z \in \mathbb{D} : \operatorname{Im}(z) > 0\}).$$

We say that ∂D is *piecewise analytic* if it is locally analytic except perhaps at a finite number of points. If ∂D is locally analytic at z, $\partial D'$ is locally analytic at w, and $f : D \to D'$ is a conformal transformation with $f(z) = w$ (in the sense that $f(z_n) \to w$ if $z_n \to z$), then the Schwarz reflection principle (see, e.g., [**1**, Theorem 4.24, Thoerem 6.4]) shows that f can be extended to be a conformal transformation of $D \cup \mathcal{B}(z, \epsilon)$ for some $\epsilon > 0$.

If ∂D is piecewise analytic, then $\operatorname{hm}(z, D; \cdot)$ is absolutely continuous with respect to one-dimensional Lebesgue measure (length)[1]. The density of $\operatorname{hm}(z, D; \cdot)$ with respect to length is called the *Poisson kernel* and will be denoted $H_D(z, w)$. Equation (2.3) becomes

$$(2.5) \qquad u(z) = \int_{\partial D} F(w)\, H_D(z, w)\, |dw|.$$

The Poisson kernel is characterized by the following. If $w \in \partial D$ and ∂D is locally analytic at w:

- $H_D(\cdot, w)$ is a harmonic function in D;
- If $z_n \in D$ with $z_n \to w$, then $H_D(z_n, \cdot)$ approaches the delta function at w.

EXAMPLE 2.16. Suppose $D = r\mathbb{D} = \{|z| < r\}$. Then

$$(2.6) \qquad H_{r\mathbb{D}}(z, w) = \frac{1}{2\pi r} \frac{r^2 - |z|^2}{|w - z|^2} = \frac{1}{2\pi r} \operatorname{Re} \frac{w + z}{w - z}, \quad |z| < r,\ |w| = r.$$

EXERCISE 2.17. Show that for every positive integer k there is a positive number $c(k)$ such that if D is a domain, $z \in D$, $u : D \to \mathbb{R}$ is harmonic, and j is a nonnegative integer no larger than k, then
$$|\partial_x^j \partial_y^{k-j} u(z)| \leq c(k)\, \operatorname{dist}(z, \partial D)^{-k}\, \|u\|_\infty.$$

(Hint: one may assume that $z = 0$ and $D = \mathbb{D}$. Use (2.5) and (2.6).)

---

[1] We omit the proof of this fact. It is obvious for $\mathbb{D}$ and conformal invariance establishes it at locally analytic boundary points.

REMARK 2.18. Suppose u is a harmonic function in $\mathbb{D}_+ := \mathbb{D} \cap \{x+iy : y > 0\}$, continuous on $\overline{\mathbb{D}}_+$, whose boundary value is 0 on $\mathbb{R} \cap \partial \mathbb{D}_+$. Then we can extend u to a harmonic function on $\mathbb{D}$ by $u(\bar{z}) = -u(z)$. To see this, note that there is a unique harmonic function v on $\mathbb{D}$ with this boundary value, and symmetry shows that v must be zero on $\mathbb{R}$. Hence $v = u$ on $\mathbb{D}_+$.

The following follows immediately from Theorem 2.2.

PROPOSITION 2.19. *Suppose $f : D \to D'$ is a conformal transformation that is continuous and one-to-one on $\overline{D}$. Then if $z \in D, V \subset \partial D$,*
$$\mathrm{hm}(f(z), D'; f(V)) = \mathrm{hm}(z, D; V).$$

REMARK 2.20. We will sometimes write the conclusion of this proposition as
$$f \circ \mathrm{hm}(z, D) = \mathrm{hm}(f(z), D').$$

COROLLARY 2.21. *Suppose $f : D \to D'$ is a conformal transformation and $u : D' \to \mathbb{R}$ is a harmonic function. Then $v(z) = u(f(z))$ is a harmonic function on D.*

EXERCISE 2.22. Suppose $f : D \to D'$ is a conformal transformation, $z \in D$, $w \in \partial D$, ∂D is locally analytic at w, and $\partial D'$ is locally analytic at $f(w)$. Show that
$$H_{D'}(f(z), f(w)) = |f'(w)|^{-1} H_D(z, w).$$

EXERCISE 2.23. Let $D = \mathbb{H} = \{x + iy : y > 0\}$. Show that

(2.7) $$H_{\mathbb{H}}(x + iy, x') = \frac{1}{\pi} \frac{y}{(x - x')^2 + y^2}, \quad -\infty < x, x' < \infty, \ 0 < y < \infty.$$

(Hint: Use Exercise 2.22. The function $f(z) = i(1+z)/(1-z)$ transforms $\mathbb{D}$ onto $\mathbb{H}$.) Use this to show that for $z \in \mathbb{H}$,

(2.8) $$\mathbf{P}^z\{B_{\tau_\mathbb{H}} \leq 0\} = \frac{1}{\pi} \arg(z).$$

EXAMPLE 2.24. Let D be the half-infinite strip $\{x + iy : x > 0, 0 < y < \pi\}$, and $w = iq$ with $q \in (0, \pi)$. Then $H_D(\cdot, iq)$ is the unique harmonic function in D, vanishing as $\mathrm{Re}(z) \to \infty$, with boundary value the delta function at iq. Using separation of variables (see, e.g., [8]), we can write this function explicitly:
$$H_D(x + iy, iq) = \frac{2}{\pi} \sum_{n=1}^{\infty} e^{-nx} \sin(nq) \sin(ny).$$

From this series it is easy to check that there is a constant c such that for all $x \geq 1$ and all $0 < y < \pi$,

(2.9) $$\left| H_D(x + iy, iq) - \frac{2}{\pi} e^{-x} \sin q \sin y \right| \leq c\, e^{-2x} \sin q \sin y.$$

We can write this as

(2.10)
$$H_D(x + iy, iq) = \frac{2}{\pi} e^{-x} \sin q \sin y\, [1 + O(e^{-x})], \quad 1 \leq x < \infty, \ 0 < y, q < \pi,$$

where $O(e^{-x})$ denotes an error term which can depend on x, q, y but is bounded in absolute value by $c\, e^{-x}$ for some c independent of x, y, q. Also, for $x > 0$,
$$\int_0^\pi H_D(x+iy, iq) \sin y\, dy = e^{-x} \sin q$$
The probability density $(1/2) \sin q$, $0 < q < \pi$, is the density of the hitting measure of $[0, i\pi]$ by "Brownian motion in D starting at $x = \infty$ conditioned to leave D at $[0, i\pi]$". More precisely, let
$$\overline{H}_D(x+iy, iq) := \frac{H_D(x+iy, iq)}{\int_0^\pi H_D(x+iy, iq')\, dq'}$$
be the conditional density of B_{τ_D} given that $B_{\tau_D} \in [0, i\pi]$. Then there is a constant c such

(2.11) $\quad |\overline{H}_D(x+iy, iq) - \dfrac{1}{2} \sin q| \leq c\, e^{-x} \sin q, \quad 1 \leq x < \infty,\ 0 < y, q < \pi.$

In fact (as pointed out in the next example), conformal mapping can be used to give an exact expression for $H_D(x+iy, iq)$. However, the approximations derived here will be all that we need, and the method of series expansion has the advantage that it works in dimensions other than two.

EXAMPLE 2.25. Let
$$D = \{x+iy : x^2 + y^2 > 1, y > 0\}.$$
This is the image of the half-infinite strip above under the map $z \mapsto e^z$. Then (2.10) can be rewritten

(2.12) $\quad H_D(z, e^{i\theta}) = \dfrac{2}{\pi} \dfrac{\operatorname{Im}(z)}{|z|^2} \sin\theta\, [1 + O(|z|^{-1})].$

In fact, the map $z \mapsto z + z^{-1}$ maps D conformally onto $\mathbb{H}$ sending the upper half circle onto $[-2, 2]$. Using Exercises 2.22 and 2.23, we can find $H_D(z, e^{i\theta})$ exactly. For the purposes of this book, the expression in (2.12) will suffice.

PROPOSITION 2.26 (Harnack principle). *Suppose D is a domain and $u: D \to (0, \infty)$ is a positive harmonic function.*
(i) If $D = \mathbb{D}$ and $r < 1$, then
$$\frac{1-r}{1+r} u(0) \leq u(z) \leq \frac{1+r}{1-r} u(0), \quad |z| \leq r.$$
(ii) For every compact $K \subset D$ there is a $C(D, K) < \infty$, independent of u, such that
$$u(z) \leq C(D, K)\, u(w), \quad z, w \in K.$$

PROOF. The first assertion follows from (2.5) and (2.6). For the second assertion we will prove more. Call $z, w \in D$ adjacent if
$$|z - w| < \frac{1}{2} \max\{\operatorname{dist}(z, \partial D), \operatorname{dist}(w, \partial D)\}.$$
By (i), $u(z) \leq 3\, u(w)$ for adjacent points. Let $d = d_D$ denote the corresponding graph distance on D, i.e., $d(z, w)$ is the smallest integer k such that there is a finite sequence $z = z_0, z_1, \ldots, z_k = w$ such that z_j is adjacent to z_{j-1} for $j = 1, \ldots, k$. Then,
$$u(z) \leq 3^{d(z,w)}\, u(w).$$

If we fix some $z \in D$ and let $U_k = \{w \in D : d(z, w) \leq k\}$, then $U_1, U_2, \ldots$ is an increasing sequence of open sets whose union is D. By compactness, $K \subset U_k$ for some k. □

EXERCISE 2.27. Suppose $r < 1$ and in the preceding proof we called z, w r-adjacent in D if

$$|z - w| < r \max\{\text{dist}(z, \partial D), \text{dist}(w, \partial D)\}.$$

Let $d_r = d_{r,D}$ denote the corresponding graph distance. Show that for every $\epsilon > 0$ there is a $C_\epsilon < \infty$ such that for all z, w, D and all $\epsilon \leq r \leq s \leq 1 - \epsilon$, $d_s(z, w) \leq d_r(z, w) \leq C_\epsilon d_s(z, w)$. In particular, if $u : D \to (0, \infty)$ is a positive harmonic function, then for all z, w, D and all $\epsilon \leq r \leq 1 - \epsilon$,

$$u(z) \leq \left(\frac{1+\epsilon}{1-\epsilon}\right)^{C_\epsilon d_r(z,w)} u(w).$$

PROPOSITION 2.28. *For every $r \in (0, 1)$, there exists a $c_r > 0$ such that if $D = \mathbb{D}_+ = \{x + iy : x^2 + y^2 < 1 : y > 0\}$ and $u : D \to (0, \infty)$ is a harmonic function such that for each $x \in (-1, 1)$, $\lim_{\epsilon \to 0+} u(x + i\epsilon) = 0$, then for all $z = x + iy \in D \cap r\mathbb{D}$,*

(2.13) $$u(x + iy) \geq c_r\, u(i/2)\, y.$$

In particular, $\partial_y u(x) \geq c_r u(i/2) > 0$ for $-r < x < r$.

PROOF. We can extend u to a harmonic function on $\mathbb{D}$ by $u(\bar{z}) = -u(z)$; this establishes the existence of $\partial_y u(x)$. Let $\delta = (1-r)/2$. Using the "gambler's ruin" estimate[2] on the imaginary part, it is easy to see that there is a $\rho_r > 0$ such that the probability that a Brownian motion starting at $x + iy \in D \cap r\mathbb{D}$ reaches $\{w \in D : \text{dist}(w, \partial D) > \delta\}$ without leaving D is bounded above by $\rho_r y$. Hence $u_y(x + iy) \geq \rho_r \inf\{u(w) : \text{dist}(w, \partial D) > \delta\} \geq c_r\, u(i/2)$, where the last inequality uses the Harnack principle. □

PROPOSITION 2.29. *Suppose u_n is a sequence of harmonic functions on a domain D that are locally bounded, i.e., for each $z \in D$ there exist C, ϵ such that $|u_n(w)| \leq C$ for $|w - z| \leq \epsilon$. Then there exists a subsequence u_{n_k} and a harmonic function u on D such that for each $z \in D$, $u_{n_k}(z) \to u(z)$.*

Similarly, if f_n is a sequence of analytic functions on D that are locally bounded, then there exists a subsequence f_{n_k} and an analytic function f such that for each $z \in D$, $f_{n_k}(z) \to f(z)$.

PROOF. Let Q be a countable dense subset of D. Since for each $z \in Q$, $\{u_n(z)\}$ is bounded, the Cantor diagonalization argument shows that there is a u and a subsequence $\{n_k\}$ such that $u_{n_k}(z) \to u(z)$ for all $z \in Q$. Using Exercise 2.17, we can see that $\partial_x u_n$ and $\partial_y u_n$ are uniformly bounded on compact sets, and from this we can extend u to D by continuity and $u_{n_k}(z) \to u(z)$ for all $z \in D$. It is easy to check that u satisfies the spherical mean value property (at least locally) and hence is harmonic. A similar proof works for analytic functions. □

---

[2]The gambler's ruin estimate states that if B_t is a standard one-dimensional Brownian motion with $B_0 = x \in (0, 1)$, then the probability that it hits 1 before 0 is x.

2.4. Green's function

For this section, let D be a regular domain, and, as before, let τ_D be the first time that Brownian motion leaves the domain. If $z, w \in D$ and $t > 0$, let $p_D(t, z, w)$ be the density (in w) of $B_{t \wedge \tau_D}$ assuming $B_0 = z$, i.e.,

(2.14) $$p_D(t, z, w) = \lim_{\epsilon \to 0+} (\pi \epsilon^2)^{-1} \mathbf{P}^z \{|B_t - w| \leq \epsilon; t < \tau_D\}.$$

We let $p_D(0, z, \cdot)$ be the delta function at z and we set $p_D(t, z, w) = 0$ if either z or w is not in D. Then the function $p_D(t, z, w)$ satisfies the following:

- $\dot{p}_D(t, z, w) = \frac{1}{2} \Delta_w p_D(t, z, w), \quad t > 0, \ z, w \in D,$
- $p_D(t, z, w) = 0$ if $z \notin D$ or $w \notin D$,
- As $t \to 0+$, $p_D(t, z, \cdot)$ approaches the delta function at z.
-
$$p_D(s + t, z, w) = \int_{\mathbb{C}} p_D(s, z, z') \, p_D(t, z', w) \, dA(z'),$$

where A denotes area.

Let $p(t, z, w) = (2\pi t)^{-1} \exp\{-|z - w|^2/2t\}$ be the transition density for Brownian motion in all of $\mathbb{C}$. Clearly $p(t, z, w) \geq p_D(t, z, w)$. Note that

$$\sup\{p(t, z, w) : 0 \leq t < \infty, |z - w| \geq r\} = (\pi e r^2)^{-1},$$

and hence

$$p_D(t, z, w) \geq p(t, z, w) - \sup\{p(s, z', w) : 0 \leq s \leq t, |z' - w| \geq \text{dist}(w, \partial D)\}$$
(2.15) $$\geq p(t, z, w) - (\pi e \, \text{dist}(w, \partial D)^2)^{-1}.$$

LEMMA 2.30. *Suppose D is a regular domain. If $z, w \in D$ and $t > 0$, then*

$$\lim_{\epsilon \to 0+} \epsilon^{-2} \mathbf{P}^z \{|B_t - w| \leq \epsilon; \, 0 < \text{dist}(B[0, t], \partial D) \leq \epsilon\} = 0.$$

PROOF. Assume that $z = 0$ and choose $r > 0$ so that $\text{dist}(w, \partial D) > 3r$. Then if $\epsilon < r$, $\text{dist}(z', \partial D) \leq r$, and $s > 0$,

$$\mathbf{P}^{z'}\{|B_s - w| \leq \epsilon\} \leq \pi \epsilon^2 \sup p(s', z', w') \leq \epsilon^2/(r^2 e),$$

where the supremum is over all $s' > 0$ and w' with $|z' - w'| \geq r$. If $\epsilon \in (0, r/2)$, let σ_ϵ be the first time t that $\text{dist}(B_t, \partial D) \leq \epsilon$ and let $\eta_\epsilon = \eta_{\epsilon, r}$ be the first time after σ that $|B_t - B_{\sigma_\epsilon}| > r/2$. Then,

$$\mathbf{P}^0\{|B_t - w| \leq \epsilon; \, 0 < \text{dist}(B[0, t], \partial D) \leq \epsilon\} \leq \mathbf{P}^0\{\eta_\epsilon < \tau \wedge t\} \, [\epsilon^2/(r^2 e)].$$

However, it is straightforward to use Exercise 2.4 to show that

$$\lim_{\epsilon \to 0+} \mathbf{P}^0\{\eta_\epsilon < \tau \wedge t\} = 0.$$

□

PROPOSITION 2.31. *If D is a regular domain and $z, w \in D$ and $t > 0$,*

(2.16) $$p_D(t, z, w) = p_D(t, w, z).$$

PROOF. Assume that $z = 0$. Let $B_s, 0 \leq s \leq t$ be a Brownian motion starting at 0 and let

$$\tilde{B}_s = w - (B_t - B_{t-s}), \quad 0 \leq s \leq t,$$

which is a Brownian motion starting at w. The event $\{|B_t - w| \leq \epsilon\}$ is the same as the event $\{|\tilde{B}_t| \leq \epsilon\}$. Note that the set $\tilde{B}[0, t]$ is a translation of the set $B[0, t]$ by a number of absolute value at most ϵ. Hence the symmetric difference of the events $\{B[0,t] \subset D\}$ and $\{\tilde{B}[0,t] \subset D\}$ is contained in

$$\{0 < \text{dist}(B[0,t], \partial D) \leq \epsilon\} \cup \{0 < \text{dist}(\tilde{B}[0,t], \partial D) \leq \epsilon\}.$$

Hence, by Lemma 2.30, the probability of the symmetric difference of the events

$$\{|B_t - w| \leq \epsilon, B[0,t] \subset D\}, \quad \{|\tilde{B}_t| \leq \epsilon, \tilde{B}[0,t] \subset D\}$$

is $o(\epsilon^2)$ as $\epsilon \to 0+$. The proposition follows from (2.14). $\square$

LEMMA 2.32. *Suppose D is a regular domain. Then for every $z, w \in D$, there exists a $c = c(z, w, D) < \infty$ such that for $t > 1$,*

$$\mathbf{P}^z\{\tau_D > t\}, \mathbf{P}^w\{\tau_D > t\} \leq c\, (\log t)^{-1},$$

(2.17) $$p_D(t, z, w) \leq c\, t^{-1} (\log t)^{-2}.$$

PROOF. Assume that 0 is a regular point of the boundary and that $|z| = 1$ (if $|z| = a > 0$, then $\mathbf{P}^z\{\tau_D > t\} = \mathbf{P}^{z/a}\{\tau_{a^{-1}D} > t/a^2\}$.) Let T_R be the first time that the Brownian motion reaches the circle of radius R. Then,

$$\mathbf{P}\{T_{t^{1/4}} > t\} \leq \mathbf{P}^z\{|B_t| \leq t^{1/4}\} = \int_{|z'| < t^{1/4}} p(t, z, z')\, dA(z') \leq 2 t^{-1/2}.$$

Hence, it suffices to show that

(2.18) $$\mathbf{P}^z\{\tau_D > T_R\} \leq c\, (\log R)^{-1}.$$

Since 0 is a regular point, it is easy to show that there is a $\rho = \rho(D) > 0$ such that the probability that a Brownian motion starting on the unit circle leaves D before reaching the circle of radius 2 is at least ρ. Let q_R be the supremum of $\mathbf{P}^{z'}\{\tau_D > T_R\}$ where the supremum is over $|z'| = 1$. By Exercise 2.13, the probability that a Brownian motion starting at the circle of radius 2 reaches the circle of radius R before hitting the circle of radius 1 is $\log 2 / \log R$. Hence, using the strong Markov property, we get the following inequality,

$$q_R \leq (1-\rho)\left[\frac{\log 2}{\log R} + (1 - \frac{\log 2}{\log R}) q_R\right] \leq (1-\rho)\left[\frac{\log 2}{\log R} + q_R\right],$$

which gives $q_R \leq (1-\rho) \log 2 / (\rho \log R)$.

To prove (2.17), write

$$p_D(t, z, w) = \int_{\mathbb{C}} \int_{\mathbb{C}} p_D(t/3, z, z')\, p_D(t/3, z', w')\, p_D(t/3, w', w)\, dA(z')\, dA(w').$$

Since $p_D(t/3, z', w') \leq p(t/3, 0, 0) \leq c/t$, we have (using (2.16))

$$\begin{aligned}p_D(t, z, w) &\leq c\, t^{-1} \left[\int_{\mathbb{C}} p_D(t/3, z, z')\, dA(z')\right]\left[\int_{\mathbb{C}} p_D(t/3, w, w')\, dA(w')\right]\\ &= c\, t^{-1}\, \mathbf{P}^z\{\tau_D > t/3\}\, \mathbf{P}^w\{\tau_D > t/3\}.\end{aligned}$$

$\square$

The *Green's function* for D [3] is defined by

$$G_D(z, w) = \pi \int_0^\infty p_D(t, z, w)\, dt.$$

---

[3] More precisely, the Green's function for Brownian motion stopped at ∂D or the Green's function for the Laplacian with Dirichlet boundary conditions on D.

Note that $G_D(z,w) = 0$ if $z \notin D$ or $w \notin D$. If $z = w$ the integral is infinite and we set $G_D(z,z) = \infty$. However, if $z \neq w$, then the integral is finite. The convergence as $t \to 0$ follows from $p_D(t,z,w) \leq p(t,z,w)$ and the convergence as $t \to \infty$ follows from (2.17). Using (2.16), we see that $G_D(z,w) = G_D(w,z)$. The multiplicative factor π is chosen for convenience.

LEMMA 2.33. *Suppose D is a regular domain and $z \in D$. There is a $c = c(z,D) < \infty$, such that for every $w \in D$, $|w - z| < 1$,*
$$\left| G_D(z,w) - \pi \int_0^1 p(t,z,w)\, dt \right| \leq c.$$

PROOF. An examination of the proof of Lemma 2.32 shows that for fixed z, D, the constant can be chosen uniformly for all $|z - w| \leq 1$. Therefore
$$\int_1^\infty p_D(t,z,w)\, dt \leq c_1,$$
where $c_1 = c_1(D,z) < \infty$. Also, (2.15) shows that
$$p(t,z,w) - p_D(t,z,w) \leq c_2(D,z)$$
for $|z - w| < \text{dist}(z, \partial D)/2$ and $p(t,z,w) \leq c_3(D,z)$ for $|z - w| \geq \text{dist}(z, \partial D)/2$. $\square$

PROPOSITION 2.34. *$G_D(z,\cdot)$ is the unique harmonic function on $D \setminus \{z\}$ such that $G_D(z,w) \to 0$ as $w \to \partial D$ and $G_D(z,w) = -\log|z - w| + O(1)$ as $w \to z$.*

PROOF. The asymptotics of $G_D(z,w)$ as $w \to z$ follow from Lemma 2.33 and the estimate
$$\int_0^1 p(t,z,w)\, dt = \int_0^1 \frac{1}{2\pi t} e^{-|z-w|^2/2t}\, dt = -\frac{1}{\pi} \log|z - w| + O(1).$$
Harmonicity of $G_D(z,\cdot)$ follows from
$$\frac{1}{2} \Delta_w G_D(z,w) = \int_0^\infty \dot{p}_D(t,z,w)\, dt$$
and $p_D(0+,z,w) = p_D(\infty,z,w) = 0$. Alternatively, one can show that it satisfies the mean value property. To show uniqueness, suppose ϕ is a harmonic function in $D \setminus \{z\}$ such that ϕ is continuous on $\overline{D} \setminus \{z\}$; $\phi \equiv 0$ on ∂D and as $w \to z$,
$$\phi(w) = -\log|w - z| + O(1).$$
Let $D_{z,\epsilon} = D \setminus \overline{B}(z,\epsilon)$. By applying Proposition 2.11 to $D_{z,\epsilon}$, we get
$$\phi(w) = \mathbf{P}^w\{\tau_{D_{z,\epsilon}} < \tau_D\} [-\log \epsilon + O(1)].$$
By letting $\epsilon \to 0+$, we get that
$$\phi(w) = \lim_{\epsilon \to 0+} -(\log \epsilon)\, \mathbf{P}^w\{\tau_{D_{z,\epsilon}} < \tau_D\}.$$
This establishes uniqueness. $\square$

REMARK 2.35. The proof shows that
$$(2.19) \qquad G_D(z,w) = \lim_{\epsilon \to 0+} -(\log \epsilon)\, \mathbf{P}^w\{\tau_{D_{z,\epsilon}} < \tau_D\}.$$
In particular, the limit on the right exists. (One can give a direct proof of the existence of this limit.)

PROPOSITION 2.36. *If D is a regular domain and z, w are distinct points in D,*
$$G_D(z,w) = \mathbf{E}^w[\log|B_{\tau_D} - z|] - \log|z - w|.$$

PROOF. If D is bounded, then
$$\phi(w) = G_D(z,w) + \log|z - w|,$$
is a bounded harmonic function on $D \setminus \{z\}$ that is continuous on $\partial D \setminus \{z\}$. Hence, by Proposition 2.10,
$$\phi(w) = \mathbf{E}^w[\phi(B_{\tau_D})] = \mathbf{E}^w[\log|B_{\tau_D} - z|].$$
If D is unbounded, find an increasing sequence of bounded domains D_n whose union is D and use the dominated convergence theorem. Note that the estimate $\mathbf{E}^w[\log|B_{\tau_D} - z|] < \infty$ follows from (2.18). □

PROPOSITION 2.37. *Suppose D, D' are regular domains and $f : D \to D'$ is a conformal transformation. Then*
$$G_D(z,w) = G_{D'}(f(z), f(w)).$$

PROOF. By Corollary 2.21, $w \mapsto G_{D'}(f(z), f(w))$ is a harmonic function on $D \setminus \{z\}$. As $w \to z$, $|f(w) - f(z)| = |f'(z)||w - z| + O(|w - z|^2)$, and hence
$$G_{D'}(f(z), f(w)) = -\log|w - z| + O(1).$$
□

EXAMPLE 2.38. Let $D = \mathbb{D}$. Then $G_\mathbb{D}(0, z) = G_\mathbb{D}(z, 0) = -\log|z|$. If $w \in \mathbb{D}$, then
$$f(\zeta) = \frac{\zeta - w}{1 - \overline{w}\zeta}$$
is a conformal transformation of $\mathbb{D}$ onto itself with $f(w) = 0$. hence
$$G_\mathbb{D}(w, z) = G_\mathbb{D}(f(w), f(z)) = \log\frac{|1 - \overline{w}z|}{|w - z|}.$$

EXAMPLE 2.39. Let $D = \mathbb{H}, x, x' \in \mathbb{R}, y, y' > 0$. Using the Möbius transformation $f(z) = (z - x')/y'$, we get
$$G_\mathbb{H}(x + iy, x' + iy') = G_\mathbb{H}(\frac{x - x' + iy}{y'}, i).$$
The linear fractional transformation $g(z) = (i-z)/(i+z)$ takes $\mathbb{H}$ onto $\mathbb{D}$ and hence
$$G_\mathbb{H}(x + iy, i) = G_\mathbb{D}(g(x + iy), g(i)) = -\log|g(x + iy)| = \frac{1}{2}\log\frac{x^2 + (y + 1)^2}{x^2 + (y - 1)^2}.$$

REMARK 2.40. If $z \in D, w \in \partial D$ and ∂D is locally analytic at w, then
$$(2.20) \qquad 2\pi H_D(z, w) = \lim_{\epsilon \to 0+} \epsilon^{-1} G_D(z, w + \epsilon \mathbf{n}),$$
where $\mathbf{n}$ denotes the inward unit normal at w. The factor 2π comes from our definitions of H_D, G_D; for example, $H_\mathbb{D}(0, 1) = 1/2\pi$, but $G_\mathbb{D}(0, 1 - \epsilon) = -\log(1 - \epsilon) \sim \epsilon$ as $\epsilon \to 0+$.

REMARK 2.41. The assumption that D is regular is only for convenience. If D is a domain such that ∂D has at least one regular point, we can define
$$p_D(t,z,w) = \lim_{n\to\infty} p_{D_n}(t,z,w), \quad G_D(z,w) = \lim_{n\to\infty} G_{D_n}(z,w),$$
where D_n is an increasing sequence of regular domains whose union is D. The argument in Lemma 2.32 shows that the limit is finite.

CHAPTER 3

Conformal mappings

3.1. Simply connected domains

A domain $D \subset \mathbb{C}$ is *simply connected* if $\hat{\mathbb{C}} \setminus D$ is a connected subset of the Riemann sphere $\hat{\mathbb{C}}$. Equivalently, D is simply connected if and only if the region bounded by every simple closed curve $\gamma : [a, b] \to D$ is contained in D, i.e., that the "winding number" about each $z \notin D$ of γ is zero. In particular, if f is a holomorphic function in D and γ is a closed C^1 curve in D, then

$$\int_\gamma f(z)\, dz = 0. \tag{3.1}$$

For any such f and any fixed $z_0 \in D$, we can define the antiderivative

$$F(w) = \int_\gamma f(z)\, dz,$$

where the integral is over any C^1 curve in D from z_0 to w; (3.1) shows that the value is independent of the choice of γ, and it is easy to see that $F'(w) = f(w)$.

LEMMA 3.1. *If D is a simply connected domain and f is a holomorphic function from D to $\mathbb{C}\setminus\{0\}$, then there exists a holomorphic function g on D such that $f = e^g$.*

PROOF. Fix $z_0 \in D$, and let $g(z_0)$ be any complex number with $\exp\{g(z_0)\} = f(z_0)$. For other $w \in D$, let

$$g(w) = g(z_0) + \int_\gamma \frac{f'(z)}{f(z)}\, dz,$$

where γ is a curve from z_0 to w. Then $g'(z) = f'(z)/f(z)$ and $[fe^{-g}]'(z) = 0$. □

REMARK 3.2. The function g in the lemma is unique up to translations by integer multiples of $2\pi i$. We can write

$$g(z) = \log f(z) = \log |f(z)| + i \arg f(z).$$

The content of the proposition is that we can find a continuous version of $\arg f$ in simply connected domains. Lemma 3.1 is not true for non-simply connected domains. For example, there is no continuous function g on the annulus $\{r_1 < |z| < r_2\}$ such that $z = e^{g(z)}$.

REMARK 3.3. If $a \in \mathbb{C} \setminus \{0\}$, then $h_a := \exp\{g/a\}$ satisfies $[h_a]^a = f$. Hence we can find roots of non-zero holomorphic functions on simply connected domains.

PROPOSITION 3.4. *If D is a simply connected domain and $u : D \to \mathbb{R}$ is a harmonic function, then there is a harmonic function $v : D \to \mathbb{R}$ such that $f = u + iv$ is analytic in D. Moreover, v is unique up to an additive constant.*

PROOF. If $u \equiv 0$, then the Cauchy-Riemann equations imply that v is constant; this gives uniqueness up to an additive constant. To establish existence, choose $z_0 \in D$ and define
$$v(w) = \int_\gamma \Phi \cdot dz,$$
where Φ is the vector field $(-\partial_y u, \partial_x u)$ and γ is a curve connecting z_0 and w. Since u is harmonic, Green's theorem and simple connectedness of D imply that the integral is zero over all closed curves, and hence v is well defined. It is easy to check that (u, v) satisfy the Cauchy-Riemann equations. $\square$

LEMMA 3.5 (Hurwitz). *Suppose f_n is a sequence of one-to-one analytic functions on a domain D converging to the analytic function f. Then either f is constant or it is one-to-one.*

PROOF. Assume f is not constant. Let z_0, w_0 be distinct points in D and assume $f(z_0) = f(w_0)$; without loss of generality assume that $w_0 = 0$. Let $g_n(z) = f_n(z) - f_n(z_0), g(z) = f(z) - f(z_0)$. Then $g(0) = 0$, and there is an $\epsilon > 0$ such that $g(w) \neq 0$ for $0 < |w| < 2\epsilon$. Let γ denote the circle of radius ϵ about 0 oriented counterclockwise. Then,
$$\frac{1}{2\pi i} \int_\gamma \frac{g'(z)}{g(z)} \, dz = \lim_{n \to \infty} \frac{1}{2\pi i} \int_\gamma \frac{g_n'(z)}{g_n(z)} \, dz = 0.$$
(All of the integrals in the limit are zero since g_n is nonzero on $D \setminus \{z_0\}$.) However, by writing $g(z) = z^n h(z)$ with $h(0) \neq 0$, the left-hand side can be seen to be the degree of the zero of g at 0. Hence, $g(0) \neq 0$, which is a contradiction. $\square$

THEOREM 3.6 (Riemann mapping theorem). *Let D a be simply connected domain other than $\mathbb{C}$ and $w \in D$. Then there exists a unique conformal transformation $f : D \to \mathbb{D}$ with $f(w) = 0, f'(w) > 0$.*

PROOF. Uniqueness is easy: if f, f_1 both satisfy the conclusion of the theorem, then $h = f \circ f_1^{-1}$ is a conformal transformation of $\mathbb{D}$ onto $\mathbb{D}$ with $h(0) = 0, h'(0) > 0$. This implies that h is the identity (see the remark after the Schwarz lemma, Lemma 2.1). Existence for $D = \mathbb{D}$ follows from the fact that there exists a Möbius transformation f satisfying $f(w) = 0, f'(w) > 0$.

Let $\mathcal{G}$ be the set of conformal transformations $f : D \to f(D)$ with $f(w) = 0, f'(w) > 0$, and $f(D) \subset \mathbb{D}$. If $f \in \mathcal{G}$, then the Schwarz lemma applied to $f^*(z) = f(w + z \operatorname{dist}(w, \partial D))$ tells us that $f'(w) \leq [\operatorname{dist}(w, \partial D)]^{-1}$. We will now show that $\mathcal{G}$ is non-empty. Let $w_0 \in \mathbb{C} \setminus D$. Then $(z - w_0)^{-1}$ is a non-zero analytic function in D; hence, since D is simply connected, there is an analytic function g on D with $g(z)^2 = (z - w_0)^{-1}$. The open mapping theorem tells us that for some $\epsilon > 0, \mathcal{B}(g(w), \epsilon) \subset g(D)$. Since g^2 is one-to-one, so is g and $\mathcal{B}(-g(w), \epsilon) \cap g(D) = \emptyset$. Hence, $\hat{f}(z) := \epsilon/(g(z) + g(w))$ is a conformal transformation of D onto a subset of $\mathbb{D}$. By composing $\hat{f}$ with a Möbius transformation, we can find an $f \in \mathcal{G}$.

Let $M = \sup\{f'(w) : f \in \mathcal{G}\} \leq [\operatorname{dist}(w, \partial D)]^{-1} < \infty$, and let $f_1, f_2, \ldots$ be a sequence of functions in $\mathcal{G}$ with $f_j'(w) \to M$. By Proposition 2.29, we can find a subsequence that converges to an analytic function f with $f(w) = 0, f(D) \subset \mathbb{D}$. Also (using, say, the Cauchy integral formula) $f'(w) = M$ which implies that f is not constant and, using Lemma 3.5, that f is one-to-one.

We will show that $f(D) = \mathbb{D}$. Suppose not, and let $z_0 \in \mathbb{D} \setminus f(D)$. Let

$$h(z) = \frac{z - z_0}{1 - \bar{z}_0 z}$$

be the Möbius transformation in $\mathbb{D}$ with $h(z_0) = 0, h'(z_0) > 0, h(0) = -z_0$. Since $h \circ f$ is one-to-one and non-zero, there is a one-to-one analytic function g on D with $g^2 = h \circ f$. Let $\hat{h}$ be the Möbius transformation such that $\hat{h}(g(w)) = 0$ and $\hat{h}'(g(w)) g'(w) > 0$. Then $\hat{f} := \hat{h} \circ g \in \mathcal{G}$. The explicit form of the Möbius transformations (see (2.2)) gives $|h'(0)| = 1 - |z_0|^2$ and

$$|\hat{h}'(g(w))| = \frac{1}{1 - |g(w)|^2} = \frac{1}{1 - |h(0)|} = \frac{1}{1 - |z_0|}.$$

Hence

$$\hat{f}'(w) = |\hat{h}'(g(w))| \, |g'(w)| = \frac{1}{1 - |z_0|} \frac{|h'(0)| \, f'(w)}{2 \, |g(w)|} = \frac{1 + |z_0|}{2 \sqrt{|z_0|}} f'(w) > f'(w),$$

which contradicts the maximality of $f'(w)$. $\square$

We will often consider the inverse of the function described above, i.e., if D is a simply connected domain and w a specified point in D, there is a unique conformal transformation $f : \mathbb{D} \to D$ with $f(0) = w, f'(0) > 0$. Suppose D is bounded. It is natural to ask when f can be extended to a continuous map from $\overline{\mathbb{D}}$ to $\overline{D}$. We have already noted in §2.3 that if $z \in \partial \mathbb{D}$ and ∂D is locally analytic at $f(z)$, then we can extend f to a conformal transformation of $\mathbb{D} \cup \mathcal{B}(z, \epsilon)$ for some $\epsilon > 0$. If ∂D is analytic (i.e., locally analytic at all $z \in \partial D$), then a compactness argument shows that f can be extended to a conformal transformation of $(1 + r)\mathbb{D}$ for some $r > 0$.

In general, one cannot extend the map f to a continuous function on $\overline{\mathbb{D}}$. For example, if

$$D = \{x + iy : -1 < x < 1, 0 < y < 2\} \setminus \bigcup_{n=1}^{\infty} \{x + \frac{i}{n} : 0 \leq x < 1\},$$

there is no continuous extension. There is a topological characterization of the domains for which one can extend f. A closed set $K \subset \mathbb{C}$ is *locally connected* if for every $\epsilon > 0$, there is a $\delta > 0$ such that if $z, w \in K$ with $|z - w| < \delta$, then there exists a connected $K_1 \subset K$ with $z, w \in K_1$ and $\text{diam}(K_1) < \epsilon$. (Here, $\text{diam}(K_1) = \sup\{|z - w| : z, w \in K_1\}$.) We refer the reader to [**71**, Theorem 2.1] for a proof of the following.

PROPOSITION 3.7. *Let f be a conformal transformation of $\mathbb{D}$ onto a bounded domain D. Then f has a continuous extension to $\overline{\mathbb{D}}$ if and only if $\mathbb{C} \setminus D$ is locally connected.*

EXAMPLE 3.8. Suppose $\gamma^1, \ldots, \gamma^k$ are curves (images of $[0, 1]$), not necessarily simple. Then (see [**70**, pp. 88-89]), $K := \gamma^1[0, 1] \cup \cdots \cup \gamma^k[0, 1]$ is locally connected. The open set $\mathbb{C} \setminus K$ consists of a finite or countably infinite number of connected components, each of which is simply connected (as a subset of $\hat{\mathbb{C}}$). If D is one of these components, then $\mathbb{C} \setminus D$ is locally connected (if $z, w \in \mathbb{C} \setminus D$, either $[z, w] \subset \mathbb{C} \setminus D$ or we can let z', w' be the first and last points on $[z, w]$ in $\overline{D}$. Then we can connect z and w by the union of $[z, z'], [w', w]$ and a connected set connecting $[z', w']$ in K). In Chapter 8 we consider domains D as above where some or all of the curves γ^j are Brownian paths.

If $f : \mathbb{D} \to D$ is a conformal transformation and D is bounded with $\mathbb{C} \setminus D$ locally connected, then we can parametrize ∂D by $\gamma(t) = f(e^{2\pi it})$. Conversely, if ∂D is a curve, then ∂D is locally connected and (using the argument in the previous paragraph) $\mathbb{C} \setminus \partial D$ is locally connected; hence, f gives a parametrization of ∂D. This parametrization might not be one-to-one. For example, if D is the split disk $D = \mathbb{D} \setminus [-1, 0]$, then the points in $(-1, 0)$ are hit twice by f. Recall that a closed curve $\gamma : [a, b] \to \mathbb{C}$ is called a *Jordan curve* if it is one-to-one on $[a, b)$, i.e., if γ is a homeomorphism of (the circle) $[a, b]$ with a and b identified. A bounded domain D is called a *Jordan domain* if ∂D is a Jordan curve. Jordan domains are simply connected.

PROPOSITION 3.9. *Suppose D is a bounded domain with $\mathbb{C} \setminus D$ locally connected. Suppose $f : \mathbb{D} \to D$ is a conformal transformation, extended to a continuous map from $\overline{\mathbb{D}}$ onto $\overline{D}$. Then D is a Jordan domain if and only if f is one-to-one on $\partial \mathbb{D}$.*

PROOF. If f is one-to-one, then f gives a parametrization of ∂D as a Jordan curve. For a proof of the other direction see [**71**, §2.3]. □

PROPOSITION 3.10. *If D, D' are Jordan domains and z_1, z_2, z_3 and z_1', z_2', z_3' are points on $\partial D, \partial D'$, respectively, oriented counterclockwise, then there is a unique conformal transformation $f : D \to D'$, that can be extended to a homeomorphism from $\overline{D}$ to $\overline{D}'$ such that $f(z_1) = z_1', f(z_2) = z_2', f(z_3) = z_3'$.*

PROOF. By the Riemann mapping theorem, it suffices to prove the result when $D = D' = \mathbb{D}$. This can be done directly using the form of the Möbius transformations of $\mathbb{D}$, see (2.2). □

If D, D' are Jordan domains, then we can specify uniquely a conformal map $f : D \to D'$ if we specify where three boundary points go or where one interior point and one boundary point go. Roughly speaking, there are three real "degrees of freedom" in choosing f. Specifying where a boundary point goes uses one real degree of freedom while specifying an interior point uses one complex degree (i.e., two real degrees) of freedom.

3.2. Univalent functions

We say that a function $f : D \to \mathbb{C}$ is *univalent* if it is analytic and one-to-one on D. Let $\mathcal{A}$ denote the collection of simply connected domains D other than $\mathbb{C}$ containing the origin. If $A \in \mathcal{A}$, let inrad(A) denote the *inradius (with respect to the origin)*, i.e., inrad$(A) = \text{dist}(0, \partial A)$. Let $\mathcal{S}^*$ denote the set of univalent functions f on $\mathbb{D}$ with $f(0) = 0$ and $f'(0) > 0$. The Riemann mapping theorem says that there is a one-to-one correspondence between $\mathcal{S}^*$ and $\mathcal{A}$ given by $f \longleftrightarrow f(\mathbb{D})$. In other words, the study of simply connected domains reduces to the study of univalent functions on $\mathbb{D}$.

We let $\mathcal{S}$ denote the set of $f \in \mathcal{S}^*$ with $f'(0) = 1$, and we let $\mathcal{A}_1$ be the corresponding subset of $\mathcal{A}$. Any $f \in \mathcal{S}$ has an expansion at 0,

$$f(z) = z + \sum_{n=2}^{\infty} a_n z^n.$$

3.2. UNIVALENT FUNCTIONS

An important function in $\mathcal{S}$ is the *Koebe function* f_{Koebe} defined by

$$f_{\text{Koebe}}(z) = \frac{z}{(1-z)^2} = \sum_{n=1}^{\infty} n\, z^n.$$

EXERCISE 3.11. *Suppose $f(z) = a_1 z + a_2 z^2 + \cdots \in \mathcal{S}^*$. Show that*

$$\text{area}[f(\mathbb{D})] = \pi \sum_{n=1}^{\infty} n\, |a_n|^2.$$

EXERCISE 3.12. *Show that f_{Koebe} is one-to-one on $\mathbb{D}$ and $f_{\text{Koebe}}(\mathbb{D}) = \mathbb{C} \setminus (-\infty, -1/4]$. Hint:*

$$f_{\text{Koebe}}(z) = \frac{1}{4}\left(\frac{1+z}{1-z}\right)^2 - \frac{1}{4}.$$

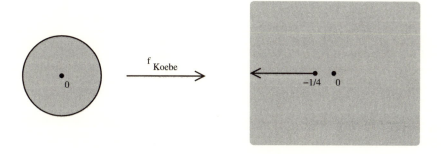

FIGURE 3.1. The Koebe function f_{Koebe}

A *compact hull* K is a compact, connected subset of $\mathbb{C}$ larger than a single point such that $\mathbb{C} \setminus K$ is connected. For any compact hull, there is a unique conformal transformation $F_K : \mathbb{C} \setminus \overline{\mathbb{D}} \to \mathbb{C} \setminus K$ such that $\lim_{z \to \infty} F_K(z)/z > 0$. In fact, if $0 \in K$, $F_K(z) = 1/f_K(1/z)$ where f_K is the conformal transformation of $\mathbb{D}$ with $f_K(0) = 0$, $f'_K(0) > 0$ onto the image of $\mathbb{C} \setminus K$ under the map $z \mapsto 1/z$. We define the *(logarithmic) capacity*, $\text{cap}(K)$, by $\text{cap}(K) = -\log f'_K(0) = \log [\lim_{z \to \infty} F_K(z)/z]$.

Let $\mathcal{H}^*$ be the set of compact hulls and let $\mathcal{H}$ be the set of compact hulls containing the origin. Let $\mathcal{H}_0^*, \mathcal{H}_0$ be the set of hulls K in $\mathcal{H}^*, \mathcal{H}$, respectively, with $\text{cap}(K) = 0$. If $K \in \mathcal{H}_0^*$, then F_K has a Laurent expansion

$$F_K(z) = z + b_0 + \sum_{n=1}^{\infty} \frac{b_n}{z^n}.$$

Also, $f_K \in \mathcal{S}$ if and only if $K \in \mathcal{H}_0$.

PROPOSITION 3.13 (Area Theorem). *If $K \in \mathcal{H}_0^*$, then*

$$\text{area}(K) = \pi \left[1 - \sum_{n=1}^{\infty} n\, |b_n|^2\right].$$

In particular, $\sum_{n=1}^{\infty} n\, |b_n|^2 \leq 1$.

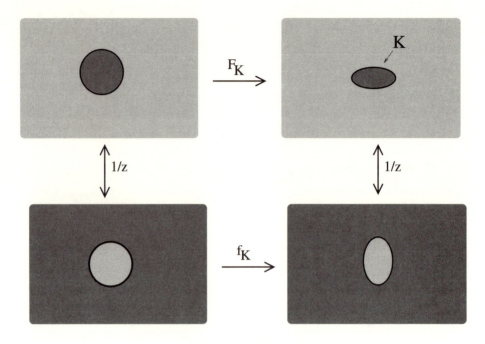

FIGURE 3.2. The maps F_K, f_K. The domain is the region shaded lighter.

PROOF. If γ is any smooth, simple closed curve, oriented counterclockwise, bounding a region D, then (2.1) and Green's theorem give
$$\int_\gamma \bar{z}\, dz = \int_\gamma (x, y) \cdot dz + i \int_\gamma (-y, x) \cdot dz = 2i\,\text{area}(D).$$
Suppose $r > 1$ and K_r denotes the region bounded by the image of the circle of radius r under F_K. Then
$$\text{area}(K) = \lim_{r \to 1+} \text{area}(K_r) = \frac{1}{2i} \int_{\partial K_r} \bar{z}\, dz = \frac{1}{2i} \int_0^{2\pi} \overline{F_K(re^{i\theta})} \, [ire^{i\theta} F_k'(re^{i\theta})]\, d\theta.$$
But,
$$\overline{F_K(re^{i\theta})} = re^{-i\theta} + b_0 + \sum_{n=1}^\infty \overline{b_n}\, r^{-n}\, e^{i\theta n},$$
$$re^{i\theta} F_K'(re^{i\theta}) = re^{i\theta} - \sum_{n=1}^\infty n\, b_n\, r^{-n}\, e^{-i\theta n}.$$
For integers m,
$$\int_0^{2\pi} e^{i\theta m}\, d\theta = \begin{cases} 0, & m \neq 0, \\ 2\pi, & m = 0, \end{cases}$$
and, hence,
$$\text{area}(K) = \lim_{r \to 1+} \pi \left[r^2 - \sum_{n=1}^\infty n\, |b_n|^2\, r^{-2n} \right].$$
□

3.2. UNIVALENT FUNCTIONS

EXAMPLE 3.14. If $K = [-2, 2]$, then
$$F_K(z) = z + \frac{1}{z}.$$
This can be verified using Exercise 3.12 and the relation
$$f_{\text{Koebe}}(z) = \frac{1}{z + z^{-1} - 2}.$$
In this example, $|b_1| = 1$. The Area theorem implies that $|b_1| \leq 1$ with equality possible only if all the higher coefficients vanish.

LEMMA 3.15. *If $f \in \mathcal{S}$, then there is an odd function $h \in \mathcal{S}$ such that for each $z \in \mathbb{D}$, $h(z)^2 = f(z^2)$.*

PROOF. Since $f(z)/z$ is a non-zero analytic function on $\mathbb{D}$, there is an analytic function g on $\mathbb{D}$ with $g(z)^2 = f(z)/z$ (see Remark 3.3). Let $h(z) = zg(z^2)$ which is clearly an odd function with $h(z)^2 = f(z^2)$, $h(0) = 0$, and $h'(0) = 1$. If $h(z_1) = h(z_2)$, then univalence of f implies $z_1^2 = z_2^2$, which then implies $z_1 = z_2$ (since h is odd). Therefore, $h \in \mathcal{S}$. □

PROPOSITION 3.16 (Bieberbach). *If $f \in \mathcal{S}$, then $|a_2| \leq 2$.*

PROOF. Suppose $f(z) = z + a_2 z^2 + \cdots \in \mathcal{S}$ and let h be as in Lemma 3.15. Then
$$h(z) = z + \frac{a_2}{2} z^2 + \cdots.$$
Let $g(z) = 1/h(1/z)$, and note that g has an expansion
$$g(z) = z - \frac{a_2}{2z} + \cdots.$$
Therefore, the Area Theorem implies that $|a_2| \leq 2$. □

THEOREM 3.17 (Koebe 1/4 Theorem). *If $f \in \mathcal{S}$, then $\mathcal{B}(0, 1/4) \subset f(\mathbb{D})$.*

PROOF. Suppose $f(z) = z + a_2 z^2 + \cdots \in \mathcal{S}$. By Proposition 3.16, $|a_2| \leq 2$. Suppose $z_0 \notin f(D)$. Let
$$\hat{f}(z) := \frac{z_0 f(z)}{z_0 - f(z)} = z + (a_2 + \frac{1}{z_0}) z^2 + \cdots.$$
Note that $\hat{f}$ is one-to-one since it is a composition of f and a linear fractional transformation, and hence, $\hat{f} \in \mathcal{S}$. By Proposition 3.16, $|a_2 + (1/z_0)| \leq 2$, and, since $|a_2| \leq 2$, $|z_0| \geq 1/4$. □

REMARK 3.18. The Koebe function f_{Koebe} shows that this theorem is sharp. By examining the argument one can see that if $f \in \mathcal{S}$ is not a rotation of f_{Koebe}, then there exists an $r > 1/4$ such that $\mathcal{B}(0, r) \subset f(\mathbb{D})$.

COROLLARY 3.19. *Suppose $f : D \to D'$ is a conformal transformation with $f(z) = z'$. Then*
$$\frac{d'}{4d} \leq |f'(z)| \leq \frac{4 d'}{d},$$
where $d = \text{dist}(z, \partial D), d' = \text{dist}(z', \partial D')$.

PROOF. Assume without loss of generality that $z = z' = 0$. Then,
$$\tilde{f}(w) := \frac{f(dw)}{d\, f'(0)} \in \mathcal{S}.$$
By the Koebe 1/4 Theorem, $\mathcal{B}(0, 1/4) \subset \tilde{f}(\mathbb{D})$, and hence $d' \geq (1/4)\,|f'(0)|\,d$. This gives the upper bound, and the lower bound is obtained by considering f^{-1}. □

PROPOSITION 3.20. *If $f \in \mathcal{S}$ and $z \in \mathbb{D}$,*
$$\left| \frac{z\, f''(z)}{f'(z)} - \frac{2\,|z|^2}{1 - |z|^2} \right| \leq \frac{4\,|z|}{1 - |z|^2}.$$

PROOF. Let $T_z(w) = (w + z)/(1 + \bar{z} w)$ be the Möbius transformation with $T_z(0) = z$ and $T'_z(0) = 1 - |z|^2$. Then $\hat{f} \in \mathcal{S}$ where
$$\hat{f}(w) = \frac{f(T_z(w)) - f(z)}{f'(z)\,(1 - |z|^2)}.$$
Using
$$T_z(w) = z + (1 - |z|^2)\,w - \bar{x}\,(1 - |z|^2)\,w^2 + \cdots,$$
$$f(T_z(w)) = f(z) + f'(z)\,[T_z(w) - z] + \frac{f''(z)}{2}\,[T_z(w) - z]^2 + \cdots,$$
we get
$$\hat{f}(w) = w + \left[\frac{f''(z)\,(1 - |z|^2)}{2\,f'(z)} - \bar{z}\right] w^2 + \cdots$$
Therefore, by Proposition 3.16,
$$\left| \frac{f''(z)\,(1 - |z|^2)}{2\,f'(z)} - \bar{z} \right| \leq 2.$$
Multiplying both sides by $2|z|/(1 - |z|^2)$ gives the result. □

THEOREM 3.21 (Distortion Theorem). *If $f \in \mathcal{S}$ and $z \in \mathbb{D}$,*
$$\frac{1 - |z|}{(1 + |z|)^3} \leq |f'(z)| \leq \frac{1 + |z|}{(1 - |z|)^3}.$$

PROOF. It suffices to prove the estimate for $z = x \in (0, 1)$, since if $z = re^{i\theta}$, we can consider $\hat{f}_\theta(w) = e^{-i\theta} f(we^{i\theta})$. Since $f'(z)$ is nonzero and $f'(0) = 1$, by Lemma 3.1 we can find an analytic function h with $h(0) = 0$ and $e^h = f'$; in particular, $\operatorname{Re} h = \log |f'|$. Note that
$$x\, \partial_x \operatorname{Re}[h(x)] = \operatorname{Re}\left[\frac{x\, f''(x)}{f'(x)} \right],$$
and hence, by Proposition 3.20,
$$\frac{2x - 4}{1 - x^2} \leq \partial_x \log |f'(x)| \leq \frac{2x + 4}{1 - x^2}.$$
Since $\log |f'(0)| = 0$, we can integrate this and then exponentiate to get the theorem. □

REMARK 3.22. The assumption that f is one-to-one in $\mathbb{D}$ is important in the Distortion Theorem. For example, if
$$f_n(z) = n^{-1}\left[e^{nz} - 1\right],$$
then f_n is locally one-to-one on $\mathbb{D}$ with $f_n(0) = 0$, $f_n'(0) = 1$. However, $f_n'(1/n) = e$, so f_n does not satisfy the conclusion of the Distortion Theorem for large n.

THEOREM 3.23 (Growth Theorem). *If $f \in \mathcal{S}$ and $z \in \mathbb{D}$,*
$$\frac{|z|}{(1+|z|)^2} \leq |f(z)| \leq \frac{|z|}{(1-|z|)^2}.$$

PROOF. As in Theorem 3.21, it suffices to prove the upper bound for $z = x \in [0, 1)$. Then by the Distortion Theorem,
$$|f(x)| = \left|\int_0^x f'(s)\,ds\right| \leq \int_0^x |f'(s)|\,ds \leq \int_0^x \frac{1+s}{(1-s)^3}\,ds = \frac{x}{(1-x)^2}.$$
The lower bound is immediate if $|f(z)| \geq 1/4$. We will prove the lower bound when $f(z) = x \in [0, 1/4)$; by considering $\tilde{f}_\theta$ as in Theorem 3.21, this will establish the result for all $|f(z)| < 1/4$. By the Koebe 1/4 Theorem, the line segment $[0, x]$ is in $f(\mathbb{D})$; let $\gamma(t) = f^{-1}(t), 0 \leq t \leq x$. Then,
$$x = \int_0^x f'(\gamma(s))\,\gamma'(s)\,ds = \int_0^x |f'(\gamma(s))|\,|\gamma'(s)|\,ds = \int_\gamma |f'(w)|\,|dw|.$$
The second equality uses the fact that the integral is increasing and hence that $f'(\gamma(s))\,\gamma'(s)$ is nonnegative. Using the Distortion Theorem,
$$\int_\gamma |f'(w)|\,|dw| \geq \int_0^{|z|} \frac{1-r}{(1+r)^3}\,dr = \frac{|z|}{(1+|z|)^2}.$$
$\square$

REMARK 3.24. The inequalities in both the Distortion Theorem and the Growth Theorem are sharp. The examples to show this are rotations of the Koebe function f_{Koebe}. See [30] for more details.

COROLLARY 3.25. *If $f : D \to D'$ is a conformal transformation with $f(z) = z'$, then for all $r \in (0, 1)$ and all $|w - z| \leq r\,\text{dist}(z, \partial D)$,*
$$|f(w) - z'| \leq \frac{4\,|w - z|}{(1-r)^2}\frac{\text{dist}(z', \partial D')}{\text{dist}(z, \partial D)}.$$

PROOF. By translation and scaling, we may assume $z = z' = 0$, $\text{dist}(z, \partial D) = 1$, and $f'(0) = 1$, i.e., $f \in \mathcal{S}$. Then Theorem 3.23 gives $|f(w)| \leq |w|/(1-r)^2$ and Theorem 3.17 implies $\text{dist}(0, \partial D') \geq 1/4$. $\square$

PROPOSITION 3.26. *For each $0 < r < 1$, there is a $C_r < \infty$ such that if $f \in \mathcal{S}$ and $|z| \leq r$, then $|f(z) - z| \leq C_r|z|^2$. (In fact, the optimal C_r is $(2-r)/(1-r)^2$.)*

PROOF. If we combine Proposition 3.20 and the Distortion Theorem, we can obtain a uniform bound on $|f''(z)|$ over all $f \in \mathcal{S}$ and $|z| \leq r$. This gives the first statement. The parenthetical statement, which we will not use, follows from the

Bieberbach conjecture proved by de Branges [**15**], which states that $|a_n| \leq n$ for all n. Therefore,

$$|f(z) - z| \leq |z|^2 \sum_{n=2}^{\infty} |a_n|\, |z|^{n-2} \leq |z|^2 \sum_{n=2}^{\infty} n\, r^{n-2} = \frac{2-r}{(1-r)^2}\, |z|^2.$$

The Koebe function f_{Koebe} can be used to see that no smaller value of C_r works. □

REMARK 3.27. Recall from Exercise 2.27, that if $0 < r < 1$, we call z and w r-adjacent in D if

$$|z - w| < r\, \max\{\operatorname{dist}(z, \partial D), \operatorname{dist}(w, \partial D)\},$$

i.e., if $z \in \mathcal{B}(w, r\operatorname{dist}(w, \partial D))$ or vice versa. Let $d_{r,D}$ denote the corresponding graph distance in D. As noted in the exercise, if $0 < r < s < 1$, then $d_{s,D}(z,w) \leq d_{r,D}(z,w) \leq C_{r,s}\, d_{s,D}(z,w)$. Suppose $f : D \to D'$ is a conformal transformation. It is not hard to give examples to show that $d_{r,D'}(f(z), f(w))$ does not have to equal $d_{r,D}(z,w)$. The Growth Theorem and the Koebe 1/4 Theorem, however, tell us that if z, w are $(1/8)$-adjacent in D, then $f(z)$ and $f(w)$ are r-adjacent in D' where $r = [(1/8)/(7/8)^2]/[1/4] = 32/49$. In particular (see Exercise 2.27), there is a constant c such that

(3.2) $$d_{1/2, D'}(f(z), f(w)) \leq c\, d_{1/2, D}(z, w).$$

The *hyperbolic metric* ρ is defined on $\mathbb{D}$ by

$$\rho(z, w) = \inf \int_\gamma \frac{|dz'|}{1 - |z'|^2},$$

where the infimum is over all curves γ from z to w. The definition makes it clear that this is a metric, and it can be checked that if T is a Möbius transformation of $\mathbb{D}$, then $\rho(T(z), T(w)) = \rho(z, w)$. We can therefore define $\rho_D(z, w)$ for simply connected D other than $\mathbb{C}$ to be $\rho(f(z), f(w))$, where $f : D \to \mathbb{D}$ is a conformal transformation. The invariance of ρ under Möbius transformations shows that this is well defined. If $z = 0$, the infimum is obtained by the line segment from 0 to w; hence,

$$\rho(0, w) = \frac{1}{2} \log \frac{1 + |w|}{1 - |w|} \sim \frac{1}{2} \log \frac{1}{1 - |w|}, \quad |w| \to 1-.$$

It is easy to check that $d_{1/2, \mathbb{D}}(0, w) \sim c\, \log[1/(1 - |w|)]$ as $|w| \to 1-$. Hence, using (3.2), we see that $d_{1/2, D}$ and ρ_D are essentially comparable metrics. To be more precise, there exist c_1, c_2, independent of D, with

$$c_1\, \rho_D(z, w) \leq d_{1/2, D}(z, w) \leq c_2\, \rho_D(z, w),$$

provided that $\rho_D(z, w) \geq 1$.

3.3. Capacity

In the previous section we defined the capacity of a compact hull K by $F_K(z) \sim e^{\operatorname{cap}(K)} z$, $z \to \infty$, where F_K is the conformal transformation of $\mathbb{C} \setminus \overline{\mathbb{D}}$ onto $\mathbb{C} \setminus K$ such that $\lim_{z \to \infty} F_K(z)/z > 0$. If $w \in \mathbb{C}, a > 0$, then

$$F_{K+w}(z) = F_K(z) + w, \quad F_{aK}(z) = a\, F_K(z),$$

and hence, $\operatorname{cap}(K + w) = \operatorname{cap}(K), \operatorname{cap}(aK) = \operatorname{cap}(A) + \log a$. Also $\operatorname{cap}(\overline{\mathbb{D}}) = 0$. Here we will discuss some properties and equivalent definitions of capacity. Let

3.3. CAPACITY

$g_K = F_K^{-1}$, and, as before, let $f_K(z) = 1/F_K(1/z)$ so that $f_K'(0) = e^{-\text{cap}(K)}$. In particular, $f_K \in \mathcal{S}$ if and only if $K \in \mathcal{H}_0$. For any hull K, let $\text{rad}(K) = \sup\{|z| : z \in K\}$, i.e., $\text{rad}(K)$ is the radius of the smallest closed disk about the origin containing K.

PROPOSITION 3.28. *If $K_1, K_2 \in \mathcal{H}$ with $K_1 \subset K_2$, then $\text{cap}(K_1) \leq \text{cap}(K_2)$. The inequality is strict unless $K_1 = K_2$.*

PROOF. The map $h := f_{K_1}^{-1} \circ f_{K_2}$ takes $\mathbb{D}$ into $\mathbb{D}$ with $h(0) = 0$. By the Schwarz lemma, $h'(0) \leq 1$ and hence $f_{K_1}'(0) \geq f_{K_2}'(0)$. If $K_1 \neq K_2$, then h is not onto and hence $h'(0) < 1, f_{K_1}'(0) > f_{K_2}'(0)$. □

PROPOSITION 3.29. *If $K \in \mathcal{H}_0$, then $1 \leq \text{rad}(K) \leq 4$. Also, $[-4, 0] \in \mathcal{H}_0$.*

PROOF. Since $\text{cap}(\overline{\mathbb{D}}) = 0$, the previous proposition shows $\text{cap}(K) < 0$ for all K with $\text{rad}(K) < 1$. Suppose $K \in \mathcal{H}_0$. By the Koebe $1/4$ Theorem, $f_K(\mathbb{D})$ contains $\mathcal{B}(0, 1/4)$. Hence, $F_K(\mathbb{C} \setminus \mathbb{D})$ contains $\{|z| > 4\}$ which implies $\text{rad}(K) \leq 4$. The final statement follows since $f_{[-4,0]} = f_{\text{Koebe}}$. □

If
$$f_K(z) = z + a_2 z^2 + a_3 z^3 + \cdots,$$
and we let $w = 1/z$, then
$$F_K(w) = \frac{1}{f_K(1/w)} = w + b_0 + b_1 w^{-1} + \cdots = w - a_2 + (a_2^2 - a_3) w^{-1} + \cdots$$
In particular, if $K \in \mathcal{H}_0$, $|b_0| \leq 2$. The estimate in the next proposition is a similar uniform estimate.

PROPOSITION 3.30. *There exists a $c < \infty$ such that for all $K \in \mathcal{H}_0$ and all $|z| > 1$,*
$$|F_K(z) - z| \leq c.$$

PROOF. It suffices to find a constant c that works for all $K \in \mathcal{H}_0$ and all $|z| > 2$, since that would imply $|F_K(z)| \leq c + 2$ for $1 < |z| \leq 2$. Let $w = 1/z$. Then by Theorem 3.23 and Proposition 3.26,
$$|F_K(z) - z| = \left| \frac{1}{f_K(w)} - \frac{1}{w} \right| = \frac{|w - f_K(w)|}{|w| |f_K(w)|} \leq \frac{9}{4} \frac{|w - f_K(w)|}{|w|^2} \leq \frac{9}{4} C_{1/2}.$$
□

If $K \in \mathcal{H}^*$, let
$$\phi_K(z) = \begin{cases} \log |g_K(z)|, & z \notin K \\ 0, & z \in K. \end{cases}$$
Then ϕ_K is the unique harmonic function on $\mathbb{C} \setminus K$ with boundary value 0 on K and such that $\phi_K(z) \sim \log|z|$ as $z \to \infty$. Uniqueness can be seen by noting that for every $\epsilon > 0$, the function $\phi_K(z) - \log|z|$ is bounded on $\{|w| \geq \epsilon\}$; harmonic on $\{|w| > \epsilon\} \setminus K$; and has boundary value $\phi_K(z) - \log|z|$ for $|z| = \epsilon$ or $z \in K \cap \{|w| > \epsilon\}$, Proposition 2.11 gives the uniqueness of such a function. If D is the image of $\mathbb{C} \setminus K$ under the map $z \mapsto 1/z$, then $\phi_K(z) = G_D(1/z, 0)$, where G_D denotes the Green's function as in §2.4.

EXERCISE 3.31. *Suppose* $K \in \mathcal{H}$, *and suppose* ϕ *is a positive harmonic function on* $\mathbb{C} \setminus K$ *with boundary value* 0 *on* K *and such that for some* c, $\phi(z) \leq c \log|z|$ *for all* $z \in \mathbb{C} \setminus K$. *Then there exists an* $\alpha > 0$ *such that* $\phi(z) = \alpha \, \phi_K(z)$. *(Hint: It suffices to prove the result for* $K = \overline{\mathbb{D}}$ *(why?). Prove that if* $K = \overline{\mathbb{D}}$, ϕ *must be radially symmetric, and then use this to find the function.)*

PROPOSITION 3.32. *There exists a* $c < \infty$, *such that if* $K \in \mathcal{H}$ *and* $|z| > 4 \, e^{\mathrm{cap}(K)}$,
$$|\phi_K(z) - \log|z| + \mathrm{cap}(K)| \leq c \, e^{\mathrm{cap}(K)} \, |z|^{-1}.$$

PROOF. From the scaling relations $\phi_{rK}(z) = \phi_K(z/r)$ and $\mathrm{cap}(rK) = \mathrm{cap}(K) + \log r$, we can see that if suffices to prove the result for $K \in \mathcal{H}_0$. Assume $K \in \mathcal{H}_0$ and $|z| > 4$. By Proposition 3.29, $z \in \mathbb{C} \setminus K$. Proposition 3.30 says that there is a c' such that $|g_K(z) - z| \leq c'$. Hence,
$$\left| \log|g_K(z)| - \log|z| \right| = \left| \log\left[1 + \frac{|g_K(z)| - |z|}{|z|}\right] \right| \leq \frac{c}{|z|}.$$
(The final inequality holds provided $|z| \geq 2c'$; however, the left hand side is bounded uniformly for $4 \leq |z| \leq 2c'$.) □

COROLLARY 3.33. *If* $K \in \mathcal{H}$, B *is a complex Brownian motion, and* τ *is the first time that* B *reaches* K, *then*
$$(3.3) \qquad \phi_K(z) = \log|z| - \mathbf{E}^z[\log|B_\tau|].$$
In particular, there exists a $c < \infty$, *such that if* $|z| > 4 \, e^{\mathrm{cap}(K)}$,
$$|\, \mathbf{E}^z[\log|B_\tau|] - \mathrm{cap}(K) \,| \leq c \, e^{\mathrm{cap}(K)} \, |z|^{-1}.$$

PROOF. For every $\epsilon > 0$, let $\tau_\epsilon = \tau \wedge \min\{t : |B_t| \leq \epsilon\}$. Let $h(z) = \log|z| - \phi_K(z)$, which is a bounded harmonic function on $\mathbb{C} \setminus (K \cup \overline{B}(0, \epsilon))$ and hence by the optional sampling theorem, $h(z) = \mathbf{E}^z[h(B_{\tau_\epsilon})]$. Note that there is a $c' = c'(K, z)$ such that $|h(B_{\tau_\epsilon})| \leq c' + |\log Y|$ where $Y = \min\{|B_t| : 0 \leq t \leq \tau\}$. Moreover, there is a $\rho < 1$ such that for all $r < \mathrm{rad}(K)$,
$$\mathbf{P}\{Y \leq r/2 \mid Y \leq r\} \leq \rho.$$
In fact, we can choose $1 - \rho$ to be the probability that a Brownian motion starting on the unit circle makes a closed loop in the annulus $\{1/2 < |w| < 1\}$ before reaching the circle of radius $1/2$. Hence $\mathbf{P}\{Y < 2^{-m} \mathrm{rad}(K)\} \leq \rho^m$, from which $\mathbf{E}[|\log Y|] < \infty$ follows easily. Therefore, by dominated convergence, $h(z) = \mathbf{E}^z[h(B_\tau)]$ which gives (3.3). The last assertion then follows from Proposition 3.32. □

COROLLARY 3.34. *Suppose* $K \in \mathcal{H}$ *and* $r > 4 \, e^{\mathrm{cap}(K)}$. *Let* B *be a Brownian motion started uniformly on the circle of radius* r *about the origin and let* τ *be as in Corollary 3.33. Then* $\mathbf{E}[\log|B_\tau|] = \mathrm{cap}(K)$.

PROOF. Let $M(r)$ be the average value of $\psi(z) = \mathbf{E}^z[\log|B_\tau|]$ on $\overline{B}(0, r)$. By the strong Markov property, $M(r)$ is constant for $r > \mathrm{rad}(K)$, and Corollary 3.33 shows that $\lim_{r \to \infty} M(r) = \mathrm{cap}(K)$. □

REMARK 3.35. We can define $\mathrm{cap}(K)$ for any compact set K by $\mathrm{cap}(K) = \mathbf{E}[\log|B_\tau|]$, where B_0 is started uniformly on any circle that surrounds K. Monotonicity follows from (2.4).

3.4. Half-plane capacity

Let $\mathbb{H} = \{x + iy : y > 0\}$ be the upper half plane. We will call a bounded subset $A \subset \mathbb{H}$ a *compact $\mathbb{H}$-hull* if $A = \mathbb{H} \cap \overline{A}$ and $\mathbb{H} \setminus A$ is simply connected. Let $\mathcal{Q}$ denote the set of compact $\mathbb{H}$-hulls. For each $A \in \mathcal{Q}$, we let A^* be the closure of $\{z : z \in A \text{ or } \overline{z} \in A\}$. If A is connected, then $A^* \in \mathcal{H}^*$.

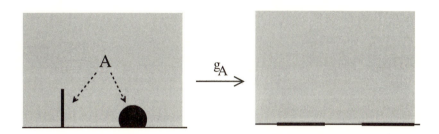

PROPOSITION 3.36. *For each $A \in \mathcal{Q}$, there is a unique conformal transformation $g_A : \mathbb{H} \setminus A \to \mathbb{H}$ such that*
$$\lim_{z \to \infty} [g_A(z) - z] = 0.$$

PROOF. The Riemann mapping theorem tells us that there are many conformal transformations $g : \mathbb{H} \setminus A \to \mathbb{H}$ with $|g(z)| \to \infty$ as $z \to \infty$. Suppose that A is contained in $\mathcal{B}(0, r)$. The Schwarz reflection principle [**75**, Theorem 11.14] shows that g can be extended to a conformal transformation of $\{|z| > r\}$ with $g(\overline{z}) = \overline{g(z)}$. The function $f(z) = 1/g(1/z)$ can be expanded about the origin; this gives an expansion for g near infinity:
$$g(z) = b_{-1} z + b_0 + b_1 z^{-1} + \cdots.$$

Since reals are sent to reals, it is easy to see that $b_j \in \mathbb{R}$, and since points in $\mathbb{H}$ are sent to points in $\mathbb{H}$, $b_{-1} > 0$. The conformal transformations of $\mathbb{H}$ onto $\mathbb{H}$ that send infinity to infinity are of the form $z \mapsto az + b$ where $a > 0, b \in \mathbb{R}$. Hence, there is a unique choice of g, which we denote by g_A, such that $b_{-1} = 1$ and $b_0 = 0$. $\square$

DEFINITION 3.37. If $A \in \mathcal{Q}$, the *half-plane capacity (from infinity)*, hcap(A), is defined by
$$\text{hcap}(A) = \lim_{z \to \infty} z \, [g_A(z) - z].$$

In other words,
$$g_A(z) = z + \frac{\text{hcap}(A)}{z} + O(\frac{1}{|z|^2}), \quad z \to \infty.$$

If $r > 0, x \in \mathbb{R}$, and $A \in \mathcal{Q}$, then it is easy to check that
$$g_{rA}(z) = r \, g_A(z/r), \quad g_{A+x}(z) = g_A(z - x) + x;$$
hence,
$$\text{hcap}(rA) = r^2 \, \text{hcap}(A), \quad \text{hcap}(A + x) = \text{hcap}(A).$$

EXAMPLE 3.38. If $A = \overline{\mathbb{D}} \cap \mathbb{H}$, then
$$g_A(z) = z + \frac{1}{z}.$$
If $A' = (0, i]$ [1], then
$$g_{A'}(z) = \sqrt{z^2 + 1} = z + \frac{1}{2z} + \cdots.$$
Hence, $\mathrm{hcap}(\overline{\mathbb{D}} \cap \mathbb{H}) = 1, \mathrm{hcap}((0, i]) = 1/2$.

EXAMPLE 3.39. Suppose $0 < \alpha < 1$ and let A be the line segment from the origin to $\alpha^\alpha (1-\alpha)^{1-\alpha} e^{i\alpha\pi}$. Then
$$g_A^{-1}(z) = [z + (1-\alpha)]^\alpha \, [z - \alpha]^{1-\alpha} = z - \frac{\alpha(1-\alpha)}{2z} + \cdots,$$
and hence $\mathrm{hcap}(A) = \alpha(1-\alpha)/2$. Note that g_A maps $\alpha^\alpha (1-\alpha)^{1-\alpha} e^{i\alpha\pi}$ to $2\alpha - 1$. Using the scaling rule, we see that

$$\mathrm{hcap}[\{re^{i\alpha\pi} : 0 < r \leq 1\}] = \frac{1}{2} \alpha^{1-2\alpha}(1-\alpha)^{2\alpha-1}. \tag{3.4}$$

Note that this tends to zero as $\alpha \to 0+$.

EXERCISE 3.40. Suppose $A \in \mathcal{Q}$ and z_n is a sequence of points in $\mathbb{H} \setminus A$ such that $z_n \to z$ with $z \in A$. Show that $\mathrm{Im}[g_A(z_n)] \to 0$. Show by example that it is not always true that $g_A(z_n)$ converges.

PROPOSITION 3.41. Suppose $A \in \mathcal{Q}$, B_t is a Brownian motion, and let $\tau = \tau_{\mathbb{H} \setminus A}$ be the smallest t with $B_t \in \mathbb{R} \cup A$. Then for all $z \in \mathbb{H} \setminus A$,

$$\mathrm{Im}(z) = \mathrm{Im}[g_A(z)] + \mathbf{E}^z[\mathrm{Im}(B_\tau)]. \tag{3.5}$$

Also,

$$\mathrm{hcap}(A) = \lim_{y \to \infty} y \, \mathbf{E}^{iy}[\mathrm{Im}(B_\tau)], \tag{3.6}$$

and if $\mathrm{rad}(A) < 1$,

$$\mathrm{hcap}(A) = \frac{2}{\pi} \int_0^\pi \mathbf{E}^{e^{i\theta}}[\mathrm{Im}(B_\tau)] \sin\theta \, d\theta. \tag{3.7}$$

PROOF. Note that $\phi(z) = \mathrm{Im}[z - g_A(z)]$ is a bounded harmonic function on $\mathbb{H} \setminus A$. Hence (3.5) follows from the optional sampling theorem since $\mathrm{Im}[g_A(w)] = 0$ for $w \in \mathbb{R} \cup A$. Also, (3.6) follows immediately from (3.5) and the definition of hcap. Now assume $\mathrm{rad}(A) < 1$. For $y > 1$, let $p(iy, \cdot)$ denote the density of $B_{\tau_{\mathbb{H} \setminus \mathbb{D}}}$ assuming $B_0 = iy$. The strong Markov property gives

$$\mathbf{E}^{iy}[\mathrm{Im}(B_\tau)] = \int_0^\pi \mathbf{E}^{e^{i\theta}}[\mathrm{Im}(B_\tau)] \, p(iy, e^{i\theta}) \, d\theta.$$

But (2.12) gives

$$p(iy, e^{i\theta}) = y^{-1} \frac{2}{\pi} \sin\theta \, [1 + O(y^{-1})], \quad 0 < \theta < \pi,$$

from which (3.7) follows. □

---

[1] If w, z are complex numbers we write $[w, z], (w, z], (w, z)$ for the closed, half-open, and open line segments connecting w, z.

We could have used (3.6) to define hcap. In fact, this definition does not require $\mathbb{H} \setminus A$ to be simply connected so we will define hcap(A) for bounded subsets of $\mathbb{H}$ by (3.6). If $A \subset \{|z| < 1\}$, then hcap(A) also satisfies (3.7). Using (2.4), we can see that hcap(A) is monotone in A.

If $A, A' \in \mathcal{Q}$, $A \subset A'$, then $g_{A'} = g_{g_A(A' \setminus A)} \circ g_A$, and hence

$$(3.8) \quad \text{hcap}(A') = \text{hcap}(A) + \text{hcap}(g_A(A' \setminus A)).$$

This also demonstrates monotonicity. In particular,

$$(3.9) \quad \text{hcap}(A) \leq \text{hcap}(\text{rad}(A)\,(\mathbb{H} \cap \overline{\mathbb{D}})) = [\text{rad}(A)]^2 \, \text{hcap}(\mathbb{H} \cap \overline{\mathbb{D}}) = [\text{rad}(A)]^2,$$

In contrast to cap, there is no similar lower bound for connected A; see (3.4). The next proposition shows that hcap satisfies an inequality that is characteristic of "capacities".

PROPOSITION 3.42. *Suppose A_1, A_2 are bounded subsets of $\mathbb{H}$ with $A_j = \overline{A_j} \cap \mathbb{H}$. Then*

$$(3.10) \quad \text{hcap}(A_1) + \text{hcap}(A_2) \geq \text{hcap}(A_1 \cup A_2) + \text{hcap}(A_1 \cap A_2).$$

PROOF. Let $\tau_j = \tau_{\mathbb{H} \setminus A_j}, \tau = \tau_{\mathbb{H} \setminus (A_1 \cup A_2)}, \eta = \tau_{\mathbb{H} \setminus (A_1 \cap A_2)}$. Then for y large,

$$\mathbf{E}^{iy}[\text{Im}(B_{\tau_1})] + \mathbf{E}^{iy}[\text{Im}(B_{\tau_2})] =$$

$$\mathbf{E}^{iy}[\text{Im}(B_\tau)] + \mathbf{E}^{iy}[\text{Im}(B_{\tau_1}); \tau_2 < \tau_1] + \mathbf{E}^{iy}[\text{Im}(B_{\tau_2}); \tau_1 \leq \tau_2].$$

Using (2.4), we get

$$\mathbf{E}^{iy}[\text{Im}(B_{\tau_1}); \tau_2 < \tau_1] + \mathbf{E}^{iy}[\text{Im}(B_{\tau_2}); \tau_1 \leq \tau_2]$$
$$\geq \mathbf{E}^{iy}[\text{Im}(B_\eta); \tau_2 < \tau_1] + \mathbf{E}^{iy}[\text{Im}(B_\eta); \tau_1 \leq \tau_2]$$
$$= \mathbf{E}^{iy}[\text{Im}(B_\eta)].$$

If we multiply by y and let $y \to \infty$ we get (3.10). □

We now consider the real part of g_A. If $x \in \mathbb{R}$ with $|x| > \text{rad}(A)$, then $g_A(x)$ is well defined since the boundary of $\mathbb{H} \setminus A$ is straight near x. If $\partial(\mathbb{H} \setminus A)$ is a simple curve, then g_A is well defined on all of $\partial(\mathbb{H} \setminus A)$.

PROPOSITION 3.43. *Suppose $A \in \mathcal{Q}$. If $x > \text{rad}(A)$,*

$$g_A(x) = \lim_{y \to \infty} \pi y \left[\frac{1}{2} - \mathbf{P}^{iy}\{B_\tau \in [x, \infty)\} \right].$$

If $x < -\text{rad}(A)$,

$$g_A(x) = \lim_{y \to \infty} \pi y \left[\mathbf{P}^{iy}\{B_\tau \in (-\infty, x]\} - \frac{1}{2} \right].$$

Here $\tau = \tau_{\mathbb{H} \setminus A}$ is as in Proposition 3.41.

PROOF. We will prove the first equality; the second is done similarly. Let $x > \text{rad}(A)$. If $A = \emptyset$ (so that $g_A(z) = z$), then (see Exercise 2.23),

$$\lim_{y \to \infty} \pi y \left[\frac{1}{2} - \mathbf{P}^{iy}\{B_\tau \in [x, \infty)\} \right] = \lim_{y \to \infty} \pi y \, \mathbf{P}^{iy}\{B_\tau \in [0, x]\}$$

$$(3.11) \qquad\qquad\qquad\qquad = \lim_{y \to \infty} \pi y \int_0^{x/y} \frac{1}{\pi(1 + s^2)} \, ds = x.$$

For $A \neq \emptyset$, write $g_A = u_A + iv_A$. By conformal invariance of Brownian motion, if σ is the first time that B_t reaches $\mathbb{R}$,

$$\mathbf{P}^{iy}\{B_\tau \in [x, \infty)\} = \mathbf{P}^{g_A(iy)}\{B_\sigma \in [g_A(x), \infty)\}$$
$$= \mathbf{P}^{iv_A(iy)}\{B_\sigma \in [g_A(x) - u_A(iy), \infty)\}.$$

But as $y \to \infty$, $g_A(iy) = iy - (i/y) + O(y^{-2})$; in particular, $v_A(iy) \sim y$ and $y\, u_A(iy) \to 0$. Hence the proposition follows from (3.11). $\square$

COROLLARY 3.44. *Suppose $A \in \mathcal{Q}$. If $\text{rad}(A) \leq 1$ and $x > 1$,*

$$x \leq g_A(x) \leq x + \frac{1}{x}.$$

If $\text{rad}(A) = 1$ and $x < -1$,

$$x + \frac{1}{x} \leq g_A(x) \leq x.$$

Moreover, for any A,

(3.12) $$|g_A(z) - z| \leq 3\,\text{rad}(A), \quad z \in \mathbb{H} \setminus A.$$

PROOF. Assume $x > 0$ ($x < 0$ is the same). From Proposition 3.43, we see that $g_A(x)$ is monotone in A. Hence, if $\text{rad}(A) \leq 1$ and $x > 1$, $g_A(x)$ is maximized when $A = \mathbb{D} \cap \mathbb{H}$ for which $g_A(x) = x + (1/x)$. By scaling it suffices to prove the last inequality when $\text{rad}(A) = 1$. In this case, $-2 \leq g_A(-1) \leq g_A(1) \leq 2$ and hence $|g_A(z) - z| \leq 3$ for $z \in \partial(\mathbb{H} \setminus A)$. (Here we are assuming that g_A is defined up to the boundary — if this is not the case, we can approximate A by a sequence A_n of nice sets.) Since $|g_A(z) - z| \leq 3$ on $\partial(\mathbb{H} \setminus A)$, the maximal principle implies this is true on $\mathbb{H} \setminus A$. $\square$

REMARK 3.45. The constant 3 in the corollary is optimal. For example, if we let A_θ be the hull bounded by the line segment $[-1, e^{i\theta}]$ and the circular arc $[e^{i\theta}, e^{i\pi}]$, then as $\theta \to 0+$, $g_{A_\theta}(-1+\theta) \to 2$ (since as $\theta \to 0+$, A_θ approaches $\overline{\mathbb{D}} \cap \mathbb{H}$).

We know that

$$g_A(z) = z + \frac{\text{hcap}(A)}{z} + O(\frac{1}{|z|^2}), \quad z \to \infty,$$

where the $O(\cdot)$ term depends on A. The next proposition derives a uniform estimate on the error in terms of $\text{hcap}(A)$ and $\text{rad}(A)$.

PROPOSITION 3.46. *There is a $c < \infty$ such that for all $A \in \mathcal{Q}$ and $|z| \geq 2\,\text{rad}(A)$,*

$$\left| z - g_A(z) + \frac{\text{hcap}(A)}{z} \right| \leq c\,\frac{\text{rad}(A)\,\text{hcap}(A)}{|z|^2}.$$

PROOF. By scaling we may assume that $\text{rad}(A) = 1$. Let $h(z) = z - g_A(z) + z^{-1}\text{hcap}(A)$ and let

$$v(z) = \text{Im}[h(z)] = \text{Im}[z - g_A(z)] - \frac{\text{Im}(z)}{|z|^2}\,\text{hcap}(A).$$

Let σ be the first time that a Brownian motion B_t reaches $\partial \mathbb{D} \cup \mathbb{R}$ and let $p(z, e^{i\theta})$ denote the density of B_σ on $\partial \mathbb{D}$. Then, by (3.5),

$$\text{Im}[z - g_A(z)] = \int_0^\pi \mathbf{E}^{e^{i\theta}}[\text{Im}(B_\tau)]\, p(z, e^{i\theta})\, d\theta.$$

But (2.12) tells us that

$$p(z, e^{i\theta}) = \frac{\text{Im}(z)}{|z|^2} \frac{2}{\pi} \sin\theta \, [1 + O(\frac{1}{|z|})].$$

Using (3.7), we then get $|v(z)| \leq c \, \text{hcap}(A) \, \text{Im}(z)/|z|^3$ for $|z| \geq 2$. From this (see Exercise 2.17 and the remark following), we get that

$$|\partial_x v(z)| \leq c \, \text{hcap}(A)/|z|^3, \quad |\partial_y v(z)| \leq c \, \text{hcap}(A)/|z|^3,$$

and hence $|h'(z)| \leq c \, \text{hcap}(A)/|z|^3$ for $|z| \geq 2$. Since $h(iy) \to 0$ as $y \to \infty$, for $y \geq 2$,

$$|h(iy)| = \left| \int_y^\infty h'(iy_1) \, dy_1 \right| \leq \int_y^\infty |h'(iy_1)| \, dy_1 \leq c \, \frac{\text{hcap}(A)}{y^2}.$$

Similarly, by integrating along the circle of radius $r \geq 2$, $|h(re^{i\theta})| \leq |h(ir)| + c \, \text{hcap}(A) \, r^{-2}$. □

REMARK 3.47. This proof uses a useful technique to show that an analytic function $f = u + iv$ is close to the zero function when we know that v is close to zero. If $f(z) = 0$ at one point (in the above case, that point is infinity) and $|f'|$ is small, then f is small. But to estimate $|f'|$ we need only estimate the partial derivatives of v, and these in turn can be estimated in terms of the magnitude of v. For another example, see Proposition 3.51 at the end of this section.

REMARK 3.48. As before, let τ be the first time that B_t reaches $\mathbb{R} \cup A$ and let $\sigma = \tau_{\mathbb{C} \setminus r\mathbb{D}}$ be the first time that B_t reaches the circle of radius r where $r \geq \text{rad}(A)$. Suppose F is a bounded nonnegative function on $\mathbb{R} \cup A$ that vanishes on $\mathbb{R}$ and let $v(z) = \mathbf{E}^z[F(B_\tau)]$ be the harmonic function in $\mathbb{H} \setminus A$ with boundary value F. Then for $|z| > r$,

$$v(z) = \mathbf{E}^z[F(B_\tau)] = \mathbf{P}^z\{\sigma < \tau\} \, \mathbf{E}^z[F(B_\tau) \mid \sigma < \tau],$$

We can consider (2.12) as a combination of two estimates:

$$\mathbf{P}^z\{\sigma < \tau\} = \frac{r \, \text{Im}(z)}{|z|^2} [1 + O(\frac{r}{|z|})] = r \, \text{Im}(-\frac{1}{z}) \, [1 + O(\frac{r}{|z|})],$$

and if $|w| \geq |z|$,

$$\mathbf{E}^w[F(B_\tau) \mid \sigma < \tau] = \mathbf{E}^z[F(B_\tau) \mid \sigma < \tau] \, [1 + O(\frac{r}{|z|})].$$

REMARK 3.49. If $A_t, t > 0$, is a parametrized collection of sets, where $A_t \in \mathcal{Q}$ with $\text{hcap}(A_t) = t$ and $\text{rad}(A_t) \to 0$ as $t \to 0+$, then Proposition 3.46 implies for every $z \in \mathbb{H}$,

$$\lim_{t \to 0+} \frac{g_{A_t}(z) - z}{t} = \frac{1}{z}.$$

REMARK 3.50. The notion of half-plane capacity we have defined here is the one that arises in considering Loewner chains. However, there is another natural definition of a capacity from infinity in $\mathbb{H}$,

(3.13) $$\text{cap}_{\mathbb{H}}(A) = \lim_{y \to \infty} y \, \mathbf{P}^{iy}\{B_\tau \in A\}.$$

Using conformal invariance and Exercise 2.23, we can see that this is the same as
$$\operatorname{cap}_{\mathbb{H}}(A) = \lim_{y \to \infty} y\, \mathbf{P}^{iy}\{B_\sigma \in g_A(A)\} = \frac{1}{\pi}\operatorname{length}(g_A(A)),$$
where σ is the first time that the Brownian motion reaches $\mathbb{R}$. This capacity satisfies the different scaling rule $\operatorname{cap}_{\mathbb{H}}(rA) = r\operatorname{cap}_{\mathbb{H}}(A)$. It is not difficult (using an argument as in Exercise 2.7) to show that there exist constants c_1, c_2 such that for all connected $A \in \mathcal{Q}$,

(3.14) $$c_1 \operatorname{diam}(A) \le \operatorname{cap}_{\mathbb{H}}(A) \le c_2 \operatorname{diam}(A).$$

Also, we leave it to the reader to verify the relation
$$\operatorname{cap}_{\mathbb{H}}(A_1) + \operatorname{cap}_{\mathbb{H}}(A_2) \ge \operatorname{cap}_{\mathbb{H}}(A_1 \cup A_2) + \operatorname{cap}_{\mathbb{H}}(A_1 \cap A_2).$$

Suppose $A \in \mathcal{Q}, 0 \in \overline{A}$, and $\overline{A}$ is connected, i.e., $\overline{A} \setminus A$ is a compact interval of $\mathbb{R}$. Let $\hat{A} = \{z : z \in \overline{A} \text{ or } \overline{z} \in A\}$. Then $\hat{A} \in \mathcal{H}$. Let $\hat{g} : \mathbb{C} \setminus \hat{A} \to \mathbb{C} \setminus \overline{\mathbb{D}}$ be the conformal transformation with
$$\hat{g}(z) \sim e^{-\operatorname{cap}(\hat{A})}\, z, \quad z \to \infty.$$

Here cap denotes logarithmic capacity as in §3.3. Then $g^* := \hat{g} + (1/\hat{g})$ maps $\mathbb{H} \setminus A$ onto $\mathbb{H}$ with $g^*(A) = [-2, 2]$. But
$$g_A(z) = e^{\operatorname{cap}(\hat{A})}\, g^*(z) + r,$$
for some $r \in \mathbb{R}$. Therefore, $\operatorname{length}[g_A(A)] = 4\, e^{\operatorname{cap}(\hat{A})}$ and $\operatorname{cap}_{\mathbb{H}}(A) = (4/\pi)e^{\operatorname{cap}(\hat{A})}$. There is an expression for $\operatorname{hcap}(A)$ that is similar to (3.13) but replaces B with an "$\mathbb{H}$-excursion"; see Corollary 5.16.

PROPOSITION 3.51. *There is a $c < \infty$ such that if $A \in \mathcal{Q}$ and $\epsilon = \sup\{\operatorname{Im}(z) : z \in A\}$, then for all z with $\operatorname{Im}(z) > 2\epsilon$,*
$$|g_A(z) - z| \le c\epsilon \left[1 + \log^+\left(\frac{\operatorname{diam}(A)}{|z| \wedge \operatorname{diam}(A)}\right) \right].$$

PROOF. By scaling we may assume $\operatorname{diam}(A) = 1$, and by translation invariance it suffices to prove the result for $z = iy$. Let $h(z) = g_A(z) - z$, $v(z) = \operatorname{Im}[h(z)]$. Note that $|v(iy)| \le \epsilon \mathbf{P}^{iy}\{B_\tau \in A\} \le c\epsilon [1 \wedge y^{-1}]$. Therefore, $|\partial_x v(iy)|, |\partial_y v(iy)| \le c\epsilon\, y^{-1}\, [1 \wedge y^{-1}]$, and hence similar bounds hold for $h'(iy)$. Since $h(iy) \to 0$ as $y \to \infty$,
$$|h(iy)| = \left|\int_y^\infty h'(ir)\, dr\right| \le c\epsilon \int_y^\infty r^{-1}\, [1 \wedge r^{-1}]\, dr \le c\epsilon\, [1 + \log^+(1 \vee y^{-1})].$$
□

LEMMA 3.52. *There exists a $c > 0$ such that if $-\infty < x_1 < x_2 < \infty, 0 < r \le (x_2 - x_1)/2$ and $A = \{x + iy : x_1 \le x \le x_2, 0 < y \le r\}$, then for all $z \in \mathbb{H}$,*
$$|g_A^{-1}(z) - z| \le c\, r\, \log\left(\frac{x_2 - x_1}{r}\right).$$

PROOF. By the scaling and translation rules, it suffices to prove the result when $x_1 = -1, x_2 = 1, r \le 1$, and for this this we can give the conformal map explicitly, see, e.g. [**42**, §13.5], and the estimate can be derived from this. However, we will sketch a proof of the estimate without using the explicit form of the map.

By the maximal principle, it suffices to establish the estimate for $x \in \mathbb{R}$. We will discuss one case here, the estimate

(3.15) $$g_A(1) - 1 \leq cr \log(2/r).$$

Let B_t be a Brownian motion started at iy ($y > r$), and let $\tau = \tau_{\mathbb{H}\setminus A}$ as before and $\sigma = \sigma_r = \inf\{t : \text{Im}(B_t) = r\}$. By symmetry, $g_A(ir) = 0$, and hence, by conformal invariance

$$g_A(1) = \lim_{y \to \infty} \pi y \, \mathbf{P}^{iy}\{B_\tau \in A_+\}.$$

Here $A_+ = A \cap \{\text{Re}(z) \geq 0\}$. If $p(iy, \cdot)$ denotes the density of $\text{Re}[B_\sigma]$ given $B_0 = iy$, we know that

$$p(iy, x) \sim \frac{1}{\pi y}, \quad y \to \infty.$$

Hence, the strong Markov property gives

$$g_A(1) = \int_{-\infty}^{\infty} \theta(x) \, dx.$$

where $\theta(x) = \theta(x, r) = \mathbf{P}^{x+ri}\{B_\tau \in A_+\}$. We now use standard estimates (e.g., gambler's ruin) for Brownian motion to see that

$$\theta(x) = 0, \quad -1 \leq x < 0,$$

$$\theta(x) = 1, \quad 0 \leq x \leq 1,$$

$$\theta(x) \leq c \left(1 \wedge \frac{r}{x-1}\right), \quad 1 < x \leq 2,$$

$$\theta(x) \leq c \frac{r}{x^2}, \quad x \leq -1 \text{ or } x \geq 2.$$

This gives (3.15).
□

REMARK 3.53. This lemma used special properties of the rectangle. If we had considered instead the triangle with sides $[-(1/2), -(1/2)+ir]$, $[-(1/2)+ir, (1/2)+ir]$, $[-(1/2), (1/2)+ir]$, then as $r \to 0+$, $g_A(-(1/2)+r) \to 1/2$, and hence is not $O(r)$.

REMARK 3.54. With the aid of the lemma, we can see that we can approximate any nonempty $A \in \mathcal{Q}$ by $A_n \in \mathcal{Q}$ whose boundary are simple curves. To be more precise, let $I = I_A = [x_1, x_2]$ be the smallest closed interval containing $\overline{g_A(A)}$, i.e., the smallest closed interval containing all limit points of $g_A(z_n)$ for sequences $z_n \in \mathbb{H} \setminus A$ approaching points in A. For each n, let $V_n = \{x + iy : x_1 \leq x \leq x_2, 0 < y \leq 1/n\}$ and let $A_n = g_A^{-1}(V_n) \cup A$. Since $g_{A_n} = g_{V_n} \circ g_A$, Lemma 3.52 tells us that $|g_A^{-1}(z) - g_{A_n}^{-1}(z)| \leq c_A \log n/n$ for $z \in \mathbb{H}$. Let J be the smallest closed interval (perhaps one point) containing $\overline{A} \cap \mathbb{R}$. Then J is also the smallest closed interval containing A_n and

$$\bigcap_{n \geq 1} \overline{A_n} = A \cup J.$$

3.5. Transformations on $\mathbb{D}$

Suppose $A \subset \mathbb{D}$ such that $A = \overline{A} \cap \mathbb{D}$ and $D_A := \mathbb{D} \setminus A$ is a simply connected domain containing the origin. Let $\tilde{g}_A$ be the unique conformal transformation of D_A onto $\mathbb{D}$ with $\tilde{g}_A(0) = 0, \tilde{g}'_A(0) > 0$. The Schwarz lemma applied to $\tilde{g}_A^{-1}$ shows that $\tilde{g}'_A(0) \geq 1$. If $\text{diam}(A)$ is small, then near A, $\mathbb{D} \setminus A$ looks similar to the upper half plane with a compact $\mathbb{H}$-hull removed. We will derive a proposition analogous to Proposition 3.46 for $\tilde{g}_A$. There is an essential difference. In the last section we consider the map g_A which has infinity, which is a boundary point for $\mathbb{H} \setminus A$, as a fixed point. Here $\tilde{g}_A$ fixes the origin, which is an interior point of $\mathbb{D} \setminus A$. However, if $\text{diam}(A)$ is small, this should make only a slight difference. We would like to use the logarithm (or more precisely $-i \log$) to map the disk to the upper half plane sending the origin to infinity. However log is not a conformal transformation (or even well defined) on $\mathbb{D} \setminus \{0\}$. We can do this locally near the unit circle and this is the basis for our argument.

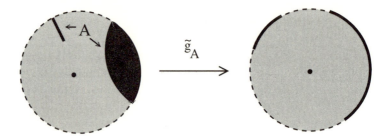

Since $\tilde{g}_A(z)/z$ is a bounded, non-zero analytic function on D_A with value $\tilde{g}'_A(0)$ at the origin, we can define $h_A(z) = \log(\tilde{g}_A(z)/z)$ where the branch of the logarithm is chosen so that $h_A(0) = \log \tilde{g}'_A(0) \geq 0$. Since $\phi(z) := \log|\tilde{g}_A(z)| - \log|z|$ is a bounded harmonic function for $z \in D_A \setminus \{0\}$ with boundary value $-\log|z|$ on ∂D_A,

$$\text{Re}[h_A(z)] = \log|\tilde{g}_A(z)| - \log|z| = -\mathbf{E}^z[\log|B_\sigma|], \quad z \in D_A,$$

where $\sigma = \sigma_A$ is the first time that the Brownian motion B_t reaches $A \cup \partial \mathbb{D}$. Let $\text{rad}_\theta(A)$ denote the radius of A with respect to $e^{i\theta}$, i.e., the smallest r such that $A \subset \overline{\mathcal{B}}(e^{i\theta}, r)$.

PROPOSITION 3.55. *There is a $c < \infty$ such that if A is as in the first paragraph of this section, $\theta \in [0, 2\pi)$, and $z \in \{w \in \mathbb{D} : |w - e^{i\theta}| \geq c\, \text{rad}_\theta(A)\}$,*

$$\left| h_A(z) - \frac{e^{i\theta} + z}{e^{i\theta} - z} \log \tilde{g}'_A(0) \right| \leq c\, \frac{\text{rad}_\theta(A) \log \tilde{g}'_A(0)}{|e^{i\theta} - z|^2}.$$

PROOF. Since $\tilde{g}_{e^{i\theta} A}(z) = e^{i\theta} \tilde{g}_A(e^{-i\theta} z)$, it suffices to prove the result when $\theta = 0$. Let $r = \text{rad}_0(A)$, and we may assume $r < 1/4$, since we can always choose $c \geq 8$. Let

$$\Phi(z) = i\, \frac{1-z}{1+z}$$

which is the linear fractional transformation of $\mathbb{D}$ onto $\mathbb{H}$ with $\Phi(0) = i, \Phi(1) = 0$, $\Phi(-1) = \infty$, and let $A' = \Phi(A) \in \mathcal{Q}$. Note that $\text{rad}(A') \leq 2r < 1/2$ since $|\Phi'(z)| < 2$ for $|z - 1| < 1$. Let $\hat{h}(w) = i\, h \circ \Phi^{-1}(w)$ and $F(w) = -\log|\Phi^{-1}(w)|$.

3.5. TRANSFORMATIONS ON $\mathbb{D}$

Then $\hat{h}(w) = \hat{u}(w) + i\hat{v}(w)$ is an analytic function on $D' := \mathbb{H} \setminus A'$ satisfying $\hat{v}(w) = \mathbf{E}^w[F(B_{\tau_{D'}})]$. Also, $\hat{h}(i) = ih(0) = i\alpha$ where $\alpha = \log \tilde{g}'_A(0) \geq 0$. As in the proof of Proposition 3.46, we have for $|w| \geq 4r$,

$$\begin{aligned}\hat{v}(w) &= \hat{v}(i) \frac{\mathbf{P}^w\{\eta \leq \sigma\}}{\mathbf{P}^i\{\eta \leq \sigma\}} \left[1 + rO(\frac{1}{|w| \wedge 1})\right] \\ &= \alpha \operatorname{Im}[-\frac{1}{w}] \left[1 + rO(\frac{1}{|w| \wedge 1})\right],\end{aligned}$$

where $\eta = \eta_{2r}$ is the first time that the Brownian motion reaches the circle of radius $2r$. Writing $h_A = u + iv$, this gives for $|z - 1| \geq c'r$,

$$u(z) = \alpha \operatorname{Re}[\frac{1+z}{1-z}] \left[1 + rO(\frac{1}{|z-1|})\right].$$

Let

$$h_1(z) = h_A(z) - \alpha \frac{1+z}{1-z}.$$

Then for $|z - 1| \geq c'r$,

$$|\operatorname{Re} h_1(z)| \leq c\, r\, \alpha\, |1-z|^{-2}.$$

Then, as in Proposition 3.46,

$$|h'_1(z)| \leq c\, r\, \alpha\, |1-z|^{-3}.$$

(Again, we estimate the derivative by estimating the partials of $\operatorname{Re} h_1$, and for $|z|$ near 1 with $|z - 1| \geq cr$ we need to use a reflection argument.) Since $h_1(0) = 0$, integrating again gives $|h_1(z)| \leq c\, r\, \alpha\, |1-z|^{-2}$. $\square$

REMARK 3.56. We can write the conclusion of this proposition as

$$h_A(z) = \log \tilde{g}'_a(0) \left[\frac{e^{i\theta} + z}{e^{i\theta} - z} + O(\frac{\operatorname{rad}_A(\theta)}{|e^{i\theta} - z|^2})\right].$$

By exponentiating, we see that as $\operatorname{rad}_\theta(A) \to 0$,

$$\tilde{g}_A(z) = z\left[1 + \log \tilde{g}'_A(0)\left(\frac{e^{i\theta} + z}{e^{i\theta} - z} + O(\frac{\operatorname{rad}_A(\theta)}{|e^{i\theta} - z|^2})\right)\right].$$

If $A_t, t > 0$ are given with $\tilde{g}'_{A_t}(0) = e^t$ and $\operatorname{rad}_\theta(A_t) \to 0$ as $t \to 0+$, then for all $z \in \mathbb{D}$,

$$\lim_{t \to 0+} \frac{\tilde{g}_{A_t}(z) - z}{t} = z\, \frac{e^{i\theta} + z}{e^{i\theta} - z}.$$

PROPOSITION 3.57. *There is a $c < \infty$ such that if $\epsilon > 0$, A, h_A are as in the beginning of this section, $A \subset \{w \in \mathbb{D} : |w| > 1 - \epsilon\}$ and $|z| \leq 1 - 2\epsilon$,*

$$|h_A(z)| \leq c\epsilon\, [1 - \log(1 - |z|)].$$

PROOF. Since $\tilde{g}'_A(0) \in [0, (1-\epsilon)^{-1}]$, $h_A(0) = \log g'_A(0) \leq c\epsilon$. Let $u(z) = \operatorname{Re}[h_A(z)]$. Since $|u(z)| = -\mathbf{E}^z[\log|B_\sigma|] \leq c\epsilon$, $|\partial_x u(z)|, |\partial_y u(z)| \leq c\, \epsilon/(1-|z|)$ for $|z| \leq 1 - 2\epsilon$ (see Exercise 2.17). Therefore,

$$|h(z)| \leq |h(0)| + \int_0^{|z|} |h'(rz/|z|)|\, dr$$

$$\leq c\left[\epsilon + \int_0^{|z|} \frac{\epsilon}{1-r}\, dr\right] = c\epsilon\,[1 - \log(1-|z|)].$$

$\square$

PROPOSITION 3.58. *There is a c such that if A and Φ are as in the proof of Proposition 3.55 with $\mathrm{rad}_0(A) < 1/4$, then*

$$|\log \tilde{g}_A'(0) - 2\,\mathrm{hcap}[\Phi(A)]| \leq c\,\mathrm{hcap}[\Phi(A)]\,\mathrm{rad}_0(A).$$

PROOF. Let $r = \mathrm{rad}_0(A), \alpha = \mathrm{hcap}[\Phi(A)]$. Note that $u_A := \Phi^{-1} \circ g_{\Phi(A)} \circ \Phi$ is a conformal transformation of D_A onto $\mathbb{D}$ with $u_A(0) = w := \Phi^{-1}(g_{\Phi(A)}(i))$. Hence, if $T_w = (z-w)/(1-\overline{w}z)$ is the Möbius transformation with $T_w(w) = 0$, $T_w'(0) > 0$,

$$g_A'(0) = |(T_w \circ u_A)'(0)| = |T_w'(w)|\,|(\Phi^{-1})'(g_{\Phi(A)}(i))|\,|g_{\Phi(A)}'(i)|\,|\Phi'(0)|.$$

Simple calculation gives

$$\Phi'(0) = 2i, \quad (\Phi^{-1})'(z) = \frac{2z}{(i+z)^2}, \quad T_w'(w) = \frac{1}{1-|w|^2} = 1 + O(|w|^2).$$

From Proposition 3.46 we see that $g_{\Phi(A)}(i) = (1-\alpha)\,i + O(\alpha r)$. Differentiating the expression in Proposition 3.46 we get $g_A'(i) = 1 + \alpha + O(\alpha r)$, and hence

$$|(\Phi)^{-1}(g_{\Phi_A}(i))| = (1/2)\,[1 + \alpha + O(\alpha r)].$$

Combining all of these estimates gives $g_A'(0) = 1 + 2\alpha + O(\alpha r)$. $\square$

REMARK 3.59. The factor of 2 in the proposition comes from the fact that $|\Phi'(1)| = 1/2$. More generally, if A_n is a sequence of such sets with

$$r_n := \mathrm{rad}_0(A_n) \to 0,$$

and ϕ is a conformal map in a neighborhood of 1 locally mapping $\partial \mathbb{D}$ to $\mathbb{R}$ and $\mathbb{D}$ into $\mathbb{H}$, then

$$\log \tilde{g}_{A_n}'(0) = |\phi'(1)|^{-1}\,\mathrm{hcap}[\phi(A_n)]\,[1 + O(r_n)].$$

3.6. Carathéodory convergence

DEFINITION 3.60. A sequence of functions f_n on a domain D converges to f *uniformly on compact sets*, written $f_n \xrightarrow{u.c.} f$, if for each compact $K \subset D$, $f_n \longrightarrow f$ uniformly on K.

PROPOSITION 3.61. *Suppose $f_n \in \mathcal{S}$. Then there is an $f \in \mathcal{S}$ and a subsequence f_{n_j} such that $f_{n_j} \xrightarrow{u.c.} f$.*

PROOF. The Growth and Distortion Theorems give uniform bounds on $|f_n|$ and $|f_n'|$ on every closed disk of radius $r < 1$. Hence, the existence of a a subsequence f_{n_j} and an analytic function such that f_{n_j} converges to f and f_{n_j}' converges to f' pointwise follows from Proposition 2.29. These uniform estimates also show that $f_{n_j} \xrightarrow{u.c.} f$. Clearly $f(0) = 0, f'(0) = 1$ and univalence of f follows from Lemma 3.5. $\square$

DEFINITION 3.62. Suppose D_n is a sequence of domains in $\mathcal{A}$, and let f_n be the conformal transformation of $\mathbb{D}$ onto D_n with $f_n(0) = 0, f_n'(0) > 0$. We define *convergence in the Carathéodory sense*, denoted $\xrightarrow{Cara}$ as follows:

- $D_n \xrightarrow{Cara} \mathbb{C}$ if $f_n'(0) \to \infty$;
- $D_n \xrightarrow{Cara} \{0\}$ if $f_n'(0) \to 0$ (which implies that $f_n \xrightarrow{u.c.} 0$);
- $D_n \xrightarrow{Cara} D \in \mathcal{A}$ if $f_n \xrightarrow{u.c.} f$ where f is the conformal transformation of $\mathbb{D}$ onto D with $f(0) = 0, f'(0) > 0$.

If D_n is a sequence of simply connected domains containing z_n we say that D_n converges to D in the Carathéodory sense with respect to z_n and z, written $(D_n, z_n) \xrightarrow{Cara} (D, z)$, if $D_n - z_n \xrightarrow{Cara} D - z$.

If $D_n \in \mathcal{A}$, then there is a subsequence $\{n_j\}$ such that $f_{n_j}'(0)$ converges (possibly to 0 or ∞). Using Proposition 3.61, we see that there is a subsubsequence, which we also denote D_{n_j}, and a $D \in \mathcal{A} \cup \{\{0\}, \mathbb{C}\}$ such that $D_{n_j} \xrightarrow{Cara} D$. There is an equivalent, more topological, definition of $\xrightarrow{Cara}$ that we give as a proposition. We will only state it in the case that $D_n, D \in \mathcal{A}$.

PROPOSITION 3.63. *Suppose $D_n, D \in \mathcal{A}$. Then $D_n \xrightarrow{Cara} D$ if and only if the following holds. For each subsequence $\{n_j\}$, call the kernel of the subsequence the largest domain $\tilde{D}$ containing the origin such that for all compact $K \subset \tilde{D}$, $K \subset D_{n_j}$ for all but finitely many j. Then the kernel of every subsequence is D.*

PROOF. Suppose $D_n \xrightarrow{Cara} D$; then $f_n \xrightarrow{u.c.} f$, where $f_n, f \in \mathcal{S}^*$ with $f_n(\mathbb{D}) = D_n, f(\mathbb{D}) = D$. Let $r > 0$. Since $f_n \xrightarrow{u.c.} f$, for all n sufficiently large, $f((1-2r)\overline{\mathbb{D}}) \subset f_n((1-r)\mathbb{D})$. Any compact $K \subset D$ is contained in $f((1-2r)\overline{\mathbb{D}})$ for some $r > 0$, so $K \subset D_n$ for all sufficiently large n. Hence, the kernel of every subsequence contains D. Now suppose for some subsequence, which we write as just D_n, the kernel is $\tilde{D}$, which we know contains D. Let K_m be an increasing sequence of compact sets whose union is $\tilde{D}$. For each m, there is an N_m such that $K_m \subset D_n$ for all $n \geq N_m$. Since $\{f_n^{-1}\}_{n \geq N_m}$ is a uniformly bounded sequence of univalent functions on K_m, we can use a diagonalization argument to find a subsequence, which we also write as f_n^{-1}, that converges to an analytic function $\hat{g}$ on $\tilde{D} = \cup_m K_m$. Note that $\hat{g} = f^{-1}$ on D. Since $\hat{g}$ is not constant, Lemma 3.5 implies that $\hat{g}$ is univalent. But $\hat{g}(\tilde{D}) = \mathbb{D} = f^{-1}(D)$. Hence $\tilde{D} = D$.

Conversely, assume that the kernel of every subsequence is $D \in \mathcal{A}$ and choose r_1, r_2 with $0 < r_1 < \text{dist}(0, \partial D) < r_2 < \infty$. Then for all n sufficiently large, $\overline{\mathcal{B}}(0, r_1) \subset D_n, \overline{\mathcal{B}}(0, r_2) \not\subset D_n$. Using Corollary 3.19, we see that $f_n'(0)$ is bounded away from zero and infinity. We can find a subsequence that converges to f uniformly on compact sets; in fact, from the argument in the previous paragraph we can see that $f(\mathbb{D}) = D$. Since all subsequences contain subsubsequences that converge to f, $f_n \xrightarrow{u.c.} f$. □

REMARK 3.64. It follows that if $D_n, D \in \mathcal{A}$ and $0 < r < 1$, then $f_n(r\overline{\mathbb{D}}) \to f(r\overline{\mathbb{D}})$ in the Hausdorff metric.[2] However, the domains can be much different. For

---

[2]The distance between two compact subsets A, B is
$$\max \{ \sup\{\text{dist}(a, B) : a \in A\}, \sup\{\text{dist}(b, A) : b \in B\} \}.$$

example, if
$$D_n = \mathbb{C} \setminus \left([1, \infty) \cup \{e^{i\theta} : 0 \leq \theta \leq 2\pi - \frac{1}{n}\} \right),$$
then $D_n \xrightarrow{Cara} \mathbb{D}$. This can be seen using Proposition 3.63, since the kernel of every subsequence contains $\mathbb{D}$ but contains no point in $\partial \mathbb{D}$.

REMARK 3.65. Recall that we call ∂D *analytic* if it is locally analytic at every point. A bounded Jordan domain D is analytic if and only if for every conformal transformation $f : \mathbb{D} \to D$ there is an $\epsilon > 0$ such that f can be extended to a conformal transformation of $(1 + \epsilon)\mathbb{D}$. Any simply connected domain can be approximated in the Carathéodory sense by analytic domains. In fact, if $D \in \mathcal{A}$, f is the corresponding Riemann map, and $r_n \uparrow 1$, then the domains $D_n := f(r_n \mathbb{D})$ are analytic domains increasing to D with $D_n \xrightarrow{Cara} D$. Note that $f_{D_n}(z) = f(r_n z)$.

DEFINITION 3.66. Suppose D is a domain and $K \subset \partial D$ is closed. A sequence of functions f_n converges to f *uniformly away from* K if for every $\epsilon > 0$, $f_n \longrightarrow f$ uniformly on $D \setminus \{\text{dist}(z, K) < \epsilon\}$.

DEFINITION 3.67. Suppose $A_n, A \in \mathcal{Q}$ and g_{A_n}, g_A are as defined in §3.4. Then $A_n \xrightarrow{Cara} A$ if $g_{A_n}^{-1}$ converges to g_A^{-1} uniformly away from $\mathbb{R}$.

PROPOSITION 3.68. *Suppose $A_n \in \mathcal{Q}$ with $\text{rad}(A_n)$ uniformly bounded. Then there exists an $A \in \mathcal{Q}$ (possibly empty) and a subsequence A_{n_j} such that $A_{n_j} \xrightarrow{Cara} A$.*

PROOF. Let $f_n = g_{A_n}^{-1}$. Assume, for ease, that $\text{rad}(A_n) \leq 1$ for each n and hence $\text{hcap}(A_n) \leq 1$ for each n. By taking a subsequence if necessary, we may assume $\text{hcap}(A_n) \to \alpha \in [0, 1]$. Since $\text{rad}(A_n) \leq 1$, (3.12) tells us that $h_n(z) := f_n(z) - z$ is a uniformly bounded sequence of analytic functions on $\mathbb{H}$. Hence there exists a subsequence, which we will denote by just h_n and an analytic function h such that $h_n \xrightarrow{u.c.} h$. Let $f(z) = h(z) + z$ so that $f_n \xrightarrow{u.c.} f$. Since each f_n is univalent and the limit function f is not constant, we know that f is a conformal transformation of $\mathbb{H}$ onto a subdomain of $\mathbb{H}$. Proposition 3.46 shows that there is a c such that for all n and all $|z| \geq 1$,
$$|f_n(z) - (z - \frac{\text{hcap}(A_n)}{z})| \leq \frac{c}{|z|^2},$$
and hence
$$|f(z) - (z - \frac{\alpha}{z})| \leq \frac{c}{|z|^2}.$$
Hence, $f(\mathbb{H}) = \mathbb{H} \setminus A$ where $A \in \mathcal{Q}$ with $\text{hcap}(A) = \alpha$. □

3.7. Extremal distance

If D, D' are bounded Jordan domains each with three specified boundary points, z_1, z_2, z_3 and z_1', z_2', z_3', respectively, ordered counterclockwise, then there is a unique conformal transformation from D to D' that can be extended continuously to $\overline{D}$ that maps z_j to z_j'. In this section we will consider simply connected domains with *four* specified boundary points.

Let $\mathcal{R}_L$ denote the $L \times \pi$ rectangle
$$\mathcal{R}_L = \{x + iy : 0 < x < L, 0 < y < \pi\},$$

with the four points being the vertices of the rectangle. Let $\partial_1 = [0, \pi i], \partial_2 = \partial_{2,L} = [L, L+\pi i]$ be the vertical boundaries, and let $\partial_3 = \partial_{3,L} = [0, L], \partial_4 = \partial_{4,L} = [i\pi, L + i\pi]$ be the horizontal boundaries. Let $\tau = \tau_{\mathcal{R}_L}$ and for each $z \in \mathcal{R}_L$, let

$$f_1(z) = f_{1,L}(z) = 2 \min\{\mathbf{P}^z\{B_\tau \in \partial_1\}, \mathbf{P}^z\{B_\tau \in \partial_2\}\},$$

$$f_2(z) = f_{2,L}(z) = 2 \min\{\mathbf{P}^z\{B_\tau \in \partial_3\}, \mathbf{P}^z\{B_\tau \in \partial_4\}\},$$

and let $\Theta(\mathcal{R}_L; \partial_1, \partial_2) = \sup\{f_1(z) : z \in D\}$.

PROPOSITION 3.69. *Let $\Theta(L) = \Theta(\mathcal{R}_L; \partial_1, \partial_2)$. The supremum in the definition of $\Theta(L)$ is obtained when z is the center $(L/2) + i(\pi/2)$. Also, $\Theta(L)$ is a continuous, strictly decreasing function of L with $\Theta(\pi^2/L) + \Theta(L) = 1$. Moreover, as $L \to \infty$,*

(3.16) $$\Theta(L) = (8/\pi) e^{-L/2} + O(e^{-L}).$$

PROOF. Let z_0 be a point at which the supremum is obtained. Then the two probabilities must be equal (if not, we could move z_0 in the real direction and increase the minimum), and hence $\text{Re}[z_0] = L/2$. On the line $\text{Re}[z_0] = L/2$, $f_1(z) = h(z) := \mathbf{P}^z\{B_\tau \in \partial_1\} + \mathbf{P}^z\{B_\tau \in \partial_2\}$. Note that $h((L/2)+iy) = \mathbf{P}\{\sigma_1 < \sigma_2\}$, where σ_1, σ_2 are independent random variables, the first having the distribution of the first exit of a one-dimensional Brownian motion from $[0, L]$ starting at the midpoint, and the second the distribution of the first exit of a one-dimensional Brownian motion from $[0, \pi]$ starting at y. However, for all $t > 0$, $\mathbf{P}^{\pi/2}\{\sigma_2 \geq t\} \geq \mathbf{P}^y\{\sigma_2 \geq t\}$, with equality only if $y = \pi/2$, since we can split the exit time starting at $\pi/2$ into the time to reach $\{y, \pi - y\}$ plus the time after that to leave the interval. Hence h is maximized for $y = \pi/2$. Strict decrease in L is immediate. Using separation of variables, we can find $\Theta(L)$ explicitly,

$$\Theta(L) = 2\mathbf{P}^{(L/2)+i(\pi/2)}\{B_\tau \in \partial_2\} = \sum_{n>0,\text{odd}} (-1)^{(n-1)/2} \frac{8 \sinh(nL/2)}{n\pi \sinh(nL)}.$$

This gives continuity and (3.16). Finally, the relation $\Theta(\pi^2/L) + \Theta(L) = 1$ follows from symmetry and the trivial equality

$$\sum_{j=1}^{4} \mathbf{P}^{(L/2)+i(\pi/2)}\{B_\tau \in \partial_j\} = 1.$$

□

A Jordan domain[3] D with four boundary points z_1, z_2, z_3, z_4 ordered counterclockwise is called a *conformal rectangle*. Let A_1 be the arc in ∂D between z_1, z_2 and A_2 the arc between z_3, z_4. The quantity $\Theta(D; A_1, A_2)$ is defined in the same way as $\Theta(\mathcal{R}_L; \partial_1, \partial_2)$, and it is easy to see that this is a conformal invariant; moreover the "center" of the rectangle, i.e., the point at which the supremum is obtained, is also conformally invariant. We define the π-*extremal distance* $L(A_1, A_2; D)$ or $L(D; A_1, A_2)$ between A_1 and A_2 to be the L such that $\Theta(D; A_1, A_2) = \Theta(\mathcal{R}_L; \partial_1, \partial_2)$. In fact, it is easy to see that there is a conformal transformation $f : D \to \mathcal{R}_L$ such that $f(z_1) = \pi i, f(z_2) = 0, f(z_3) = L, f(z_4) = L + \pi i$. If A_3, A_4 are the complementary arcs in ∂D to A_1, A_2, then $L(A_3, A_4; D) = \pi^2/L$.

---

[3]It is easy to modify the definition for domains whose boundaries are closed (not necessarily simple) curves in $\hat{\mathbb{C}}$.

The π-extremal distance is an example of a more general quantity defined on families of curves. Suppose D is an open subset of $\mathbb{C}$ (not necessarily connected) and Γ is a collection of piecewise C^1 curves $\gamma : [a,b] \to \mathbb{C}$ with $\gamma(a,b) \subset D$. (We will say that γ is in D if $\gamma(a,b) \subset D$.) Then the *module* of Γ, mod(Γ), is defined by

$$\mathrm{mod}(\Gamma) = \inf_\rho \int_D \rho(z)^2 \, dA(z),$$

where A denotes area and the infimum is over all nonnegative Borel functions ρ with

(3.17) $$\int_\gamma \rho(z) \, |dz| \geq 1, \quad \gamma \in \Gamma.$$

Note that the particular parametrization of γ is not relevant. The *extremal length* or *extremal distance* of Γ is defined to be $1/\mathrm{mod}(\Gamma)$ and the π-extremal distance of Γ is π times the extremal distance. (We will show below that this agrees with our previous definition.)

REMARK 3.70. We choose to use π-extremal distance rather than extremal distance throughout this book because it makes the formulas nicer. In particular, it avoids factors of π and $1/\pi$ in exponents in expressions such as (3.16).

EXAMPLE 3.71. Let $D = \mathcal{R}_L$ and let Γ be the collection of curves γ in $\mathcal{R}_L$ with $\gamma(a) \in \partial_1, \gamma(b) \in \partial_2$. If $\rho \equiv 1/L$ on $\mathcal{R}_L$, then ρ satisfies (3.17). Hence $\mathrm{mod}(\Gamma) \leq \pi/L$. However, if ρ satisfies (3.17), then by Hölder's inequality,

$$\int_{\mathcal{R}_L} \rho(z)^2 \, dA(z) \geq \frac{1}{\pi L} \left[\int_{\mathcal{R}_L} \rho(z) \, dA(z) \right]^2$$

$$= \frac{1}{\pi L} \left[\int_0^\pi \int_0^L \rho(x+iy) \, dx \, dy \right]^2 \geq \frac{\pi}{L}.$$

Therefore, $\mathrm{mod}(\Gamma) = \pi/L$ and the π-extremal distance of Γ is L.

EXAMPLE 3.72. Let $D = \{r_1 < |z| < r_2\}$ be an annulus, and let Γ be the collection of curves γ in D with $|\gamma(a)| = r_1, |\gamma(b)| = r_2$. If ρ satisfies (3.17), then by Hölder's inequality,

$$\int_D \rho(z)^2 \, dA(z) \geq \left[\int_D |z|^{-2} \, dA(z) \right]^{-1} \left[\int_D |z|^{-1} \rho(z) \, dA(z) \right]^2$$

$$= [2\pi \log(r_2/r_1)]^{-1} \left[\int_0^{2\pi} \int_{r_1}^{r_2} \rho(re^{i\theta}) \, dr \, d\theta \right]^2$$

$$\geq 2\pi \left[\log(r_2/r_1) \right]^{-1}.$$

Hence, $\mathrm{mod}(\Gamma) \geq 2\pi \left[\log(r_2/r_1) \right]^{-1}$. By choosing $\rho(z) = [|z| \log(r_2/r_1)]^{-1}$ we can see that $\mathrm{mod}(\Gamma) \leq 2\pi \left[\log(r_2/r_1) \right]^{-1}$. The π-extremal distance of Γ is $\log(r_2/r_1)/2$.

PROPOSITION 3.73. *Suppose $f : D \to D'$ is a conformal transformation and Γ is a family of curves in D. Let $f \circ \Gamma$ be the corresponding family of curves in D'. Then*

$$\mathrm{mod}(f \circ \Gamma) = \mathrm{mod}(\Gamma).$$

3.7. EXTREMAL DISTANCE

PROOF. If $\int_{f\circ\gamma} \tilde{\rho}(z)\,|dz| \geq 1$, and $\rho(z) := |f'(z)|\,\tilde{\rho}(f(z))$, then $\int_\gamma \rho(z)\,|dz| \geq 1$. Also,

$$\int_D \rho(z)^2\,dA(z) = \int_D \tilde{\rho}(f(z))^2\,|f'(z)|^2\,dA(z) = \int_{D'} \tilde{\rho}(w)^2\,dA(w).$$

Hence, $\text{mod}(\Gamma) \leq \text{mod}(f \circ \Gamma)$. Using f^{-1}, we get the other inequality. $\square$

The first example and Proposition 3.73 show that $L(A_1, A_2; D)$ is the same as the π-extremal distance of the family of curves Γ connecting A_1 and A_2 in D. This second definition does not require that D be simply connected. We end this section by proving some simple lemmas about extremal distance that will be used in the next section.

LEMMA 3.74. *Suppose D is a domain and A_1, A_2 are arcs in ∂D. Then*

$$L(A_1, A_2; D) \geq \frac{\pi\,\text{dist}(A_1, A_2)^2}{\text{Area}(D)},$$

where $\text{dist}(A_1, A_2) = \inf\{|z - w| : z \in A_1, w \in A_2\}$.

PROOF. If Γ is the family of curves connecting A_1 and A_2 in D, then $\rho(z) := 1/\text{dist}(A_1, A_2)$ satisfies (3.17) and hence

$$\text{mod}(\Gamma) \leq \int_D \rho(z)^2\,dA(z) = \frac{\text{Area}(D)}{\text{dist}(A_1, A_2)^2}.$$

$\square$

LEMMA 3.75. *There is a $c < \infty$ such that the following holds. Suppose D is a bounded subdomain of the half-infinite strip $D' := \{x+iy : -\infty < x < 0, 0 < y < \pi\}$ whose boundary is the union of four arcs: $A_1 = [0, \pi i]$; $A_2 = \gamma[0, 1]$ where γ is a curve in D' with $\text{Im}[\gamma(0)] = 0, \text{Im}[\gamma(1)] = \pi$; $A_3 = [\gamma(0), 0]$, $A_4 = [\gamma(1), \pi i]$. Then*

$$\text{dist}(A_1, A_2) \leq L(A_1, A_2; D) \leq \text{dist}(A_1, A_2) + c.$$

PROOF. Let $l = \text{dist}(A_1, A_2) = \inf\{|\text{Re}(\gamma(s))| : 0 \leq s \leq 1\}$. The lower bound follows immediately from comparison with $\mathcal{R}_l$, and it suffices to prove the upper bound for $l \geq 2$ (if $l \leq 2$, we can consider the curve obtained by moving γ two units to the left). Suppose $s \in [0, 1]$ with $\text{Re}[\gamma(s)] = -l$. Assume, without loss of generality, that $\text{Im}[\gamma(s)] \leq \pi/2$. Consider the L-shaped domain

$$\tilde{D} = \{x + iy : -l - 1 < x < -l + 1, \frac{\pi}{2} < y < \frac{3\pi}{4}\} \cup$$

$$\{x + iy : -l - 1 < x < -l - \frac{1}{2}, 0 < y < \frac{3\pi}{4}\}.$$

It is easy to see that there is a $\rho > 0$ such that the probability that a Brownian motion starting at $-l + \theta i, \theta \in [9\pi/16, 11\pi/16]$, exits $\tilde{D}$ on the real axis is greater than ρ. If this happens, then topological considerations show that the Brownian motion hits $\gamma[0, 1]$ before leaving this domain. In other words, the probability that a Brownian motion starting at $-l + \theta i$ exits D at A_2 is greater than ρ. Using (2.9) and the strong Markov property, we see there is a constant c' such that for $j = 1, 2$, the probability that a Brownian motion starting at $-(l/2) + i(\pi/2)$ exits D at A_j is at least $c'e^{-l/2}$. This implies $\Theta(A_1, A_2; D) \geq ce^{-l/2}$ and the lemma follows from (3.16). $\square$

3.8. Beurling estimate and applications

We use extremal distance to prove the following very useful estimate. We will then use the estimate to prove a number of propositions that will be used in studying the boundary behavior of conformal maps.

THEOREM 3.76 (Beurling estimate). *There is a constant $c < \infty$ such that if $\gamma : [0,1] \to \mathbb{C}$ is a curve with $\gamma(0) = 0$ and $|\gamma(1)| = 1$, $z \in \mathbb{D}$, and B_t is a Brownian motion, then*

$$\mathbf{P}^z\{B[0, \tau_\mathbb{D}] \cap \gamma[0,1] = \emptyset\} \leq c\, |z|^{1/2}.$$

If $\gamma(t) = t$ and $z = -\epsilon$, by conformal invariance and (2.9), we see as $\epsilon \to 0+$,

$$\mathbf{P}^{-\epsilon}\{B[0, \tau_\mathbb{D}] \cap \gamma[0,1] = \emptyset\} = \mathbf{P}^{i\sqrt{\epsilon}}\{|B(\tau_{\mathbb{D}_+})| = 1\} \sim (4/\pi)\sqrt{\epsilon}.$$

Hence, the exponent in the estimate is optimal. (In fact, this probability is exactly $(4/\pi)\arctan \epsilon^{1/2}$; this can be derived by considering the map $z \mapsto -[z+z^{-1}]$ and using Exercise 2.23.) Our proof will use extremal distance.

PROOF. Let $q(\epsilon)$ be the supremum of $\mathbf{P}^z\{B[0, \tau_\mathbb{D}] \cap \gamma[0,1] = \emptyset\}$ where the supremum is over all curves γ as in the theorem and all $|z| = \epsilon$. The supremum is clearly the same if we only consider curves with $\gamma[0,1) \subset \mathbb{D}$. Note that $q(e^{-2}\epsilon) \leq (1-\delta)\, q(\epsilon)$ where δ is the probability that a Brownian motion starting on the unit circle makes a closed loop about the origin before reaching the circle of radius e^2. Therefore

(3.18)
$$\tilde{q}(\epsilon) := \sup_\gamma \sup_{|z|=\epsilon} \mathbf{P}^z\{B[0, \tau_\mathbb{D}] \cap \gamma[0,1] = \emptyset;\, B[0, \tau_\mathbb{D}] \cap (e^{-2}\epsilon)\mathbb{D} = \emptyset\} \geq \delta\, q(\epsilon),$$

and it suffices to bound $\tilde{q}(\epsilon)$. For any γ, let t_ϵ be the largest t with $|\gamma(t)| = e^{-2}\epsilon$. Then if we replace γ with the curve that goes radially from 0 to $\gamma(t_\epsilon)$ in time t_ϵ and then proceeds as γ, the probability in (3.18) is not decreased. Hence, in the definition of $\tilde{q}$ we can take the supremum over all γ of that form.

Suppose such a γ is chosen and let D' be the connected component of $\{e^{-2}\epsilon < |w| < 1\} \setminus \gamma[t_\epsilon, 1]$ containing z. It suffices to show that

$$\mathbf{P}^z\{B_{\tau_{D'}} \in \partial \mathbb{D}\} \leq c\, \epsilon^{1/2},$$

for some c independent of ϵ, γ. Let η' denote the arc of $D' \cap \{|w| = \epsilon\}$ containing z and let D'_1 be the connected component of $D' \setminus \eta'$ whose boundary intersects the unit circle. Let D, D_1, η be the images of D', D'_1, η' under a branch of the map $w \mapsto (1/2)\log w$. Note that $\text{Area}(D_1) \leq \pi\, (\hat{L}+1)$ where $\hat{L} = (1/2)|\log \epsilon|$. Therefore, by Lemma 3.74, $L(A_1, \eta; D_1) \geq \hat{L}^2/(\hat{L}+1)$. where $A_1 = \partial D_1 \cap (i\mathbb{R}) = \partial D \cap (i\mathbb{R})$.

Let f be a conformal transformation of D onto $\{x + iy : -\tilde{L} < x < 0, 0 < y < \pi\}$ that maps A_1 onto $[0, \pi i]$ and the endpoints of η to $-\tilde{L}$ and $-\tilde{L}+i\pi$. Here $\tilde{L}$ is the appropriate π-extremal distance, i.e., the unique value for which such a transformation exists. Note that $\tilde{L} \geq L(A_1, \eta; D_1)$. Let a be the infimum of $|\text{Re}(f(w))|$ over $w \in \eta$. By Lemma 3.75 and conformal invariance,

$$a \geq \tilde{L} - c \geq L(A_1, \eta, D') - c \geq \hat{L} - c_1.$$

Hence, using (2.9), the probability that a Brownian motion starting at a point on $f(\eta)$ exits $f(D)$ at $[0, \pi i]$ is bounded above by $c'e^{-\hat{L}} = c'\epsilon^{1/2}$. □

REMARK 3.77. The Beurling Projection Theorem states that if V is a closed subset of $\overline{\mathbb{D}}$ and $P(V) = \{|z| : z \in V\}$ is the radial projection of V on the positive real axis, then
$$\mathbf{P}\{B[0, \tau_{\mathbb{D}}] \cap V = \emptyset\} \leq \mathbf{P}\{B[0, \tau_{\mathbb{D}}] \cap P(V) = \emptyset\}.$$
The Beurling estimate is an immediate corollary. We will sketch the proof here. The key lemma is the following: if $V_+ = \{|x| + iy :: x + iy \in V\}$, then
$$\mathbf{P}\{B[0, \tau_{\mathbb{D}}] \cap V_+ = \emptyset\} \geq \mathbf{P}\{B[0, \tau_{\mathbb{D}}] \cap V = \emptyset\}.$$
(We leave this an exercise; see [**5**, Lemma V.4.2] for a proof.) In other words, if we reflect the entire set onto the right hand side of the line $\{\theta = \pi/2, 3\pi/2\}$, we can only increase the probability of avoiding the set. We now continue this process, reflecting across the line $\{\theta = 0, \pi\}$ to show that a maximal V with a given radial projection can be found in the quadrant $\{0 \leq \theta \leq \pi/2\}$. We now reflect along the line $\{\theta = \pi/4, 5\pi/4\}$, then the line $\{\theta = \pi/8, 9\pi/8\}$, and continue.

COROLLARY 3.78. *There is a $c > 0$ such that if $0 < r_1 < r_2 < \infty$ and $\gamma : [a, b] \to \mathbb{C}$ is a curve with $|\gamma(a)| = r_1, |\gamma(b)| = r_2$, then*
$$\mathbf{P}^z\{B[0, \tau_{r_2\mathbb{D}}] \cap \gamma[a, b] = \emptyset\} \leq c\,(r_1/r_2)^{1/2}, \quad |z| \leq r_1,$$
$$\mathbf{P}^z\{B[0, \tau_{\mathbb{C}\setminus(r_1\overline{\mathbb{D}})}] \cap \gamma[a, b] = \emptyset\} \leq c\,(r_1/r_2)^{1/2}, \quad |z| \geq r_2.$$

PROOF. By scaling we may assume $r_2 = 1$. The first inequality is an almost immediate consequence of Theorem 3.76, and the second is obtained by considering the map $w \mapsto 1/w$. □

The second inequality in the corollary is often phrased in terms of harmonic measure. Suppose K is a compact hull containing the origin. Then there is a c such that for every $\epsilon > 0$ and $|z| \geq \text{rad}(K)$,

(3.19) $$\text{hm}(z, \mathbb{C} \setminus K; \mathcal{B}(0, \epsilon\,\text{rad}(K))) \leq c\,\epsilon^{1/2}.$$

The Beurling estimate is particularly useful for the study of simply connected domains, since the boundary of a simply connected domain is connected.

PROPOSITION 3.79. *There is a $c < \infty$ such that if D is a simply connected domain in $\mathbb{C}$ other than $\mathbb{C}$, $z \in D$, and B_t is a Brownian motion, then for all $r > 0$,*

(3.20) $$\mathbf{P}^z\{\text{diam}(B[0, \tau_D]) > r\,\text{dist}(z, \partial D)\} \leq c\,r^{-1/2}.$$

PROOF. Without loss of generality we may assume that $D \in \mathcal{A}, z = 0$, and $\text{dist}(0, \partial D) = \text{inrad}(D) \in [1/2, 1]$. Suppose first that D is a bounded Jordan domain and let $r > 0$. Then either $\text{rad}(D) \leq r/4$ (in which case $\text{diam}(B[0, \tau_D]) \leq r\,\text{dist}(z, \partial D)$ w.p.1) or there is a curve in ∂D from the unit circle to the circle of radius $r/4$. In the latter case, (3.20) follows from Corollary 3.78. If D is not a bounded domain, we can write $D = \cup_{\delta > 0} D_\delta$ where $D_\delta = f((1-\delta)\mathbb{D})$ and f is a conformal transformation of $\mathbb{D}$ onto D with $f(0) = 0$. Then, $\tau_D = \lim_{\delta \to 0+} \tau_{D_\delta}$ and (3.20) holds for D if it holds for every D_δ. □

COROLLARY 3.80. *There is a $c < \infty$ such that the following holds. Suppose $A_1, A_2 \in \mathcal{Q}$ with $A_1 \subset A_2$. Suppose for some $\epsilon > 0$, $\text{dist}(z, A_1 \cup \mathbb{R}) < \epsilon$ for every $z \in \partial(\mathbb{H} \setminus A_2)$. Then,*
$$\text{hcap}(A_1) \geq \text{hcap}(A_2) - c\,\epsilon^{1/2}\,\text{rad}(A_2)^{3/2}.$$

PROOF. By scaling it suffices to prove the result when $\mathrm{rad}(A_2) = 1/2$. Let B_t be a Brownian motion starting on the unit circle with initial argument density $(1/2)\sin\theta$. Then, by (3.7), $\mathrm{hcap}(A_j) = (4/\pi)\mathbf{E}[\mathrm{Im}(B_{\tau_j})]$ where $\tau_j = \tau_{\mathbb{H}\setminus A_j}$. Note that $\tau_1 \geq \tau_2$, but (3.20) gives
$$\mathbf{P}^z\{|B_{\tau_1} - B_{\tau_2}| > r\,\epsilon\} \leq c\,r^{-1/2}.$$
Since $|\mathrm{Im}(B_{\tau_1}) - \mathrm{Im}(B_{\tau_2})| \leq 1/2$, this implies $|\mathbf{E}[\mathrm{Im}(B_{\tau_1})] - \mathbf{E}[\mathrm{Im}(B_{\tau_2})]| \leq c'\,\epsilon^{1/2}$. □

COROLLARY 3.81. *There is a $c < \infty$ such that if $0 < \epsilon, r < \infty$ and $A_1, A_2 \in \mathcal{Q}$ with $\mathrm{rad}(A_1), \mathrm{rad}(A_2) \leq r$ and*
$$\mathrm{dist}(z, A_{3-j} \cup \mathbb{R}) \leq \epsilon, \quad j = 1, 2, \quad z \in \partial(\mathbb{H} \setminus A_j),$$
then $|\mathrm{hcap}(A_1) - \mathrm{hcap}(A_2)| \leq c\,\epsilon^{1/2}\,r^{3/2}$.

PROOF. Apply the previous corollary to A_j and the hull generated by $A_1 \cup A_2$. □

PROPOSITION 3.82. *There is a $c < \infty$ such that if $A \in \mathcal{Q}$ and $\gamma : [0,1] \to \mathbb{C}$ is a curve with $\gamma(0) \in A \cup \mathbb{R}$, $\gamma(0,1] \subset \mathbb{H} \setminus A$, then*

(3.21) $$\mathrm{diam}[g_A(\gamma(0,1])] \leq c\,d^{1/2}\,r^{1/2},$$

where $d = \mathrm{diam}[\gamma(0,1]]$ and $r = \sup\{\mathrm{Im}[\gamma(t)] : 0 < t \leq 1\}$. In particular, for every $z \in \mathbb{H} \setminus A$,

(3.22) $$\mathrm{Im}[g_A(z)] \leq c\,\mathrm{dist}(z, A \cup \mathbb{R})^{1/2}\,\mathrm{Im}(z)^{1/2},$$

and the limit $\lim_{t \to 0+} g_A(\gamma(t))$ exists.

PROOF. By scaling we may assume $r = 1$. If $d \geq 1$, the estimate follows from (3.12), so we may assume that $d < 1$. Let B_t denote a Brownian motion starting at iy ($y \geq 2$), let $\tau = \tau_{\mathbb{H}\setminus A}$, $\mathcal{B} = \mathcal{B}(\gamma(0), d)$, and consider the probability that $B[0,\tau] \cap \mathcal{B} \neq \emptyset$. In order for $B[0,\tau]$ to intersect $\mathcal{B}$, it must first reach $\mathcal{B}(\gamma(0), 1)$ without leaving $\mathbb{H} \setminus A$; this probability is bounded above by c/y. Given that it has done this, Corollary 3.78 shows that the probability that the Brownian motion visits $\mathcal{B}$ before leaving $\mathbb{H} \setminus A$ is bounded above by $c\,d^{1/2}$. Hence,
$$\limsup_{y \to \infty} y\,\mathbf{P}^{iy}\{B[0,\tau] \cap \mathcal{B} \neq \emptyset\} \leq c\,d^{1/2}.$$
By conformal invariance and the behavior of g_A near infinity, this implies
$$\limsup_{y \to \infty} y\,\mathbf{P}^{iy}\{B[0,\sigma] \cap g_A(\mathcal{B}) \neq \emptyset\} \leq c\,d^{1/2},$$
where σ denotes the first time that B_t hits the real axis. Since $g_A(\gamma(0,1])$ is connected, this implies (see (3.14)) that $\mathrm{diam}[g_A(\gamma(0,1])] \leq c\,d^{1/2}$. This gives (3.21). The first extra assertion follows by taking the line segment of length $\mathrm{dist}(z, A \cup \mathbb{R}) \leq \mathrm{Im}(z)$ from z to $A \cup \mathbb{R}$. The existence of the limit follows immediately from (3.21). □

REMARK 3.83. The estimate in the last proposition is sharp. For example, if $A = (0, i]$, $0 < \epsilon < 1$, and $\gamma(t) = (1 + \epsilon t)i$, $0 \leq t \leq 1$, then
$$g_A(z) = \sqrt{z^2 + 1}, \quad g_A((1+\epsilon)i) = i\sqrt{2\epsilon + \epsilon^2},$$

and hence,
$$\text{diam}[g_A(\gamma(0,1]] \geq \sqrt{2}\,[\text{diam}(\gamma(0,1])]^{1/2}.$$

PROPOSITION 3.84. *There exists a c such that if $z \in D \in \mathcal{A}$ and $f : D \to \mathbb{D}$ is a conformal transformation with $f(0) = 0$, then*
$$1 - |f(z)| \leq c \left[\frac{\text{dist}(z, \partial D)}{\text{inrad}(D)}\right]^{1/2}.$$

PROOF. Without loss of generality, we may assume that D is an analytic Jordan domain for otherwise we can find an increasing sequence of analytic Jordan domains D_n whose union is D. By scaling, we may assume $\text{inrad}(D) = 1$, and it then suffices to prove the estimate for $d = \text{dist}(z, \partial D) < 1/4$. By Corollary 3.25 (applied to f^{-1}), we know that $f(\mathcal{B}(0, 1/2)) \supset \mathcal{B}(0, 1/16)$. Let $q(z)$ be the probability that a Brownian motion starting at z reaches $\mathcal{B}(0, 1/2)$ before leaving D. Then by Corollary 3.78, this probability is bounded above by $c'\, d^{1/2}$. Hence, by conformal invariance of Brownian motion, the probability that a Brownian motion starting at $f(z)$ reaches $\mathcal{B}(0, 1/16)$ before leaving $\mathbb{D}$ is bounded above by $c'\, d^{1/2}$. But this latter probability is exactly $-\log|f(z)|/\log 16 \geq c''(1 - |f(z)|)$. □

PROPOSITION 3.85. *There exists a $c > 0$ such that the following holds for all $D \in \mathcal{A}$. Let f be a conformal transformation of D onto $\mathbb{D}$ with $f(0) = 0$. Suppose $\gamma : [0, 1] \to \mathbb{C}$ is a simple curve with $\gamma(0) \in \partial D, \gamma(0, 1] \subset D$. Then,*
$$\text{diam}[f(\gamma(0,1])] \leq c \left[\frac{\text{diam}(\gamma(0,1])}{\text{inrad}(D)}\right]^{1/2}.$$

PROOF. By scaling we may assume $\text{inrad}(D) = 1$. By (3.19), the probability that a Brownian motion starting at the origin hits $\gamma(0, 1]$ before leaving D is bounded above by $c'\,\text{diam}(\gamma(0, 1])^{1/2}$. But (see Exercise 2.7), the probability that a Brownian motion starting at the origin hits $f(\gamma(0, 1])$ before leaving $\mathbb{D}$ is greater than $c''\,\text{diam}[f(\gamma(0, 1])]$. Hence $\text{diam}[f(\gamma(0, 1])] \leq c\,\text{diam}[\gamma(0, 1]]^{1/2}$. □

In §3.1 we discussed the problems in trying to extend a conformal map $f : \mathbb{D} \to D$ to $\partial \mathbb{D}$. This can be done if ∂D is not too bad. In the next proposition we show that a corresponding limit for f^{-1} can always be taken.

PROPOSITION 3.86. *Suppose $f : \mathbb{D} \to D$ is a conformal transformation and $\gamma : [0, 1] \to \mathbb{C}$ is a simple curve with $\gamma(0) = w_0 \in \partial D$ and $\gamma(0, 1] \subset D$. Then the limit*
$$(3.23) \qquad z_0 = \lim_{t \to 0+} f^{-1}(\gamma(t))$$

exists and is in $\partial \mathbb{D}$. Moreover, if $\hat{\gamma} : [0,1] \to \mathbb{C}$ is any simple curve with $\hat{\gamma}(0) \in \partial D, \hat{\gamma}(0,1] \subset D$ and $\lim_{t \to 0+} f^{-1}(\hat{\gamma}(t)) = z_0$, then $\hat{\gamma}(0) = w_0$.

PROOF. Without loss of generality we may assume that $D \in \mathcal{A}$, $\text{inrad}(D) = 1$, $f(0) = 0$, and $\gamma(1) = \hat{\gamma}(1) = 0$. By Proposition 3.85, $\text{diam}[f^{-1}(\gamma(0, t])] \leq c\,\text{diam}[\gamma(0, t]]^{1/2}$, and hence the limit in (3.23) exists. It is easy to see that $z_0 \in \partial \mathbb{D}$ (if $z_0 \in \mathbb{D}$, then we can see that $w_0 \in D$). For every $0 < \epsilon < |w_0|$, let t_ϵ be the smallest t with $|\gamma(t) - w_0| \geq \epsilon/2$ and let U_ϵ be the connected component of $D \cap \mathcal{B}(w_0, \epsilon)$ that contains $\gamma(0, t_\epsilon)$. Consider the (finite or countably infinite) collection of open arcs in $\partial \mathcal{B}(w_0, \epsilon) \cap D$ whose endpoints are in ∂D. We claim that there is a unique such arc A_ϵ with the property that there is a path from $\gamma(0, t_\epsilon)$ to A_ϵ staying in U_ϵ and a path from 0 to A_ϵ staying in $D \setminus U_\epsilon$. The existence of at least one can be seen by considering γ; however, existence of two different arcs would contradict the simple connectedness of D. By the same argument that establishes (3.23), we can see that $f^{-1}(A_\epsilon)$ is an arc in $\mathbb{D}$ whose endpoints are on $\partial \mathbb{D}$. Also $\text{diam}[f^{-1}(A_\epsilon)] \leq c\,\epsilon^{1/2}$.

We claim that the two endpoints of $f^{-1}(A_\epsilon)$ are distinct and neither is z_0. To see this, let V_ϵ be the connected component of $D \setminus A_\epsilon$ that contains U_ϵ. It is possible that V_ϵ contains points outside of $\mathcal{B}(w_0, \epsilon)$; however, $0 \notin V_\epsilon$. Let t_ϵ be the smallest t such that $\gamma(t) \in A_\epsilon$. Then $\gamma[0, t_\epsilon]$ divides V_ϵ into two components. In each of these components it is easy to see that one can find a point w such that with positive probability Brownian motion leaves the component at a point not in $A_\epsilon \cup \gamma[0, t_\epsilon]$. Since this is true also for the image of the component under f^{-1}, it cannot be the case that an endpoint of $f^{-1}(A_\epsilon)$ is the same as w_0.

Since any curve $\hat\gamma$ as above with $\lim_{t \to 0+} f^{-1}(\hat\gamma(t)) = z_0$ must cross through each A_ϵ, we see that w_0 is a limit point of $\hat\gamma(0, 1]$; hence $w_0 = \hat\gamma(0)$. □

PROPOSITION 3.87. *There is a constant $c < \infty$ such if $0 < \epsilon \leq 1/2$, $D = \mathbb{D} \setminus A \in \mathcal{A}$ with $A \subset \{w \in \mathbb{D} : |w| \geq 1 - \epsilon\}$, $z \in \partial \mathbb{D} \setminus \overline{A}$, and $\gamma : [0, 1] \to \mathbb{C}$ is a curve with $\gamma(0) = z$, $\gamma(0, 1] \subset D$ and $|\gamma(1)| = 1 - 2\epsilon$, then*
$$|\tilde g_A(z) - z| \leq c\,[\text{diam}(\gamma[0,1])]^{1/2},$$
where $\tilde g_A$ is as defined in §3.5.

PROOF. By Proposition 3.85, $\text{diam}[\tilde g_A(\gamma[0, 1])] \leq c\,[\text{diam}(\gamma[0,1])]^{1/2}$. In particular, $|\tilde g_A(z) - \tilde g_A(w)| \leq c\,[\text{diam}(\gamma[0,1])]^{1/2}$. where $w = \gamma(1)$. But Proposition 3.57 gives $|\tilde g_A(w) - w| \leq c\,\epsilon\log(1/\epsilon) \leq c\,[\text{diam}(\gamma[0,1])]^{1/2}$. □

3.9. Conformal annuli

We will call a domain D a *conformal annulus* if $\hat{\mathbb{C}} \setminus D$ consists of two connected components, each of which is larger than a single point.

THEOREM 3.88. *For every conformal annulus D there is a unique $r \in (0, 1)$ such that there is a conformal transformation f of D onto $A_r := \{r < |z| < 1\}$.*

PROOF. If Γ denotes the set of curves connecting the inner and outer boundary of A_r, then in Example 3.72 we showed that $\text{mod}(\Gamma) = 2\pi/(-\log r)$. Since mod is a conformal invariant, the annuli A_r for different r are not conformally invariant. This gives uniqueness.

Let K_1, K_2 be the two components of $\hat{\mathbb{C}} \setminus D$. Since $\hat{\mathbb{C}} \setminus K_2$ is simply connected it can be conformally mapped onto the unit disk. Hence we can assume that $K_2 \cap \mathbb{C} = \{|z| \geq 1\}$, and (by composing by a Möbius transformation if necessary) that $K_1 \subset \mathbb{D}$ is a compact hull containing the origin. Let $u(z) = \mathbf{P}^z\{\tau_D \in K_1\}$ be the harmonic function in D with boundary value 1_{K_1}. Let
$$\alpha_D = \int_0^{2\pi} \frac{du}{dn}(e^{i\theta})\,d\theta = -\int_C \Phi(z) \cdot dz > 0,$$
where n denotes the inward normal, $\Phi(z) = (-\partial_y u, \partial_x u)$, and C is the unit circle traversed counterclockwise. (This derivative can easily be seen to exist; see the remark after Exercise 2.17.) Let u_r, α_r be the corresponding quantity for A_r; since $u_r(z) = \log|z|/\log r$, $\alpha_r = 2\pi/\log(1/r)$. Choose r such that $\alpha_D = \alpha_r$ and write

$\alpha = \alpha_D = \alpha_r$. By using Green's Theorem, we can see that if γ is a closed curve in D, then $\int_\gamma \Phi(z) \cdot dz$ is an integer multiple of α; in fact, it is $-\alpha$ times the winding number about K_1. Fix $z_0 \in D$, and let

$$F(w) = \int_\gamma \Phi(z) \cdot dz,$$

where γ is a curve in D from z_0 to w. This is well defined up to integer multiples of α and $\exp\{i2\pi F(w)/\alpha\} = \exp\{i\,\log(1/r)\,F(w)\}$ is a single-valued function. (Single-valued is not to be confused with one-to-one.) Locally (in a simply connected neighborhood with γ restricted to curves in that neighborhood, see Proposition 3.4), the function $u + iF$ can be seen to be analytic. Hence the single-valued function $f := \exp\{(\log r)\,(u+iF)\}$ is also analytic. We claim that f is a conformal transformation of D onto $\mathbb{D}$; to show this we need to show that f is one-to-one and $f(D) = A_r$.

For every $a \in (0,1)$, let γ_a be the level curve $\{z \in D : u(z) = a\}$. It is easy to check that this is a simple curve and $D \setminus \gamma_a$ consists of two components, $\{u(z) > a\}$ and $\{u(z) < a\}$. Also (see Proposition 2.28), $\nabla u \neq 0$ on this curve. Hence f' is non-zero on D, and f is locally one-to-one. We can parametrize $\gamma_a(t), 0 \leq t \leq t_0$ smoothly (by arc length, e.g.) in a counterclockwise direction; here t_0 is the first time such that $\gamma_a(t_0) = \gamma_a(0)$. Since $du/dn > 0$, the function

$$R(t) = -\int_{\gamma_a[0,t]} \Phi(z) \cdot dz$$

is strictly increasing with $R(0) = 0, R(t_0) = \alpha$. The chain rule shows that this is also true for

$$R^*(t) = -\int_{f \circ \gamma_a[0,t]} \Phi_r(z) \cdot dz$$

where $\Phi_r = (-\partial_y u_r, \partial_x u_r)$. Hence the first t_0 such that $\gamma_a(t_0) = \gamma_a(0)$ is also the first t_0 such that $f(\gamma_A(t_0)) = f(\gamma_a(0))$. This shows that f maps γ_a one-to-one onto the circle of radius r^a. □

REMARK 3.89. Often the definition of conformal annuli is extended to allow one or both of the components of $\hat{C}$ to be single points. These cases are easy to handle using the Riemann mapping theorem.

REMARK 3.90. There are generalizations of this result to multiply connected regions where it becomes more difficult to tell if two regions are conformally equivalent. See [1, Section 6.5] or [23, Chapter 15].

CHAPTER 4

Loewner differential equation

4.1. Chordal Loewner equation

In this section we will show that to each simple curve γ starting at the origin and staying in the upper half plane, there is an associated continuous function U_t on the real line. The evolution of γ, or more precisely the evolution of the conformal transformation taking $\mathbb{H} \setminus \gamma[0, t]$ onto $\mathbb{H}$, is described by a differential equation involving U_t. Conversely, we can start with U_t, or more generally a collection of measures μ_t defined on $\mathbb{R}$, and use the differential equation to define a growing collection of hulls K_t.

Suppose $\gamma : [0, \infty) \to \mathbb{C}$ is a simple curve with $\gamma(0) \in \mathbb{R}$ and $\gamma(0, \infty) \subset \mathbb{H}$. For each $t \geq 0$, let $H_t = \mathbb{H} \setminus \gamma[0, t]$, which is a simply connected subdomain of $\mathbb{H}$. As in §3.4, let $g_t = g_{\gamma(0,t]}$ be the unique conformal transformation of H_t onto $\mathbb{H}$ such that $g_t(z) - z \to 0$ as $z \to \infty$. Then g_t has an expansion

$$g_t(z) = z + \frac{b(t)}{z} + O(\frac{1}{|z|^2}), \quad z \to \infty,$$

where $b(t) = \text{hcap}(\gamma(0,t])$. Let $f_t = g_t^{-1}$. For $s > 0$, let $\gamma^s(t) = g_s(\gamma(s+t))$. Recall (see (3.8)) that $\text{hcap}(\gamma^s(0,t]) = b(t+s) - b(s)$. Let $g_{s,t} = g_{\gamma^s(0,t-s]}$ so that $g_t = g_{s,t} \circ g_s$. The following lemma is similar to Proposition 3.82.

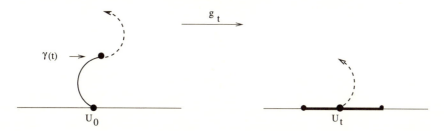

FIGURE 4.1. The map g_t

LEMMA 4.1. *There exists a constant $c < \infty$ such that if γ is a curve as above and $0 \leq s < t \leq t_0 < \infty$, then*

$$\text{diam}[g_s(\gamma(s,t])] \leq c \sqrt{\text{diam}(\gamma[0,t_0]) \, \text{osc}(\gamma, t-s, t_0)},$$

$$\|g_s - g_t\|_\infty \leq c \sqrt{\text{diam}(\gamma[0,t_0]) \, \text{osc}(\gamma, t-s, t_0)},$$

where

$$\text{osc}(\gamma, \delta, t_0) = \sup\{|\gamma(s) - \gamma(t)| : 0 \leq s, t \leq t_0; |t - s| \leq \delta\}$$

and $g_s - g_t$ is considered as a function on H_t.

PROOF. We may assume $\gamma(0) = 0$. The first assertion follows from (3.21) since, in the notation of (3.21), $d \leq \mathrm{osc}(\gamma, t - s, t_0)$ and $r \leq \mathrm{diam}(\gamma[0, t_0])$. The second assertion follows from the first and (3.12) since $\|g_s - g_t\|_\infty = \sup_z |g_{s,t}(z) - z|$. □

LEMMA 4.2. *If γ is a curve as above, then for every t, there is a unique $U_t \in \mathbb{R}$ with $g_t(\gamma(t)) = U_t$ in the sense*

(4.1) $$\lim_{z \to \gamma(t)} g_t(z) = U_t,$$

where the limit is taken over $z \in \mathbb{H} \setminus \gamma[0, t]$. Moreover,

(4.2) $$U_t = \lim_{s \to t-} g_s(\gamma(t)),$$

and $t \mapsto U_t$ is continuous.

PROOF. The previous lemma shows that the limit in (4.2) exists and gives a continuous function of t. To show that (4.1) holds, let $t > 0$. Simplicity of γ implies that there is a function $\epsilon \mapsto \delta(\epsilon)$ with $\delta(0+) = 0$ such that if $s < t$ with $|\gamma(s) - \gamma(t)| \leq \epsilon$, then $\mathrm{diam}(\gamma[s, t]) \leq \delta(\epsilon)$. If $z \in \mathcal{B}(\gamma(t), \epsilon)$, the straight line from z to $\gamma(t)$ hits $\gamma[0, t]$ first at some point, $\gamma(t')$ with $\mathrm{diam}(\gamma[t', t]) \leq \delta(\epsilon)$. The previous lemma then bounds $|g_{t'}(z) - g_{t'}(\gamma(t))| \leq c_t [\epsilon + \delta(\epsilon)]^{1/2}$, where $c_t = c_{t,\gamma}$ increases with t. □

LEMMA 4.3. *Suppose $u : [0, t_0) \to \mathbb{C}$ is a continuous function such that the right derivative*

$$u'_+(t) = \lim_{\epsilon \to 0+} \frac{u(t + \epsilon) - u(t)}{\epsilon}$$

exists everywhere and $u'_+(t)$ is a continuous function of t. Then $u'(t) = u'_+(t)$ for all $t \in (0, t_0)$.

PROOF. Without loss of generality we may assume that $u'_+(t) = 0, u(0) = 0$, for otherwise we can consider

$$u(t) - u(0) - \int_0^t u'_+(s)\, ds.$$

Let $\epsilon > 0$, and let s_0 be the supremum of all t such that $|u(s)| \leq \epsilon s$ for all $0 \leq s \leq t$. Since $u'_+(0) = 0$, $s_0 > 0$. Suppose $s_0 < t_0$. Then by continuity of u, we must have $|u(s_0)| \leq \epsilon s_0$. Since $u'_+(s_0) = 0$, there is a δ such that for $s_0 < s < s_0 + \delta$, $|u(s) - u(s_0)| \leq \epsilon(s - s_0)$ and hence $|u(s)| \leq \epsilon s$ for $0 \leq s \leq s_0 + \delta$, contradicting the maximality of s_0. Therefore $s_0 = t_0$ and $|u(t)| \leq \epsilon t$ for all t. Since this holds for all ϵ, $u(t) = 0$ for all t and $u'(t) = 0$. □

PROPOSITION 4.4. *Suppose γ is a simple curve as above such that $b(t)$ is C^1 and $b(t) \to \infty$ as $t \to \infty$. Then for $z \in \mathbb{H}$, $g_t(z)$ is the solution of the initial value problem*

(4.3) $$\dot{g}_t(z) = \frac{\dot{b}(t)}{g_t(z) - U_t}, \quad g_0(z) = z,$$

where $U_t = g_t(\gamma(t))$. If $z = \gamma(t_0)$, then this holds for $t < t_0$ and

$$U_{t_0} = \lim_{t \to t_0-} g_t(z).$$

If $z \notin \gamma(0, \infty)$, then the equation holds for all $t \geq 0$.

PROOF. Fix t_0 and let $d = \text{diam}(\gamma[0, t_0])$. Lemma 4.1 gives $\|g_s - g_t\|_\infty \leq c\delta(t-s)$, provided $0 \leq s < t \leq t_0$, where $\delta(\epsilon) = \delta(\epsilon, \gamma, t_0)$ is an increasing function with $\delta(0+) = 0$. In particular, the limit

$$U_t = \lim_{s \to t-} g_s(\gamma(t))$$

exists, and $t \mapsto U_t$ is a continuous function.

If $s \geq 0$ and $z \in H_s$, then Proposition 3.46 implies that for all sufficiently small positive ϵ,

$$g_{s+\epsilon}(z) - g_s(z) =$$

$$\frac{b(s+\epsilon) - b(s)}{g_s(z) - U_s} + \text{diam}[\gamma[s, s+\epsilon]] \, [b(s+\epsilon) - b(s)] \, O\!\left(\frac{1}{|g_s(z) - U_s|^2}\right).$$

This gives

$$\lim_{\epsilon \to 0+} \frac{g_{s+\epsilon}(z) - g_s(z)}{\epsilon} = \frac{\dot{b}(s)}{g_s(z) - U_s}.$$

Since $g_t(z)$ is continuous in t and $U_t, \dot{b}(t)$ are continuous, this implies that g_t satisfies (4.3), see Lemma 4.3. The remaining assertions follow from Theorem 4.6 below, and we will delay the proofs until then. □

REMARK 4.5. Suppose γ is a curve as above and $b(t) = \text{hcap}[\gamma(0, t])$, not necessarily C^1. Lemma 4.1, (3.8), and (3.9) tell us that b is strictly increasing and $b(t) - b(s) \leq c \, \text{diam}(\gamma[0,t]) \, \text{diam}(\gamma[s,t])$ if $s < t$; in particular, b is continuous. Therefore, we can *reparametrize γ by half-plane capacity*, i.e., let $\tilde{\gamma}(t) = \gamma(b^{-1}(2t))$. Then $\tilde{\gamma}$ is a simple curve with $\text{hcap}[\tilde{\gamma}(0, t]] = 2t$.

We started with a curve γ and found a function $t \mapsto U_t$ on the real line. We will now go in the opposite direction, and we will generalize slightly.

THEOREM 4.6. *Suppose $\mu_t, t \geq 0$, is a one parameter family of nonnegative Borel measures on $\mathbb{R}$ such that $t \mapsto \mu_t$ is continuous in the weak topology, and for each t, there is an $M_t < \infty$ such that $\sup\{\mu_s(\mathbb{R}) : 0 \leq s \leq t\} < M_t$ and $\text{supp} \, \mu_s \subset [-M_t, M_t], s \leq t$. For each $z \in \mathbb{H}$, let $g_t(z)$ denote the solution of the initial value problem*

(4.4) $$\dot{g}_t(z) = \int_{\mathbb{R}} \frac{\mu_t(du)}{g_t(z) - u}, \quad g_0(z) = z.$$

Let T_z be the supremum of all t such that the solution is well defined up to time t with $g_t(z) \in \mathbb{H}$. Let $H_t = \{z : T_z > t\}$. Then g_t is the unique conformal transformation of H_t onto $\mathbb{H}$ such that $g_t(z) - z \to 0$ as $z \to \infty$. Moreover, g_t has the expansion

$$g_t(z) = z + \frac{b(t)}{z} + O\!\left(\frac{1}{|z|^2}\right), \quad z \to \infty,$$

where

$$b(t) = \int_0^t \mu_s(\mathbb{R}) \, ds.$$

PROOF. Note that

$$\dot{g}_t(z) = \int_{\mathbb{R}} \frac{[\text{Re}[g_t(z)] - u] - i\,\text{Im}[g_t(z)]}{(\text{Re}[g_t(z)] - u)^2 + (\text{Im}[g_t(z)])^2}\,\mu_t(du). \qquad (4.5)$$

In particular, $\text{Im}[g_t(z)]$ decreases with t, and $T_z = \sup\{t : \text{Im}[g_t(z)] > 0\}$. If z, w are distinct points in H_t, and $\Delta_t(z, w) := g_t(z) - g_t(w)$, then

$$\dot{\Delta}_t(z, w) = -\Delta_t(z, w) \int_{\mathbb{R}} \frac{\mu_t(du)}{[g_t(z) - u]\,[g_t(w) - u]}.$$

Since $\Delta_0(z, w) = z - w$, this implies

$$\Delta_t(z, w) = (z - w)\,\exp\left\{ -\int_0^t \int_{\mathbb{R}} \frac{\mu_s(du)\,ds}{[g_s(z) - u]\,[g_s(w) - u]} \right\}. \qquad (4.6)$$

In particular, if $\delta = \text{Im}[g_t(z)]$ and $\text{Im}[g_t(w)] \geq \delta/2$,

$$|\Delta_t(z, w)| \leq |z - w|\,e^{2t M_t/\delta^2}. \qquad (4.7)$$

If $\epsilon < (\delta/2)\,e^{-2tM_t/\delta^2}$, then for $|z - w| < \epsilon$, $0 \leq s \leq t$, $|\Delta_s(z, w)| \leq \delta/2$ and hence (4.7) holds. This shows that $g_t(z)$ is continuous in z and then (4.6) gives

$$\begin{aligned}
g_t'(z) &= \lim_{w \to z} \frac{\Delta_t(z, w)}{z - w} \\
&= \lim_{w \to z} \exp\left\{ -\int_0^t \Big[\int_{\mathbb{R}} \frac{\mu_s(du)\,ds}{[g_s(z) - u]\,[g_s(w) - u]} \Big] \right\} \\
&= \exp\left\{ -\int_0^t \Big[\int_{\mathbb{R}} \frac{\mu_s(du)\,ds}{[g_s(z) - u]^2} \Big] \right\}.
\end{aligned}$$

Another way to get the formula for $g_t'(z)$ is to differentiate (4.4) to get

$$\dot{g}_t'(z) = -g_t'(z) \int_{\mathbb{R}} \frac{\mu_t(du)}{[g_t(z) - u]^2}, \qquad g_0'(z) = 1. \qquad (4.8)$$

(This differentiation assumes the z-differentiability of $g_t(z)$; however, we just established this fact.) Hence $g_t(z)$ is analytic in H_t and (4.6) shows that it is one-to-one on H_t. Therefore g_t is a conformal transformation with $g_t(H_t) \subset \mathbb{H}$.

To see that $g_t(H_t) = \mathbb{H}$, we start with $w \in \mathbb{H}$ and use the "inverse flow" to find a z with $g_t(z) = w$. More precisely, fix t and consider the initial value problem for $0 \leq s \leq t$,

$$\dot{h}_s(w) = -\int_{\mathbb{R}} \frac{\mu_{t-s}(du)}{h_s(w) - u}, \qquad h_0(w) = w.$$

Since $\text{Im}[h_s(w)]$ increases as s increases, the solution exists for all $0 \leq s \leq t$ and for all w. Moreover, if $h_s, 0 \leq s \leq t$, is a solution with $h_0(w) = w$, then $g_s := h_{t-s}$ satisfies (4.4) with $g_t = w$. In other words, $w = g_t(h_t(w))$.

For large z (how large depending on t and $\{\mu_s\}$),

$$\dot{g}_t(z) = \mu_t(\mathbb{R})\,g_t(z)^{-1} + \cdots = \mu_t(\mathbb{R})\,z^{-1} + \cdots,$$

so that

$$g_t(z) = z + \Big[\int_0^t \mu_s(\mathbb{R})\,ds\Big]\,z^{-1} + \cdots.$$

This establishes the last assertion. □

REMARK 4.7. If $t \mapsto \mu_t$ is piecewise continuous and right continuous, the theorem holds if we only assert that (4.4) holds as a right derivative if t is a point of discontinuity.

REMARK 4.8. This result can be stated in terms of flows along the time-varying vector field in $\mathbb{H}$,
$$V(t,z) = \int_{\mathbb{R}} \frac{\mu_t(du)}{z-u}.$$
Note that V is C^∞ in z but is only continuous in t. Equation (4.4) can be written as $\dot{g}_t(z) = V(t, g_t(z))$. The "inverse flow" h_t in the proof satisfies
$$\dot{h}_s(z) = -V(t-s, h_s(z)).$$

REMARK 4.9. Suppose μ_t is given and g_t satisfies (4.4). As before, for $s \leq t$, define $g_{s,t}$ by $g_t = g_{s,t} \circ g_s$. Then for fixed s, $g_t^* := g_{s,s+t}$ satisfies
$$(4.9) \qquad \dot{g}_t^*(z) = \int_{\mathbb{R}} \frac{\mu_{s+t}(du)}{g_t^*(z)-u},$$
with initial condition $g_0^*(z) = z$. In other words, $\tilde{g}_t := g_{s+t}$ is the solution of (4.9) with initial condition $g_0^*(z) = g_s(z)$.

An important example is $\mu_t = 2\,\delta_{U_t}$ where $t \mapsto U_t$ is a continuous function from $[0,\infty)$ to $\mathbb{R}$. Then $b(t) = 2t$, and (4.4) becomes
$$(4.10) \qquad \dot{g}_t(z) = \frac{2}{g_t(z)-U_t}, \qquad g_0(z) = z.$$
More generally, if $\mu_t = \dot{b}(t)\,\delta_{U_t}$ for some increasing C^1 function b, then
$$(4.11) \qquad \dot{g}_t(z) = \frac{\dot{b}(t)}{g_t(z)-U_t}, \qquad g_0(z) = z.$$
We will call g_t arising from (4.10) *Loewner chains* and g_t from (4.11) *generalized Loewner chains*. We will call U_t the *driving function* or the *Loewner transform*. If g_t satisfies (4.10) and $\hat{g}_t = g_{b(t)/2}$, then $\hat{g}_t$ satisfies (4.11) with driving function $\hat{U}_t = U_{b(t)/2}$.

REMARK 4.10. The equation (4.4) or the specialized form (4.10) is called the *(chordal or half-plane) Loewner (differential) equation*. The Loewner equation (4.4) is often given in terms of the inverse transformation $f_t = g_t^{-1}$. If we differentiate both sides of $f_t(g_t(z)) = z$ with respect to t, we obtain
$$(4.12) \qquad \dot{f}_t(z) = -f_t'(z) \int_{\mathbb{R}} \frac{\mu_t(du)}{z-u}, \qquad f_0(z) = z.$$
Note that f_t is a one parameter family of conformal transformations of $\mathbb{H}$ onto subsets of $\mathbb{H}$ satisfying
$$f_t(z) = z - \frac{b(t)}{z} + \cdots, \qquad z \to \infty,$$
where $b(t) = \int_0^t \mu_s(\mathbb{R})\,ds$.

EXAMPLE 4.11. If $U \equiv 0$, then the solution to (4.10) is
$$(4.13) \qquad g_t(z) = \sqrt{z^2 + 4t}.$$
If $\gamma(t) = 2\sqrt{t}\,i$, then $H_t = \mathbb{H} \setminus \gamma[0,t]$.

EXAMPLE 4.12. Suppose $0 < \alpha < 1$. The function
$$\phi(z) = [z + (1-\alpha)]^\alpha \, [z - \alpha]^{1-\alpha} = z - \frac{\alpha(1-\alpha)}{2z} + \cdots$$
maps $\mathbb{H}$ conformally onto $\mathbb{H} \setminus K$ where K is the line segment from 0 to $w_\alpha := \alpha^\alpha (1-\alpha)^{1-\alpha} e^{i(1-\alpha)\pi}$. Also, $\phi(2\alpha - 1) = w_\alpha$. Using the scaling rule, we get
$$\mathrm{hcap}([0, e^{i\alpha\pi}]) = \mathrm{hcap}([0, e^{i(1-\alpha)\pi}]) = \frac{1}{2} \alpha^{1-2\alpha} (1-\alpha)^{2\alpha - 1}.$$
Let
$$f_t(z) = \sqrt{\frac{4t}{\alpha(1-\alpha)}} \, \phi\!\left(z \sqrt{\frac{\alpha(1-\alpha)}{4t}}\right).$$
It can be checked directly that $f_t(z)$ satisfies the Loewner equation
$$\dot{f}_t(z) = -f'_t(z) \frac{2}{z - U_t},$$
where
$$U_t = c_\alpha \sqrt{t}, \quad c_\alpha = \frac{2(2\alpha - 1)}{\sqrt{\alpha(1-\alpha)}}.$$
Conversely, suppose $c > 0$ and $U_t = c\sqrt{t}$. Let $\alpha \in (1/2, 1)$ be defined by $c = 2(2\alpha - 1)/\sqrt{\alpha(1-\alpha)}$. Then if we solve (4.10) we find that $H_t = \mathbb{H} \setminus \gamma[0, t]$ where

(4.14) $$\gamma(t) = 2\sqrt{t} \left(\frac{\alpha}{1-\alpha}\right)^{\alpha - (1/2)} e^{i\pi(1-\alpha)}.$$

If $\gamma(t)$ is a simple curve with $\mathrm{hcap}(\gamma(0,t]) = 2t$ for all t then the corresponding conformal maps g_t satisfy (4.10) for some continuous $t \mapsto U_t$. However, it is not true that every Loewner chain comes from a simple curve γ. We can characterize the sets $K_t := \mathbb{H} \setminus H_t$ derived from Loewner chains as "continuously increasing hulls".

To be more precise, suppose K_t is an increasing family of hulls in $\mathcal{Q}$. Let g_t be the conformal transformation of $\mathbb{H} \setminus K_t$ onto $\mathbb{H}$ with $g_t(z) - z \to 0$ as $z \to \infty$. If $s < t$, let $K_{s,t}$ be the hull $g_s(K_t \setminus K_s) \cap \mathbb{H}$. We will say that $\{K_t\}$ is *right continuous* at t with limit U_t if $\cap_{\delta > 0} \overline{K_{t, t+\delta}}$ is the single point U_t. In particular, this implies that $\lim_{\delta \to 0+} \mathrm{diam}[K_{t, t+\delta}] = 0$. Let $b(t) = \mathrm{hcap}(K_t)$ and assume that b has a right derivative at t. If K_t is right continuous at U_t, then the proof of Proposition 3.46 shows that the right derivative
$$\lim_{\delta \to 0+} \frac{g_{t+\delta}(z) - g_t(z)}{\delta}$$
exists at each $z \in \mathbb{H} \setminus K_t$ and equals $\dot{b}(t)/(g_t(z) - U_t)$. We say that K_t is *continuously increasing* if it is right continuous at each t with limit U_t, $b(t)$ is C^1 in t, and U_t is continuous in t. By Lemma 4.3, this implies that $g_t(z)$ is continuously differentiable in t with $\dot{g}_t(z) = \dot{b}(t)/(g_t(z) - U_t)$. Hence continuously increasing hulls correspond to generalized Loewner chains. Conversely, using Lemma 4.1, it is not hard to see that the hulls generated by generalized Loewner chains are continuously increasing.

Suppose g_t is a Loewner chain with driving function U_t and let K_t be the corresponding hulls. Let $I_t = \overline{K_t} \cap \mathbb{R}$. Then it is easy to check that I_t is a compact connected interval, $[x_t^-, x_t^+]$, perhaps with $x_t^- = x_t^+$. If $x < x_t^-$ or $x > x_t^+$, then it is easy to show using Schwarz reflection that g_t extends to x and satisfies the Loewner equation

$$\dot{g}_t(x) = \frac{2}{g_t(x) - U_t}, \quad g_0(x) = x,$$

which is defined up to T_x, and $\lim_{t \to T_x-}[g_t(x) - U_t] = 0$.

Suppose K_t is a continuously increasing family of hulls as above with $U_0 = 0$. Suppose h is an analytic function defined in a neighborhood $\mathcal{N}$ of the origin of the form

$$h(z) = \alpha_1 z + \alpha_2 z^2 + \cdots, \quad \alpha_1 > 0, \quad \alpha_2, \alpha_3, \cdots \in \mathbb{R};$$

that takes reals to reals and $\mathcal{N} \cap \mathbb{H}$ into $\mathbb{H}$. Then for positive t_0 sufficiently small, $K_t^* := h(K_t), 0 \leq t \leq t_0$, is a continuously increasing family of hulls with $U_0^* = 0$. Let $b^*(t) = \mathrm{hcap}(K_t^*)$. Using Corollary 3.81 (comparing K_t^* and $\alpha_1 K_t$ to the hull generated by $K_t^* \cup (\alpha_1 K_t)$), we can see that

(4.15) $$\dot{b}^*(0) = \alpha_1^2 \dot{b}(0) = h'(0)^2 \dot{b}(0).$$

Let $g_t^*, 0 \leq t \leq t_0$, be the conformal transformations of $\mathbb{H} \setminus K_t^*$ onto $\mathbb{H}$ with $g_t^*(z) - z \to 0$ as $z \to \infty$. Then

(4.16) $$\dot{g}_0^*(z) = \frac{\dot{b}^*(0)}{z} = \frac{h'(0)^2 \dot{b}(0)}{z},$$

where the time derivative is interpreted as a right derivative.

The following lemma will be useful later. This lemma gives a uniform bound on $\mathrm{rad}(K_t)$ in terms of t and $\sup_{s \leq t} |U_s|$.

LEMMA 4.13. *If g_t is a Loewner chain with driving function U_t, then $\mathrm{rad}(K_t) \leq 4R_t$ where*

$$R_t = \max\left\{ \sqrt{t}, \ \sup\{|U_s| : 0 \leq s \leq t\} \right\}.$$

In fact, if $|z| > 4R_t$, then $|g_s(z) - z| \leq R_t$ for $0 \leq s \leq t$.

PROOF. Suppose $|z| > 4R_t$ and let σ be the first time s such that $|g_s(z) - z| \geq R_t$. If $s < t \wedge \sigma$, then $|\dot{g}_s(z)| \leq 1/R_t$. Hence

$$|z - g_s(z)| \leq s/R_t, \quad 0 \leq s \leq t \wedge \sigma.$$

Hence either $\sigma > t$ or $\sigma \geq R_t^2$. But $R_t^2 \geq t$, so $\sigma \geq t$. □

4.2. Radial Loewner equation

There is a similar Loewner equation that describes the evolution of hulls growing from the boundary of the unit disk towards the origin. If μ is a Borel measure on $\partial \mathbb{D}$ we will also write μ for the corresponding measure on $[0, 2\pi)$, i.e., if $I \subset [0, 2\pi)$, we write $\mu(I) = \mu\{e^{i\theta} : \theta \in I\}$.

THEOREM 4.14. *Suppose $\mu_t, t \geq 0$, is a one parameter family of nonnegative Borel measures on $\partial \mathbb{D}$ such that $t \mapsto \mu_t$ is continuous in the weak topology, and for each t, there is an $M_t < \infty$ such that $\sup\{\mu_s(\mathbb{R}) : s \leq t\} < M_t$. For each $z \in \mathbb{D}$, let $g_t(z)$ denote the solution of the initial value problem*

(4.17) $$\dot{g}_t(z) = g_t(z) \int_0^{2\pi} \frac{e^{i\theta} + g_t(z)}{e^{i\theta} - g_t(z)} \mu_t(d\theta), \quad g_0(z) = z.$$

Let T_z be the supremum of all t such that the solution is well defined up to time t with $g_t(z) \in \mathbb{D}$. Let $D_t = \{z : T_z > t\}$. Then g_t is the unique conformal transformation of D_t onto $\mathbb{D}$ such that $g_t(0) = 0$ and $g'_t(0) > 0$. Moreover,

$$\log g'_t(0) = \int_0^t \mu_s([0, 2\pi)) \, ds. \tag{4.18}$$

PROOF. Since $\operatorname{Re}[(e^{i\theta} + z)/(e^{i\theta} - z)] > 0$ for $z \in \mathbb{D}$, $|g_t(z)|$ increases with t. Suppose $z, w \in D_t$ and $\Delta_t(z, w) = g_t(z) - g_t(w)$. Then,

$$\dot{\Delta}_t(z, w) = \Delta_t(z, w) \int_0^{2\pi} \frac{e^{i2\theta} + (g_t(z) + g_t(w)) e^{i\theta} - g_t(z) g_t(w)}{(e^{i\theta} - g_t(z))(e^{i\theta} - g_t(w))} \mu_t(d\theta).$$

Then, as in the proof of Theorem 4.6,

$$\Delta_t(z, w) =$$

$$(z - w) \exp\left\{ \int_0^t [\int_0^{2\pi} \frac{e^{i2\theta} + (g_s(z) + g_s(w)) e^{i\theta} - g_s(z) g_s(w)}{(e^{i\theta} - g_s(z))(e^{i\theta} - g_s(w))} \mu_s(d\theta)] \, ds \right\},$$

and

$$g'_t(z) = \exp\left\{ \int_0^t [\int_0^{2\pi} \frac{e^{i2\theta} + 2 g_s(z) e^{i\theta} - g_s(z)^2}{(e^{i\theta} - g_s(z))^2} \mu_s(d\theta)] \, ds \right\}, \tag{4.19}$$

This shows that $g'_t(z)$ exists and g_t is one-to-one. Note that we can also obtain (4.19) by differentiating (4.17) with respect to z giving

$$\dot{g}'_t(z) = g'_t(z) \left[\int_0^{2\pi} \frac{e^{i\theta} + g_t(z)}{e^{i\theta} - g_t(z)} \mu_t(d\theta) + \int_0^{2\pi} \frac{2 e^{i\theta} g_t(z)}{(e^{i\theta} - g_t(z))^2} \mu_t(d\theta) \right].$$

In order to show that $g_t(D_t) = \mathbb{D}$, we consider an inverse flow. Fix $t_0 > 0$ and let h_t be defined by

$$\dot{h}_t(w) = -h_t(w) \int_0^{2\pi} \frac{e^{i\theta} + h_t(z)}{e^{i\theta} - h_t(z)} \mu_{t_0 - t}(d\theta).$$

Note that $|h_t(w)|$ decreases as t increases and

$$|\dot{h}_t(w)| \le \frac{2 |h_t(w)| M_{t_0}}{1 - |h_t(w)|} \le \frac{2 |h_t(w)| M_{t_0}}{1 - |w|}.$$

Hence $h_t(w)$ is well defined for $t \le t_0$. Also, $g_t(z) := h_{t_0 - t}(z)$ satisfies (4.17) with $g_0(z) = h_{t_0}(w)$, i.e., $g_{t_0}(h_{t_0}(w)) = w$.

Finally, note that (4.18) is (4.19) with $z = 0$. □

REMARK 4.15. The equation (4.17) is called the *(radial or disk) Loewner (differential) equation.* The radial Loewner equation is often given in terms of the inverse function $f_t = g_t^{-1}$. Note that $f_t \in \mathcal{S}^*$ with $f'_t(0) \le 1$. Differentiating $f_t(g_t(z)) = z$ with respect to t gives the partial differential equation

$$\dot{f}_t(z) = -z \, f'_t(z) \int_0^{2\pi} \frac{e^{i\theta} + z}{e^{i\theta} - z} \mu_t(d\theta), \quad f_0(z) = z. \tag{4.20}$$

4.2. RADIAL LOEWNER EQUATION

REMARK 4.16. If $e^{iz} \in D_t \setminus \{0\}$ we can define $h_t(z) = -i \log g_t(e^{iz})$ locally near z. Then,

$$\dot{h}_t(z) = -i \int_0^{2\pi} \frac{e^{i\theta} + e^{ih_t(z)}}{e^{i\theta} - e^{ih_t(z)}} \mu_t(d\theta) = \int_0^{2\pi} \cot\left[\frac{h_t(z) - \theta}{2}\right] \mu_t(d\theta).$$

If $x \in \mathbb{R}$ and $e^{ix} \notin \overline{\mathbb{D} \setminus D_t}$, then this is also valid and becomes

$$\dot{h}_t(x) = \int_0^{2\pi} \cot\left[\frac{h_t(x) - \theta}{2}\right] \mu_t(d\theta).$$

REMARK 4.17. If g_t satisfies (4.17), then $g_t^*(z) := 1/g_t(1/z)$ satisfies

(4.21) $$\dot{g}_t^*(z) = -g_t^*(z) \int_0^{2\pi} \frac{1 + g_t^*(z) e^{i\theta}}{1 - g_t^*(z) e^{i\theta}} \mu_t(d\theta).$$

Let D_t^* be the domain of g_t^*, i.e., the image of D_t under the map $z \mapsto 1/z$, and let $K_t^* = \mathbb{C} \setminus D_t^*$. Then $g_t^* = F_{K_t^*}^{-1}$, where F_K is as defined in §3.3.

We will call g_t a *radial Loewner chain* (resp., *generalized radial Loewner chain*) if it satisfies (4.17) with $\mu_t = \delta_{U_t}$ (resp., $\mu_t = b(t) \delta_{U_t}$ for continuous b) and $t \mapsto U_t$ a continuous function from $[0, \infty)$ to $\mathbb{R}$. We will call either U_t or e^{iU_t} the *driving function* or *Loewner transform* for g_t.

Suppose that g_t is a radial Loewner chain. For small t, we have a hull A_t of small radius around e^{iU_0}. Locally we can take a logarithm and consider $-i \log A_t$ as a growing hull in the half plane near U_0. To be more precise, let log denote the branch of the logarithm on $V := \mathbb{D} \setminus \{-re^{iU_0} : 0 \leq r < 1\}$ with $\log e^{iU_0} = U_0$. Choose r and t_0 sufficiently small such that for all $0 < t < t_0$ and all $z \in D_t \cap \{|w - e^{iU_0}| < r\}$, $g_t(z) \in V$. Let $\mathcal{N}$ be the image of $\mathcal{B}(e^{iU_0}, r)$ under the map $z \mapsto -i \log z$. Then the family of functions $h_t(z) = -i \log g_t(e^{iz})$ satisfy the equation

$$\dot{h}_t(z) = \cot\left[\frac{h_t(z) - U_t}{2}\right], \quad h_0(z) = z.$$

Note that when $h_t(z) - U_t$ is small, $\cot[(h_t(z) - U_t)/2]$ is approximately $2/(h_t(z) - U_t)$. Hence for points very near the growing hulls, this equation can be approximated by the chordal Loewner equation. (This approximation is the reason that the parametrization $b(t) = 2t$ is usually chosen for the chordal Loewner equation.) Let $\phi_t(z) = h_t^{-1}(z) = -i \log g_t^{-1}(e^{iz})$. Then ϕ_t satisfies

$$\dot{\phi}_t(z) = -\phi_t'(z) \cot\left[\frac{z - U_t}{2}\right], \quad \phi_0(z) = z,$$

which can be compared to (4.12).

If g_t is a generalized radial Loewner chain then the domain of g_t is $\mathbb{D} \setminus A_t$ where the A_t are hulls as in §3.5. For $s < t$, let $A_{s,t} = g_s(A_t)$. For δ small, $A_{t,t+\delta}$ is contained in the disk of radius $1/10$ about $\exp\{iU_t\}$, and hence we can let $K_{t,t+\delta}$ be the image of $A_{t,t+\delta}$ under the (locally) conformal map $z \mapsto -i \log z$. We will say that $\{A_t\}$ is a continuously growing family of radial hulls if the $K_{t,t+\delta}$ satisfy the conditions for (half-plane) continuously growing hulls as in the last section. (Note that the condition to be a half-plane continuously growing hull is really a local condition.) If $a(t) = \log g_t'(0)$, then $\dot{a}(t) = \lim_{\delta \to 0+} \operatorname{cap}(K_{t,t+\delta})/\delta$.

REMARK 4.18. Let g_t be a radial Loewner chain with the usual parametrization, $g_t'(0) = e^t$, and let $f_t = g_t^{-1}$, $d_t = \operatorname{inrad}(D_t) = \operatorname{dist}(0, A_t)$. Since $d_t^{-1} f_t$ takes $\mathbb{D}$ into

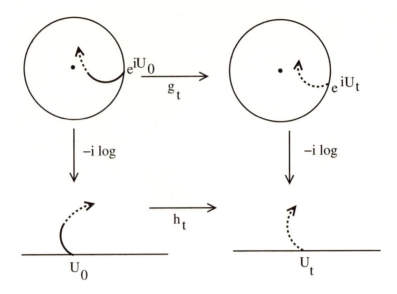

FIGURE 4.2. The maps g_t, h_t

a domain containing $\mathbb{D}$, the Schwarz lemma implies that $f'_t(0) \geq d_t$. On the other hand, Corollary 3.19 implies that $f'_t(0) \leq 4 d_t$. In other words, if $r \in (0,1)$ and σ_r denotes the supremum of all t such that $\overline{B(0,r)} \subset \mathbb{D} \setminus A_t$, then

(4.22) $$r \leq e^{-\sigma_r} \leq 4r.$$

REMARK 4.19. The radial Loewner chains give a one parameter family of conformal maps $f_t, t \geq 0$ from $\mathbb{D}$ into $\mathbb{C}$ with $f_t(0) = 0, f'_t(0) = e^{-t}$. They are obtained from a continuous function $U : [0, \infty) \to \mathbb{R}$. If $U : (-\infty, \infty) \to \mathbb{R}$ is a continuous function, we can get a one parameter family of conformal maps $f_t, -\infty < t < \infty$ satisfying (4.20) with $\mu_t = \delta_{U_t}$ and $f'_t(0) = e^{-t}$. See the next section for details.

4.3. Whole-plane Loewner equation

The radial Loewner equation describes the evolution of a hull A_t starting at the unit circle moving towards the interior. We can also think of the hull as $A_t \cup \{|z| \geq 1\}$. If we consider the image of these hulls under the inversion $z \mapsto 1/z$, then we can consider the hulls as starting at time 0 with the closed unit disk and growing toward infinity. There is nothing special about time $t = 0$; we can also consider negative times and let the hulls shrink to the point $\{0\}$. This is the basic idea of the whole-plane Loewner equation.

Suppose $K_t, -\infty < t < \infty$ is an increasing sequence of hulls in $\mathcal{H}$ as in §3.3. Let $\alpha(t) = \text{cap}(K_t)$ and suppose that $\alpha(0) = 0$, $t \mapsto \alpha(t)$ is continuous, $\alpha(t) \to -\infty$ as $t \to -\infty$, and $\alpha(t) \to \infty$ as $t \to \infty$. From Proposition 3.29, we see that the last two conditions are equivalent to saying $\cap_{t > -\infty} K_t = \{0\}$ and $\text{rad}(K_t) \to \infty$ as $t \to \infty$. Let $\tilde{K}_t$ be in $\mathcal{H}_0$ with $K_t = e^{\alpha(t)} \tilde{K}_t$. Let $F_t = F_{K_t}, \tilde{F}_t = F_{\tilde{K}_t}$ as defined in §3.3 and let $g_t = F_t^{-1}, \tilde{g}_t = \tilde{F}_t^{-1}$. Note that $F_t(z) = e^{\alpha(t)} \tilde{F}_t(z)$. If $s < t$, let $F_{s,t} = F_s^{-1} \circ F_t$ so that $F_t = F_s \circ F_{s,t}$.

4.3. WHOLE-PLANE LOEWNER EQUATION

LEMMA 4.20. *There exists a $c < \infty$ such that the following is true. Suppose $K_t, K_t^\#$ are two increasing sequences of hulls as above with $\alpha(t) = \mathrm{cap}(K_t) = \mathrm{cap}(K_t^\#)$. Let $F_s, F_s^\#, F_{s,t}, F_{s,t}^\#$ be the corresponding functions. Suppose that for some s, $F_{s,t} = F_{s,t}^\#$ for all $t \geq s$. Then for $t \geq s$ and all $|z| > 1$, $|F_t(z) - F_t^\#(z)| \leq c\, e^{\alpha(s)}$.*

PROOF. By Proposition 3.30,
$$F_t(z) = (F_s \circ F_{s,t})(z) = e^{\alpha(s)} \tilde{F}_s(F_{s,t}(z)) = e^{\alpha(s)}[F_{s,t}(z) + O(1)],$$
and similarly for $F_t^\#(z)$. □

PROPOSITION 4.21. *Suppose $\mu_t, -\infty < t < \infty$, is a continuous one parameter family of finite Borel measures on $\partial\mathbb{D}$. Let*
$$\alpha(t) = \int_0^t \mu_s([0, 2\pi))\, ds,$$
and assume that $\alpha(t) \to -\infty$ as $t \to -\infty$ and $\alpha(t) \to \infty$ as $t \to \infty$. Then there exists a unique family of conformal transformations $F_t : \mathbb{C}\setminus\overline{\mathbb{D}} \to D_t$, $-\infty < t < \infty$, such that $g_t = F_t^{-1}$ satisfies

(4.23) $$\dot{g}_t(z) = g_t(z) \int_0^{2\pi} \frac{1 + e^{i\theta} g_t(z)}{1 - e^{i\theta} g_t(z)} \mu_t(d\theta), \quad z \in D_t,$$

and for all $z \in \mathbb{C} \setminus \{0\}$, $g_t(z) \sim e^{-\alpha(t)} z$ as $t \to -\infty$.

PROOF. Let $m(t) = \mu_t[0, 2\pi)$. Fix $s \in \mathbb{R}$ and suppose first that $\mu_t(d\theta) = m(t)\, d(\theta/2\pi)$ for $t \leq s$. Then (4.23) becomes $\dot{g}_t(z) = -m(t) g_t(z)$ for $t < s$. Hence, if $-\infty < t_1 < t_2 < s$, $g_{t_2}(z) = e^{\alpha(t_1) - \alpha(t_2)} g_{t_1}(z)$; since $g_t(z) \sim e^{-\alpha(t)} z$ as $t \to -\infty$, this gives $g_t(z) = e^{-\alpha(t)} z$ for $t \leq s$.

More generally, if $s \in \mathbb{R}$, let $g_t^{(s)}(z)$ be defined by: $g_t(z) = e^{-\alpha(t)} z$ if $t \leq s$ and for $t > s$, $g_t^{(s)}(z)$ is the solution to (4.23) with initial condition $g_s^{(s)}(z) = e^{-\alpha(s)} z$. Let $F_t^{(s)}$ be the inverse of $g_s^{(s)}$. Then for $s_1 < s_2 < t$, Lemma 4.20 gives $|F_t^{(s_1)}(z) - F_t^{(s_2)}(z)| \leq e^{\alpha(s_2)} z$. This shows that the limit $F_t(z) = \lim_{s \to -\infty} F_t^{(s)}(z)$ exists as well as $g_t := F_t^{-1}$. If $s \in \mathbb{R}$, then $g_t(z), t \geq s$, is the solution to (4.23) with initial condition $g_s(z) = \lim_{s' \to -\infty} g_s^{(s')}(z)$. Note that Lemma 4.20 also gives uniqueness. □

We will call g_t a *whole-plane Loewner chain (with the standard parametrization)* if $\mu_t = \delta_{\exp\{iU_t\}}$ for a continuous function $U : (-\infty, \infty) \to \mathbb{R}$. The equation (4.23) becomes

(4.24) $$\dot{g}_t(z) = g_t(z) \frac{1 + e^{iU_t} g_t(z)}{1 - e^{iU_t} g_t(z)} = g_t(z) \frac{e^{-iU_t} + g_t(z)}{e^{-iU_t} - g_t(z)}.$$

As in the case for chordal chains, there is a one-to-one correspondence between Loewner chains and continuously growing families of hulls. We call $\{K_t\}$ a *continuously growing family of hulls in $\mathcal{H}$ (with the standard parametrization)* if $\mathrm{cap}(K_t) = t$; for each t, $\cap_{\delta > 0} g_t(K_{t+\delta}) \setminus \overline{\mathbb{D}}$ is a single point in $\partial\mathbb{D}$, say $\exp\{-iU_t\}$, and U_t can be chosen to be a continuous function of t.

If $\gamma^* : (-\infty, \infty) \to \mathbb{C}$ is a simple curve with
$$\lim_{t \to -\infty} \gamma^*(t) = 0, \quad \limsup_{t \to \infty} |\gamma^*(t)| = \infty,$$

that is parametrized so that $\operatorname{cap}(\gamma^*(-\infty, t]) = t$, then (by an argument as in Proposition 4.4) there is a $U_t : (-\infty, \infty) \to \mathbb{R}$ such that if $K_t^* = \gamma^*(-\infty, t]$ and $g_t^* = g_{K_t^*}^*$, then g_t^* satisfies (4.21) with $\mu_t = \delta_{e^{iU_t}}$. If we let $\gamma(t) = 1/\gamma^*(t)$, and let g_t be the corresponding conformal maps, then $f_t := g_t^{-1}$ satisfies (4.20) with $\mu_t = \delta_{U_t}$ and $f_t'(0) = e^{-t}$.

REMARK 4.22. The key point is that there is no real difference between the whole-plane Loewner equation and the radial Loewner equation. One gets from one to the other by the map $z \mapsto 1/z$. There *are* essential difference between these equations and the chordal Loewner equation. Radial and whole-plane Loewner equations discuss evolution from the boundary to an interior point; the chordal equation discusses evolution from the boundary to a boundary point.

DEFINITION 4.23. A simply connected domain $D \in \mathcal{A}$ is called a *slit domain* if it is of the form $D = D_t = \mathbb{C} \setminus \gamma(-\infty, t]$ where γ is a simple curve with $0 \notin \gamma(-\infty, \infty)$, $\lim_{t \to -\infty} \gamma(t) = \infty$, $\lim_{t \to \infty} \gamma(t) = 0$. Note that the curve can be parametrized so that $f_t'(0) = e^{-t}$ where $f : \mathbb{D} \to D_t$ is the conformal transformation with $f(0) = 0$, $f'(0) > 0$.

LEMMA 4.24. *The set of slit domains is dense in $\mathcal{A}$ with the topology induced by $\xrightarrow{Cara}$.*

PROOF. Since the set of bounded Jordan domains is dense (see Remark 3.65), it suffices to show that any bounded Jordan domain $D \in \mathcal{A}$ can be approximated by slit domains. We can parametrize ∂D by $\eta : [0, 1] \to \mathbb{C}$ where $\eta(0) = \eta(1) \in (0, \infty)$ and $\eta[0, 1] \cap (\eta(0), \infty) = \emptyset$, i.e., $\eta(0)$ is a point on ∂D with maximal real part. Let $D_\epsilon = \mathbb{C} \setminus [\eta[0, 1 - \epsilon] \cup (\eta(0), \infty)]$. Then each D_ϵ is a slit domain, and using Proposition 3.63 we can see that $D_\epsilon \xrightarrow{Cara} D$. □

REMARK 4.25. If $f \in \mathcal{S}$, define $a_n(f)$ by

$$f(z) = z + a_2(f) z^2 + a_3(f) z^3 + \cdots.$$

Then the map $f \mapsto a_n(f)$ is continuous from $\mathcal{S}$ (with the topology induced by $\xrightarrow{Cara}$) to $\mathbb{C}$. In particular,

$$\sup\{|a_n(f)| : f \in \mathcal{S}\} = \sup\{|a_n(f)| : f \in \mathcal{S}, f(\mathbb{D}) \text{ slit domain}\}.$$

For this reason, Loewner chains are useful for finding upper bounds on the coefficients a_n. Here we sketch an argument, originally due to Loewner, that $|a_3| \leq 3$. It suffices to show $\operatorname{Re}(a_3) \leq 3$ for every $f \in \mathcal{S}$. We will assume that we have already established that there exists some $M_3 < \infty$ such that $|a_3| \leq M_3$ for all $f \in \mathcal{S}$. It suffices to prove the results for slit domains so assume that $\gamma : (-\infty, \infty) \to \mathbb{C}$ is a simple curve with $0 \notin \gamma(-\infty, \infty)$, $\gamma(t) \to \infty$ as $t \to -\infty$, and $\gamma(t) \to 0$ as $t \to \infty$. Let f_t be the conformal transformation of $\mathbb{D}$ onto $\mathbb{C} \setminus \gamma(-\infty, t]$ with $f_t(0) = 0$, $f_t'(0) > 0$ and assume γ is parametrized so that $f_t'(0) = e^{-t}$. Then $e^t f_t \in \mathcal{S}$ (and every $f \in \mathcal{S}$ can be approximated arbitrarily close in the Carathéodory topology by functions of the form $e^t f_t$). Write

$$e^t f_t(z) = z + a_2(t) z^2 + a_3(t) z^3 + \cdots.$$

4.3. WHOLE-PLANE LOEWNER EQUATION

We will show that $\text{Re}[a_3(t)] \leq 3$. We know that there is a continuous function $U_t : (-\infty, \infty) \to \mathbb{R}$ such that f_t satisfies the Loewner equation[1]

$$\dot{f}_t(z) = -z \, f'_t(z) \, \frac{1 + \xi(t) \, z}{1 - \xi(t) \, z} \tag{4.25}$$

where $\xi(t) = e^{-iU_t}$. Then we can write

$$(e^t f_t)(z) = e^t f_t(z) + e^t \dot{f}_t(z) = e^t f_t(z) + \dot{a}_2(t) \, z^2 + \dot{a}_3(t) \, z^3 + \cdots$$

Plugging into (4.25) gives

$$e^t \dot{f}_t(z) = -z + [-2\,\xi(t) - 2\,a_2(t)] \, z^2 + [-3\,a_3(t) - 4\,\xi(t)\,a_2(t) - 2\,\xi(t)^2] \, z^3 + \cdots$$

Equating coefficients gives the differential equations

$$\dot{a}_2(t) = -2\,\xi(t) - a_2(t), \quad \dot{a}_3(t) = -2\,a_3(t) - 4\,\xi(t)\,a_2(t) - 2\,\xi(t)^2.$$

Using the fact that a_2 is uniformly bounded for $f \in S$, we can see that

$$a_2(t) = -e^{-t} \int_{-\infty}^{t} 2\,\xi(s) \, e^s \, ds, \quad |a_2(t)| \leq e^{-t} \int_{-\infty}^{t} 2\,e^s \, ds = 2.$$

This recovers Bieberbach's bound on a_2 (Proposition 3.16). Similarly, the equation for a_3 can be solved giving

$$\begin{aligned}
a_3(t) &= e^{-2t} \int_{-\infty}^{t} e^{2s} \left[-4\,\xi(s)\,a_2(s) - 2\,\xi(s)^2 \right] ds \\
&= -e^{-2t} \int_{-\infty}^{t} e^{2s} \left[4\,\xi(s) \left(-e^{-s} \int_{-\infty}^{s} 2\,\xi(r) \, e^r \, dr \right) + 2\,\xi(s)^2 \right] ds \\
&= 8 \int_{0}^{\infty} \int_{s}^{\infty} e^{-(r+s)} \, \psi(s) \, \psi(r) \, dr \, ds - 2 \int_{0}^{\infty} e^{-2s} \, \psi(s)^2 \, ds \\
&= 4 \left[\int_{0}^{\infty} e^{-s} \, \psi(s) \, ds \right]^2 - 2 \int_{0}^{\infty} e^{-2s} \, \psi(s)^2 \, ds,
\end{aligned}$$

where $\psi(s) = \xi(t - s)$. Letting $\psi(s) = e^{i\theta_s}$, we get

$$\text{Re}[a_3(t)] \leq 4 \left[\int_{0}^{\infty} e^{-s} \cos \theta_s \, ds \right]^2 + 2 \int_{0}^{\infty} e^{-2s} \left[1 - 2 \cos^2 \theta_s \right] ds.$$

But,

$$\begin{aligned}
\left[\int_{0}^{\infty} e^{-s} \cos \theta_s \, ds \right]^2 &\leq \left[\int_{0}^{\infty} e^{-s} \, ds \right] \left[\int_{0}^{\infty} e^{-s} \cos^2 \theta_s \, ds \right] \\
&= \int_{0}^{\infty} e^{-s} \cos^2 \theta_s \, ds.
\end{aligned}$$

Therefore,

$$\text{Re}[a_3(t)] \leq 1 + 4 \int_{0}^{\infty} [e^{-s} - e^{-2s}] \cos^2 \theta_s \, ds \leq 1 + 4 \int_{0}^{\infty} [e^{-s} - e^{-2s}] \, ds = 3.$$

De Branges's proof of the Bieberbach conjecture also uses Loewner chains. See [23, Chapter 17] for an exposition of this proof.

---

[1] This is the inverse equation obtained from (4.23) obtained from differentiating $f_t(g_t(z)) = z$.

4.4. Chains generated by curves

Suppose $\gamma : [0, \infty) \to \overline{\mathbb{H}}$ is a curve with $\gamma(0) \in \mathbb{R}$. At each time t, let H_t denote the unbounded component of $\mathbb{H} \setminus \gamma(0, t]$ and let $K_t = \mathbb{H} \setminus H_t$. Note that H_t is a simply connected domain and that K_t is a (not necessarily strictly) increasing family of hulls in $\mathcal{Q}$. If γ is a simple curve with $\gamma(0, \infty) \subset \mathbb{H}$, then $K_t = \gamma(0, t]$ and the hulls are strictly increasing. If γ is not a simple curve, or $\gamma(0, \infty) \cap \mathbb{R} \neq \emptyset$, then it is possible for K_t to be larger than $\overline{\cup_{s<t} K_s}$, e.g., if $\gamma(t) = e^{i\pi t}$, $0 \leq t \leq 1$, then K_s for $s < 1$ is a circular arc while K_1 is a half disk. However, if we let $\partial_t = \partial H_t \cap \mathbb{H}$, then we can see that ∂_t is contained in the closure of $\cup_{s<t} K_s$. To see this, note that if $\mathcal{B}(z, \epsilon) \cap \overline{\cup_{s<t} K_s} = \emptyset$ and $\lim_{s \to t-} \text{Im}[g_s(z)] = 0$, then the Harnack inequality implies that $\lim_{s \to t-} \text{Im}[g_s(w)] = 0$ for all $w \in \mathcal{B}(z, \epsilon)$, which in turn implies that $\mathcal{B}(z, \epsilon) \cap H_t = \emptyset$ and $\mathcal{B}(z, \epsilon) \cap \partial H_t = \emptyset$.

Using Corollary 3.80, we see that $t \mapsto \text{hcap}(K_t)$ is a continuous function. Let us assume that $\text{hcap}(K_t) \to \infty$ as $t \to \infty$. Then we can always reparametrize γ so that $\text{hcap}(K_t) = 2t$ with the proviso that γ under the new parametrization might not be continuous. For example, if

$$\gamma(t) = \begin{cases} ti, & 0 \leq t \leq 2, \\ 2i + e^{2\pi(t-2)i} - 1, & 2 \leq t \leq 3, \\ (3-t) + 2i, & t \geq 3, \end{cases}$$

then $K_t = K_3$ for $3 \leq t \leq 5$; if γ^* denotes γ reparametrized by capacity, then γ^* has a jump from $\gamma(3) = 2i$ to $\gamma(5) = -2 + 2i$. Even if the reparametrized curve is continuous it is possible that the curve does not arise from a Loewner chain. For example, if

$$\hat{\gamma}(t) = \begin{cases} ti, & 0 \leq t \leq 2, \\ 2i + e^{2\pi(t-2)i} - 1, & 2 \leq t \leq 3, \\ (t-1)i, & t \geq 3, \end{cases}$$

then the driving function U_t one would obtain from Proposition 4.4 would be discontinuous at the double point of $\hat{\gamma}$.

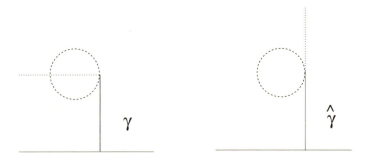

FIGURE 4.3. The curves $\gamma, \hat{\gamma}$. The curves traverse the solid line, then the dashed circle counterclockwise, and then the dotted line.

On the other hand, it is possible to obtain non-simple curves γ from Loewner chains. We will say that a Loewner chain g_t satisfying (4.4) is *generated by the curve* $\gamma : [0, \infty) \to \mathbb{C}$ if for each t, the domain of g_t, H_t, is the unbounded component of $\mathbb{H} \setminus \gamma[0, t]$. It is not true that every Loewner chain is generated by a curve (see the example below).

Suppose g_t is a Loewner chain with driving function U_t with $U_0 = 0$. Let $K_t = \overline{\mathbb{H}} \setminus H_t$ be the hulls generated by g_t and $\partial_t = \partial H_t \cap \mathbb{H}$. Let $\overline{K}_t$ denote the closure; for convenience, we set $\overline{K}_0 = \{0\}$. The Loewner equation (4.10) is valid for $z \in \overline{\mathbb{H}} \setminus \{0\}$ and $t < T_z \in (0, \infty]$ where T_z is characterized by $\lim_{t \to T_z^-} [g_t(z) - U_t] = 0$ if $T_z < \infty$. For each $t > 0$, let J_t be the points added to $\overline{K}_t$ at time t,

$$J_t = \overline{K}_t \setminus \bigcup_{s<t} \overline{K}_s.$$

We call z a *t-accessible* point if $z \in J_t$ and there exists a curve $\eta : [0,1] \to \mathbb{C}$ with $\eta(0) = z, \eta(0, 1] \subset H_t$.

LEMMA 4.26. *If $t > 0$ and z is a t-accessible point, then there is a strictly increasing sequence $s_j \uparrow t$ and a sequence of s_j-accessible points z_j with $z_j \to z$.*

PROOF. Since t-accessible points are on ∂H_t, they must be in the closure of $\bigcup_{s<t} \overline{K}_s$. Therefore, for all $\epsilon > 0$, we can find an $s < t$ with $\overline{K}_s \cap \mathcal{B}(z, \epsilon) \neq \emptyset$, and we can find an $\tilde{s}$-accessible point within distance ϵ of z by drawing a line segment from z to $K_s \cap \mathcal{B}(z, \epsilon)$. Since it is also true, that for every $s < t$, there is a δ with $\mathcal{B}(z, \delta) \subset H_s$, we can find a strictly increasing sequence s_j. □

PROPOSITION 4.27. *For each $t > 0$, there is at most one t-accessible point. Also, ∂_t is contained in the closure of the set of s-accessible points for $s \leq t$.*

PROOF. Suppose z is a t-accessible point and let η be such a curve. Since $z \in J_t$, for every $\epsilon > 0$ we can find an $s_\epsilon < t$ and an s_ϵ-accessible point z_ϵ with $|z - z_\epsilon| < \epsilon$. To see this, choose s_ϵ such that $K_{s_\epsilon} \cap \mathcal{B}(z, \epsilon) \neq \emptyset$; take a line segment from z to a point on $K_{s_\epsilon} \cap \mathcal{B}(z, \epsilon)$ and cut this segment at the first visit to K_{s_ϵ}. Let η_ϵ be the curve that starts with the reversal of this line segment followed by $\eta[0, t_\epsilon]$ where t_ϵ is the smallest $s > t$ with $|\eta(s) - \eta(t)| = \epsilon$. Then $\text{diam}[\eta_\epsilon] \leq 2\epsilon$, and hence by (3.21),

$$\text{diam}[g_{s_\epsilon}(\eta_\epsilon)] \leq c\, \epsilon^{1/2}\, M^{1/2},$$

where $M = M_t$ is chosen sufficiently large so that $\text{rad}(K_{2t}) \leq M_t$. From this we can see that

$$\lim_{r \to 0+} g_t(\eta(r)) = \lim_{s \to t-} g_s(z) = U_t.$$

Since the limit is independent of η and z, Proposition 3.86 shows that z must be unique (if it exists).

Suppose $w \in \partial_t$. Then there exist points $w' \in H_t$ arbitrarily close to w. For any such w', consider the line segment from w' to w, and let z' be the first point on this segment in $\overline{K}_t$. Let $s = T_{z'} \in (0, t]$. Then z' is an s-accessible point. Since there exist s-accessible points with $s \leq t$ arbitrarily close to w, w is contained in the closure of such points. □

EXAMPLE 4.28. Here we give an example of a Loewner chain that is not generated by a curve. Consider the "logarithmic spiral", $\gamma^*(t) = (t-1)e^{i \log|t-1|}, 0 \leq t \leq 2$, and for $t \in [0, 2] \setminus \{1\}$, let $\gamma(t) = F \circ \gamma^*(t)$ where

$$F(z) = i\left[(|z|+1)\frac{z}{|z|} + 2\right].$$

Let $\gamma(t) = (t+2)i$ for $t \geq 2$. Then $\gamma[0, 1)$ is a simple curve wrapping infinitely often around $\mathcal{B}(2i, 1)$, and $\gamma(1, 2]$ is a simple curve, not intersecting $\gamma[0, 1)$, also wrapping infinitely often around $\mathcal{B}(2i,1)$. Let $K_t = \gamma(0,t)$ for $t < 1$, $K_1 = \gamma(0,1) \cup \overline{\mathcal{B}}(2i,1)$ and for $t > 1$, $K_t = K_1 \cup \gamma(1,t]$. These hulls K_t can be obtained from a Loewner chain U_t; in fact, U_t is Hölder continuous of order $1/2$ [69]. Note that $J_1 = \overline{\mathcal{B}}(2i,1)$, but there are no 1-accessible points.

PROPOSITION 4.29. *Suppose g_t is a Loewner chain with driving function U_t and let $f_t(z) = g_t^{-1}(z), \hat{f}_t(z) = g_t^{-1}(z+U_t)$, and $V(y,t) = \hat{f}_t(iy)$. Suppose for each t the limit*

(4.26)
$$\gamma(t) = \lim_{y \to 0+} \hat{f}_t(iy)$$

exists, and the function $t \mapsto \gamma(t)$ is continuous, i.e., suppose that V is continuous on $[0,\infty) \times [0,\infty)$. Then g_t is the Loewner chain generated by the curve γ.

PROOF. The condition (4.26) and the proof of Proposition 4.27 show that $\gamma(t)$ is the only possible t-accessible point. (It is always an s-accessible point for some $s \leq t$.) Therefore the set of s-accessible points for $s \leq t$ is contained in $\gamma[0,t]$. Since $\gamma[0,t]$ is closed, Proposition 4.27 shows that ∂_t is contained in $\gamma[0,t]$. □

REMARK 4.30. Since V is clearly continuous on $(0,\infty) \times [0,\infty)$, to establish the continuity of V on $[0,\infty) \times [0,t_0]$ it suffices to find $\delta(\epsilon)$ such that $\delta(0+) = 0$ and such that

$$|V(y,t) - V(y_1,s)| \leq \delta(y + y_1 + |t-s|), \quad 0 \leq t, s \leq t_0, \ y, y_1 > 0.$$

PROPOSITION 4.31. *Suppose g_t is a Loewner chain such that for each t, $\overline{K}_t$ is locally connected. Then g_t is generated by a curve.*

PROOF. The assumption implies that the map g_t^{-1} can be extended to a continuous map from $\overline{\mathbb{H}}$ to $\overline{H_t}$. In particular, the limit $\gamma(t) = \lim_{y \to 0+} g_t^{-1}(U_t + iy)$ exists. We need to show that γ is continuous. However, we know for any Loewner chain that for all t, $\lim_{\epsilon \to 0+} \text{diam}[K_{t,t+\epsilon}] = 0$. Since g_t^{-1} is continuous on $\overline{\mathbb{H}}$, this implies $\lim_{\epsilon \to 0+} \text{diam}[K_{t+\epsilon} \setminus K_t] = 0$. □

REMARK 4.32. The converse of the proposition is true; if g_t is generated by a curve, then $\overline{K}_t$ is locally connected for all t. Note that in Example 4.28, K_1 is not locally connected, and it is not true that $\lim_{\epsilon \to 0+} \text{diam}[K_{1+\epsilon} \setminus K_1] = 0$.

The next lemma will not be used until Chapter 7, but since the proof only uses properties of the Loewner equation, we prove it here.

LEMMA 4.33. *Suppose that g_t is a Loewner chain with driving function U_t and let $f_t, \hat{f}_t, V$ be as in Proposition 4.29. Assume there exist a sequence of positive numbers $r_j \to 0$ and a c such that*

(4.27)
$$|\hat{f}'_{k2^{-2j}}(2^{-j}i)| \leq 2^j r_j, \quad k = 0, 1, \ldots, 2^{2j} - 1,$$

(4.28)
$$|U_{t+s} - U_t| \leq c\sqrt{j}\, 2^{-j}, \quad 0 \leq t \leq 1, 0 \leq s \leq 2^{-2j},$$

and

$$\lim_{j \to \infty} \sqrt{j}/\log r_j = 0.$$

Then V is continuous on $[0,1] \times [0,1]$.

PROOF. By differentiating (4.12), we get
$$\dot{f}'_t(z) = -f''_t(z)\frac{2}{z-U_t} + f'_t(z)\frac{2}{(z-U_t)^2}.$$
Proposition 3.16 implies that $|f''_t(z)| \le 2|f'_t(z)|/\text{Im}(z)$. Hence,
$$|\dot{f}'_t(z)| \le \frac{6|f'_t(z)|}{\text{Im}(z)^2}, \quad |f'_{t+s}(z)| \le \exp\left\{\frac{6s}{\text{Im}(z)}\right\}|f'_t(z)|.$$
In particular, (4.27) implies that for $k = 0, 1, \ldots, 2^{2j} - 1$,
$$(4.29) \qquad |f'_t(i2^{-j} + U_{k2^{-2j}})| \le e^6\, 2^j\, r_j, \quad k2^{-2j} \le t \le (k+1)2^{-2j}.$$
The Distortion Theorem (Theorem 3.21) tells us that if f is a univalent function on $\mathbb{D}$, then $|f'(z)| \le 12|f'(0)|$ for $|z| \le 1/2$. By iterating this, we see that if $f : \mathbb{H} \to D$ is any conformal transformation and $\text{Im}(z), \text{Im}(w) \ge y > 0$, then
$$|f'(w)| \le 144^{(|z-w|/y)+1}|f'(z)|.$$
In particular, (4.28) and (4.29) imply that there exist c, β such that
$$|\hat{f}'_t(i2^{-j})| \le c\, e^{\beta\sqrt{j}}\, 2^j\, r_j, \quad 0 \le t \le 1, j = 0, 1, 2, \ldots.$$
Using the distortion estimate again we get
(4.30)
$$|\hat{f}'_t(iy)| \le c\, e^{\beta\sqrt{j}}\, 2^j\, r_j, \quad 0 \le t \le 1,\ 2^{-j} \le y \le 2^{-j+1},\ j = 0, 1, 2, \ldots, 2^{-j}.$$
If $s \le 2^{-2j}$ and $y, y_1 \le 2^{-j}$, then
$$|\hat{f}_t(iy) - \hat{f}_{t+s}(iy_1)| \le$$
$$(4.31) \quad |\hat{f}_t(iy) - \hat{f}_t(i2^{-j})| + |\hat{f}_t(i2^{-j}) - \hat{f}_{t+s}(i2^{-j})| + |\hat{f}_{t+s}(i2^{-j}) - \hat{f}_{t+s}(iy_1)|.$$
By (4.30),
$$|\hat{f}_t(iy) - \hat{f}_t(i2^{-j})| \le R(j) := \sum_{l=j}^{\infty} c\, e^{\beta\sqrt{l}}\, r_l.$$
Our assumptions tell us that $R(j) \to 0$ as $j \to \infty$. The third term in (4.31) is handled similarly. Using (4.12) and (4.30) gives
$$|\hat{f}_t(i2^{-j}) - \hat{f}_{t+s}(i2^{-j})| \le 2s\, 2^j \sup_{t \le r \le t+s} |f'(2^{-j})| \le c\, r_j.$$
This gives the lemma (see the remark after Proposition 4.29). $\square$

For the remainder of the section, assume that g_t is a Loewner chain with driving function U_t that is generated by the path γ. As before, for $s < t$, let $g_{s,t}$ be defined by $g_t = g_{s,t} \circ g_s$. Then for fixed $s \ge 0$, the Loewner chain $g_t^{(s)} = g_{s,s+t}$ has driving function $U_t^{(s)} = U_{s+t}$ and is generated by the path $\gamma^{(s)}$ where $\gamma^{(s)}(t) = g_s(\gamma(s+t))$.

LEMMA 4.34. *Suppose g_t is generated by the curve γ. Then γ is a simple curve with $\gamma(0, \infty) \subset \mathbb{H}$ if and only if for all $s \ge 0$, $\gamma^{(s)}(0, \infty) \cap \mathbb{R} = \emptyset$.*

PROOF. The last condition for $s = 0$ is clearly equivalent to $\gamma(0, \infty) \subset \mathbb{H}$. If $s, t > 0$ then the statement $\gamma(s) = \gamma(s+t)$ is equivalent to the statement $\gamma^{(s)}(t) = g_s(\gamma(s)) = U_s \in \mathbb{R}$. $\square$

PROPOSITION 4.35. *Suppose g_t is a Loewner chain with driving function U_t that is generated by the path γ. Suppose for some $r < \sqrt{2}$ and all $s < t$.*

(4.32) $$|U_t - U_s| \leq r\sqrt{t-s}.$$

Then γ is a simple curve.

PROOF. We will show that (4.32) implies that for all $x \in \mathbb{R}\setminus\{U_0\}$ and all $t > 0$, $g_t(x) \neq U_t$. This implies that $x \in H_t^*$ for all t, where H_t^* is the symmetrization of H_t as in the Schwarz reflection principle; in particular, $x \notin \gamma(0,\infty)$. Since $U_t^{(s)}$ also satisfies (4.32), this implies that $\gamma^{(s)}(0,\infty) \cap \mathbb{R} = \emptyset$ for all $s \geq 0$ which, as we noted above, implies that γ is a simple curve. Without loss of generality, assume $U_0 = 0$, $x > 0$, and choose $\rho \in (0,1)$ such that $2\rho^3 r^{-2} - \rho > \rho^{-1} - 1$. Let t_1 be the first time that $|U_t| \geq \rho x$, let t_2 be the first time that $g_t(x) - U_t \geq \rho^{-1}x$ and let $t_0 = t_1 \wedge t_2$. By (4.32), $t_1 \geq (\rho x/r)^2$. For $t \leq t_2$, (4.10) gives $\dot{g}_t(x) \geq 2\rho/x$. If $t_0 \geq (\rho x/r)^2$, then

$$g_{(\rho x/r)^2}(x) \geq x + (2\rho^3/r^2)x \geq x\left[1 + \frac{2\rho^3}{r^2} - \rho\right] + U_{(\rho x/r)^2} > \rho^{-1}x + U_{(\rho x/r)^2}.$$

Hence $t_0 = t_2$. In particular, $g_t(x) - U_t$ always reaches $\rho^{-1}x$ before reaching 0. By iterating this argument, we see this is also true for ρ^{-k} for all k which implies that $g_t(x) - U_t$ never equals 0. □

REMARK 4.36. The assumption in the proposition can easily be weakened to the following: for every $T < \infty$, there is a $\delta > 0$ and an $r < \sqrt{2}$ such that (4.32) holds for all $0 \leq s < t \leq T$ with $t - s \leq \delta$. In particular, if U_t is Hölder α-continuous (see §1.13) for some $\alpha > 1/2$, the result holds.

REMARK 4.37. Marshall and Rohde [69] have shown that there is an $r_0 > 0$ such that if U_t satisfies (4.32), then U_t is generated by a (necessarily simple) curve. However, there are examples such as the "spiral" example that satisfy (4.32) for some $r > r_0$.

4.5. Distance to the curve

Suppose g_t is a Loewner chain generated by the curve γ. For every $z \in \mathbb{H}$, we define
$$\text{dist}(z, g) = \text{dist}[z, \mathbb{R} \cup \gamma(0, \infty)].$$
We write $\text{dist}(z, g)$ rather than $\text{dist}(z, \gamma)$ because this quantity can be defined for general solutions of the Loewner equation as in Theorem 4.6 by
$$\text{dist}(z, g) = \inf_{t < T_z} \text{dist}(z, \partial H_t).$$
While this quantity is not easily computed, the next proposition shows that it is comparable to a natural quantity defined in terms of the maps g_t.

PROPOSITION 4.38. *Suppose g_t are the conformal maps as in Theorem 4.6 and $z \in \mathbb{H}$. For $t < T_z$, let*
$$D_t(z) = \log \frac{|g_t'(z)|}{\text{Im}[g_t(z)]}.$$
Then $t \to D_t$ is increasing and the limit $D(z) := \lim_{t \to T_z-} D_t(z)$ satisfies
$$\frac{1}{4}\text{dist}(z,g) \leq e^{-D(z)} \leq 4\,\text{dist}(z,g).$$

PROOF. Assume $t < T_z$. Using (4.8), we see that
$$\begin{aligned}\partial_t[\log|g_t'(z)|] &= \partial_t \operatorname{Re}[\log g_t'(z)] \\ &= -\operatorname{Re}\int_{-\infty}^{\infty}\frac{\mu_t(du)}{(g_t(z)-u)^2} \\ &= \int_{-\infty}^{\infty}\frac{\operatorname{Im}[g_t(z)]^2 - (\operatorname{Re}[g_t(z)]-u)^2}{[\,(\operatorname{Re}[g_t(z)]-u)^2 + \operatorname{Im}[g_t(z)]^2\,]^2}\mu_t(du).\end{aligned}$$

Using (4.5), we see that
$$\partial_t[\log \operatorname{Im}[g_t(z)]] = -\int_{-\infty}^{\infty}\frac{\mu_t(du)}{(\operatorname{Re}[g_t(z)]-u)^2 + \operatorname{Im}[g_t(z)]^2}.$$

Hence,
$$(4.33) \qquad \dot{D}_t(z) = \int_{-\infty}^{\infty}\frac{2\operatorname{Im}[g_t(z)]^2 \mu_t(du)}{[\,(\operatorname{Re}[g_t(z)]-u)^2 + (\operatorname{Im}[g_t(z)])^2\,]^2},$$

from which we see that $D_t(z)$ increases with t. Corollary 3.19 gives
$$\frac{\operatorname{Im}[g_t(z)]}{4\operatorname{dist}(z,\partial H_t)} \le |g_t'(z)| \le \frac{4\operatorname{Im}[g_t(z)]}{\operatorname{dist}(z,\partial H_t)},$$

which gives the inequality. $\square$

4.6. Perturbation by conformal maps

In Chapter 6 we will discuss the locality and restriction properties for the Schramm-Loewner evolution as well as a relationship between chordal and radial *SLE*. The analysis will require understanding what happens when *SLE* is transformed by a conformal map. As a preparation for this, we will derive some formulas here which describe the evolution of the image of Loewner chains under conformal maps.

If K_t is a continuously growing collection of hulls given by solving the Loewner equation with driving function U_t, and ϕ is a conformal transformation defined near U_0, then we can consider the sets $\phi(K_t)$ for small t. In this section we derive the Loewner equation for the images. If $x \in \mathbb{R}$, we will say that an open subset $\mathcal{N}$ of $\mathbb{H}$ is an $\mathbb{H}$-neighborhood of x if $\mathcal{B}(x,\epsilon) \cap \mathbb{H} \subset \mathcal{N}$ for some $\epsilon > 0$. We say that a univalent function $\phi : \mathcal{N} \to \mathbb{H}$ is *locally real* at x_0 if for some $\epsilon > 0$,
$$\phi(z) = a_0 + a_1(z-x_0) + a_2(z-x_0)^2 + \cdots$$
with $a_0, a_1, \ldots \in \mathbb{R}$. Note that $\phi(\mathcal{N}) \subset \mathbb{H}$ implies that $a_1 > 0$.

4.6.1. Chordal. Suppose g_t is a chordal Loewner chain with driving function U_t and corresponding hulls K_t, so that
$$\dot{g}_t(z) = \frac{2}{g_t(z) - U_t}, \quad g_0(z) = z.$$
Let $x_0 = U_0$, and suppose that Φ is a locally real conformal transformation of an $\mathbb{H}$-neighborhood $\mathcal{N}$ of x_0 into $\mathbb{H}$. Let t_0 be the supremum of all t such that $K_t \subset \mathcal{N}$. Then, as mentioned at the end of §4.1, $K_t^* = \Phi(K_t), 0 \le t < t_0$, is a continuously growing family of hulls. Let g_t^* be the corresponding conformal transformations, $g_t^* : \mathbb{H} \setminus K_t^* \to \mathbb{H}$ with
$$g_t^*(z) = z + \frac{b^*(t)}{z} + \cdots, \quad z \to \infty,$$

where $b^*(t) = \text{hcap}(K_t^*)$. As before, if $s < t \le t_0$, let $K_{s,t} = g_s(K_t) \cap \mathbb{H}$, $K_{s,t}^* = g_s^*(K_t^*)$. Then $K_{s,t}^* = \Phi_s(K_{s,t})$ where $\Phi_s = \Phi \circ g_s^{-1}$. Let $\Phi_t = g_t^* \circ \Phi \circ g_t^{-1}$. Then Φ_t is a locally real conformal transformation from an $\mathbb{H}$-neighborhood of U_t into an $\mathbb{H}$-neighborhood of $U_t^* := \Phi_t(U_t)$. From (4.15) we can see that $\dot{b}(t) = 2\,\Phi_t'(U_t)^2$. Note that $\Phi_0 = \Phi$ and if $s < t$, $\Phi_t = g_{s,t}^* \circ \Phi_s \circ g_{s,t}^{-1}$, where, as before, $g_t = g_{s,t} \circ g_s$, $g_t^* = g_{s,t}^* \circ g_s$. Hence we can think of the Loewner chain as a flow Φ_t on conformal maps. The maps g_t^* satisfy the Loewner equation

$$\dot{g}_t^*(z) = \frac{\dot{b}^*(t)}{g_t^*(z) - U_t^*} = \frac{2\,\Phi_t'(U_t)^2}{g_t^*(z) - U_t^*}, \quad g_0^*(z) = z.$$

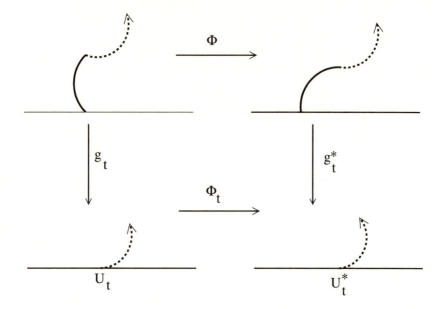

FIGURE 4.4. The maps g_t^*, Φ_t. Note that $\Phi_t = g_t^* \circ \Phi \circ g_t^{-1}$.

REMARK 4.39. If $\hat{\Phi}_t(z) = \Phi_t(z + U_t)$, then $\hat{\Phi}_t$ is a parametrized collection of conformal maps defined in a neighborhood of the origin. The Loewner equation with driving function U_t gives a flow $\hat{\Phi}_t$ on such conformal maps with initial condition $\hat{\Phi}_0(z) = \Phi(z + U_0)$.

PROPOSITION 4.40. *Under the assumptions above, the maps Φ_t satisfy*

$$(4.34) \qquad \dot{\Phi}_t(z) = 2\left[\Phi_t'(U_t)\,\frac{\Phi_t'(U_t)}{\Phi_t(z) - \Phi_t(U_t)} - \Phi_t'(z)\,\frac{1}{z - U_t}\right].$$

For each $t < t_0$, these equations are valid for z in a neighborhood of U_t. In particular, they hold at $z = U_t$ with

$$(4.35) \qquad \dot{\Phi}_t(U_t) = \lim_{z \to U_t} \dot{\Phi}_t(z) = -3\,\Phi_t''(U_t).$$

PROOF. Since the right-hand side of (4.34) is continuous in t it suffices to check right derivatives, and the argument is the same for all t so we will assume $t = 0$.

Let $f_t = g_t^{-1}$. The chain rule combined with (4.16) and (4.12) imply

$$\dot{\Phi}_0(z) = \dot{g}_0^*(\Phi \circ f_0(z)) + (g_0^*)'(\Phi \circ f_0(z))\, \Phi'(f_0(z))\, \dot{f}_0(z)$$
$$= \frac{2\, \Phi'(U_0)^2}{\Phi(z) - \Phi(U_t)} + (g_0^*)'(\Phi \circ f_0(z))\, \Phi'(f_0(z))\, [-f_0'(z)\, \frac{2}{z - U_0}],$$

which gives (4.34) at least for z in a punctured neighborhood of U_0. However, it is straightforward to take the limit in (4.35) and hence the result holds at U_0 as well. (To take the limit, assume for ease that $U_0 = 0$, $\Phi_t(0) = 0$ and write $\Phi_t(z) = a_1 z + a_2 z^2 + O(z^3)$. Then the right hand side of (4.34) is $-6a_2 + O(z)$.) □

PROPOSITION 4.41. *Under the assumptions above, the maps Φ_t satisfy*

$$(4.36) \qquad \dot{\Phi}_t'(z) = 2\left[-\frac{\Phi_t'(U_t)^2\, \Phi_t'(z)}{(\Phi_t(z) - \Phi_t(U_t))^2} + \frac{\Phi_t'(z)}{(z - U_t)^2} - \frac{\Phi_t''(z)}{z - U_t}\right].$$

For each $t < t_0$, these equations are valid in a neighborhood of U_t. In particular, they hold at $z = U_t$ with

$$(4.37) \qquad \dot{\Phi}_t'(U_t) = \lim_{z \to U_t} \dot{\Phi}_t'(z) = \frac{\Phi_t''(U_t)^2}{2\, \Phi_t'(U_t)} - \frac{4\, \Phi_t'''(U_t)}{3}.$$

PROOF. Note that (4.36) comes from differentiating (4.34) with respect to z; since the right hand side is continuous as a function of t and z, the mixed partials are equal, at least in a punctured neighborhood of U_t. Again, the limit in (4.37) is straightforward. (If we assume, for ease, that $U_t = 0$, $\Phi_t(0) = 0$ and write $\Phi_t(z) = a_1 z + a_2 z^2 + a_3 z^3 + O(z^4)$, then the right hand side of (4.36) is $(2a_2^2/a_1) - 8a_3 + O(z)$.) □

REMARK 4.42. Similar expressions can be derived for the time derivative of higher spatial derivatives, $\dot{\Phi}_t^{(n)}(U_t)$.

4.6.2. Radial. Suppose g_t is radial Loewner chain with driving function $U : [0, \infty) \to \mathbb{R}$, and let A_t be the corresponding hulls. Let $f_t = g_t^{-1}$. Suppose that Φ is a locally real conformal transformation of an $\mathbb{H}$-neighborhood of U_0 into $\mathbb{H}$. We can write $\Phi(z) = -i \log \Psi(e^{iz})$ where Ψ is a conformal transformation in a neighborhood $\mathcal{N}$ of e^{iU_0} that takes points on the unit circle to the unit circle and points in $\mathbb{D}$ to $\mathbb{D}$, and log denotes the branch of the logarithm such that $-i \log \Psi(e^{iU_0}) = \Phi(U_0)$. Let t_0 be the supremum of all t such that $A_t \subset \mathcal{N}$ and for $t < t_0$, let $A_t^* = \Psi(A_t)$. Then $A_t^*, 0 \leq t < t_0$ is a continuously growing family of radial hulls. Let g_t^* denote the generalized radial Loewner chain associated to these hulls and let U_t^* denote the driving function. Let $\Psi_t = g_t^* \circ \Psi \circ f_t$. Note that $\Psi_0 = \Psi$ and if $s < t$, $\Psi_t = g_{s,t}^* \circ \Psi_s \circ f_{s,t}$ where $g_t^* = g_{s,t}^* \circ g_s^*$, $f_t = f_s \circ f_{s,t}$. Also,

$$(4.38) \qquad \dot{g}_t^*(z) = g_t^*(z)\, |\Psi_t'(e^{iU_t})|^2\, \frac{e^{iU_t^*} + g_t^*(z)}{e^{iU_t^*} - g_t^*(z)}, \qquad g_0^*(z) = z.$$

Let $\Phi_t(z) = -i \log \Psi_t(e^{iz})$.

PROPOSITION 4.43. *Under the assumptions of this subsection, the maps Φ_t satisfy*

$$(4.39) \qquad \dot{\Phi}_t(z) = \Phi_t'(U_t)^2 \cot\left[\frac{\Phi_t(z) - \Phi_t(U_t)}{2}\right] - \Phi_t'(z) \cot\left[\frac{z - U_t}{2}\right].$$

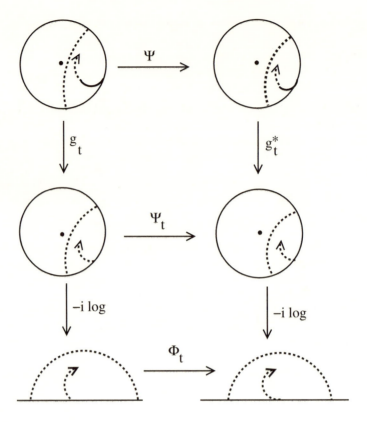

FIGURE 4.5. The maps Ψ_t, Φ_t. Note that $\Phi_t(z) = -i\log\left[g_t^* \circ \Psi \circ g_t^{-1}(e^{iz})\right]$.

For each $t < t_0$, these equations are valid in a neighborhood of U_t. In particular, they hold at $z = U_t$ with

(4.40) $$\dot\Phi_t(U_t) = \lim_{z \to U_t} \dot\Phi_t(z) = -3\,\Phi_t''(U_t).$$

PROOF. Again, we concentrate on $t = 0$, and first consider Ψ. By the chain rule,
$$\dot\Psi_0(z) = \dot g_0^*(\Psi \circ f_0(z)) + (g_0^*)'(\Psi \circ f_0(z))\,\Psi'(f_0(z))\,\dot f_0(z).$$

Hence, using (4.20) and (4.38),
$$\dot\Psi_0(z) = \Psi_0(z)\,|\Psi_0'(e^{iU_0})|^2\,\frac{e^{iU_0^*} + \Psi_0(z)}{e^{iU_0^*} - \Psi_0(z)} - z\,\Psi_0'(z)\,\frac{e^{iU_0} + z}{e^{iU_0} - z}.$$

But
$$\dot\Phi_0(z) = \frac{-i\dot\Psi_0(e^{iz})}{\Psi_0(e^{iz})}, \quad \Phi_0'(z) = \frac{e^{iz}\,\Psi_0'(e^{iz})}{\Psi_0(e^{iz})},$$

which gives (4.39). As $z \to 0$, $\cot z = z^{-1} + O(z)$, and so (4.40) follows from (4.35). □

PROPOSITION 4.44. *Under the assumptions of this subsection, the maps Φ_t satisfy*

$$\dot{\Phi}'_t(z) = -\frac{1}{2} \Phi'_t(U_t)^2 \, \Phi'_t(z) \, \csc^2\left[\frac{\Phi_t(z) - \Phi_t(U_t)}{2}\right]$$

(4.41)
$$-\Phi''_t(z) \cot\left[\frac{z-U_t}{2}\right] + \frac{1}{2} \Phi'_t(z) \csc^2\left[\frac{z-U_t}{2}\right].$$

For each $t < t_0$, these equations are valid in a neighborhood of U_t. In particular, they hold at $z = U_t$ with

(4.42) $\qquad \dot{\Phi}'_t(U_t) = \lim_{z \to U_t} \dot{\Phi}'_t(z) = \dfrac{\Phi''_t(U_t)^2}{2\,\Phi'_t(U_t)} - \dfrac{4\,\Phi'''_t(U_t)}{3} + \dfrac{\Phi'_t(U_t) - \Phi'_t(U_t)^3}{6}.$

PROOF. Differentiating (4.39) with respect to z gives (4.41). As $z \to 0$, $\csc^2 z = z^{-2} + (1/3) + O(z^2)$, so (4.42) follows from (4.37). □

4.6.3. Mapping chordal to radial. Suppose g_t is a chordal Loewner chain with driving function $U : [0, \infty) \to \mathbb{R}$ with corresponding hulls A_t and let $f_t = g_t^{-1}$. Suppose $\mathcal{N}$ is a simply connected $\mathbb{H}$-neighborhood of U_0 such that $z \mapsto e^{iz}$ is one-to-one on $\mathcal{N}$. Then $A_t^* = \exp(iA_t)$ is a continuously growing collection of hulls in the disk. Let $\tilde{g}_t$ be the unique conformal transformation of $\mathbb{D} \setminus A_t^*$ onto $\mathbb{D}$ with $\tilde{g}_t(0) = 0, \tilde{g}'_t(0) > 0$. Then $\tilde{g}_t$ satisfies the radial Loewner equation

$$\dot{\tilde{g}}_t(z) = \dot{a}(t)\, \tilde{g}_t(z)\, \frac{e^{iU_t^*} + \tilde{g}_t(z)}{e^{iU_t^*} - \tilde{g}_t(z)},$$

for some U_t^*. Here $a(t) = \log \tilde{g}'_t(0)$. Let $\Psi_t(z) = \tilde{g}_t(e^{if_t(z)})$ and $\Phi_t(z) = -i \log \Psi_t(z)$. Then $U_t^* = \Phi_t(U_t)$.

Let $A_t^\dagger := \{z + 2\pi k i : z \in A_t\}$ be the preimage of A_t^* under the exponential map above. Note that $A_t^\dagger \notin \mathcal{Q}$ since $A_t^\dagger$ is unbounded. Suppose $F : A_t^\dagger \to \mathbb{H}$ is a conformal transformation with $F(\infty) = \infty$. There is a two-parameter family of such transformations; given F, then others are of the form $bF + x$ with $b > 0, x \in \mathbb{R}$. We can choose such an F_t such that $F_t(z) = -i \log g_t^*(e^{iz})$. This specifies F_t up to an additive multiple of $2\pi i$; if we also specify $F_0(0) = 0$ and continuity of $F_t(z)$ in t, we get a unique choice. In fact, one can easily see that $F_t = \Phi_t \circ g_t$.

The following propositions are proved similarly to the propositions in the preceding two subsections.

PROPOSITION 4.45. *Under the assumptions of this subsection, the maps Φ_t satisfy*

$$\dot{\Phi}_t(z) = \Phi'_t(U_t)^2 \cot\left[\frac{\Phi_t(z) - \Phi_t(U_t)}{2}\right] - \Phi'_t(z) \frac{2}{z - U_t}.$$

For each $t < t_0$, these equations are valid in a neighborhood of U_t. In particular, they hold at $z = U_t$ with

(4.43) $\qquad \dot{\Phi}_t(U_t) = \lim_{z \to U_t} \dot{\Phi}_t(z) = -3\, \Phi''_t(U_t).$

PROPOSITION 4.46. *Under the assumptions of this subsection, the maps Φ_t satisfy*

$$\dot{\Phi}'_t(z) = -\frac{1}{2} \Phi'_t(U_t)^2 \, \Phi'_t(z) \, \csc^2\left[\frac{\Phi_t(z) - \Phi_t(U_t)}{2}\right]$$

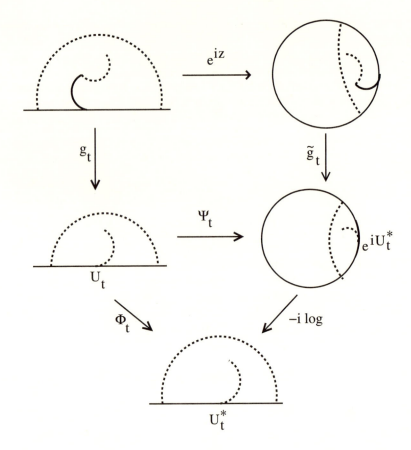

FIGURE 4.6. The maps Ψ_t, Φ_t. Note that $\Phi_t(z) = -i \log \tilde{g}_t(e^{ig_t^{-1}(z)})$.

(4.44) $$-\Phi_t''(z) \frac{2}{z - U_t} + \Phi_t'(z) \frac{2}{(z - U_t)^2}.$$

For each $t < t_0$, these equations are valid in a neighborhood of U_t. In particular, they hold at $z = U_t$ with

(4.45) $$\dot{\Phi}_t'(U_t) = \lim_{z \to U_t} \dot{\Phi}_t'(z) = \frac{\Phi_t''(U_t)^2}{2\,\Phi_t'(U_t)} - \frac{4\,\Phi_t'''(U_t)}{3} - \frac{\Phi_t'(U_t)^3}{6}.$$

4.7. Convergence of Loewner chains

Suppose $U_t^{(n)}, U_t$ are continuous functions from $[0, \infty)$ into $\mathbb{R}$ and $g_t^{(n)}, g_t$ are the corresponding chordal Loewner chains with domains $H_t^{(n)}, H_t$. Let K_t be the closure of $\mathbb{H} \setminus H_t$. We will say that the chain $g_t^{(n)}$ converges to the chain g_t in the Carathéodory sense, written $g^{(n)} \xrightarrow{Cara} g$, if for every $\epsilon > 0$ and every $T < \infty$, $g_t^{(n)}$ converges to g_t uniformly on $[0, T] \times \{z \in \mathbb{H} : \text{dist}(z, K_T) \geq \epsilon\}$.

PROPOSITION 4.47. *If $U_t^{(n)}$ converges to U_t uniformly on compact intervals $[0, T]$, then $g^{(n)} \xrightarrow{Cara} g$.*

PROOF. Fix $T < \infty$ and assume $U_t^{(n)}$ converges to U_t uniformly on $[0, T]$. Let $\epsilon > 0$. For every $\delta > 0$, let
$$V_\delta = V_{\delta,T} = \{z \in \mathbb{H} : |g_t(z) - U_t| > \delta \text{ for } 0 \leq t \leq T\}.$$
Then V_δ is an open subset of H_T. Using Lemma 4.13, we can see that there is an $R = R_T$ (depending on the function U_t) such that for all $z \in \mathbb{H}$ with $|z| > R$ and all $\delta < 1$, $z \in V_\delta$. Also, each $z \in H_T$ is in V_δ for some $\delta > 0$. By a compactness argument, we can find a $\delta > 0$ such that
$$\{z \in \mathbb{H} : |z| \leq R, \text{dist}(z, K_T) \geq \epsilon\} \subset V_\delta.$$
Hence, it suffices to show that $g_t^{(n)}$ converges uniformly to g_t on V_δ. Recall that $g_0^n(z) = g_0(z) = z$. Let $z \in V_\delta$ and let $\sigma = \sigma_{n,\delta,z}$ be the first time s such that $|g_s^{(n)}(z) - g_s(z)| \geq \delta/4$. Let $C = 4\delta^{-2}$ and choose positive $\alpha < \delta/4$ such that $\alpha(e^{CT} - 1) < \delta/4$. Suppose n is chosen sufficiently large so that $|U_t - U_t^{(n)}| \leq \alpha$ for $0 \leq t \leq T$. If $0 \leq t \leq \sigma \wedge T$,
$$|g_t^{(n)}(z) - U_t^{(n)}| \geq |g_t(z) - U_t| - |g_t^{(n)}(z) - g_t(z)| - |U_t - U_t^{(n)}| \geq \delta/2.$$
Let $h(t) = g_t(z) - g_t^{(n)}(z)$. Then if $0 < t < \sigma \wedge T$,
$$|\dot{h}(t)| = \left| \frac{2}{g_t(z) - U_t} - \frac{2}{g_t^n(z) - U_t^{(n)}} \right|$$
$$\leq 2 \frac{|g_t(z) - g_t^{(n)}(z)| + |U_t - U_t^{(n)}|}{|g_t(z) - U_t||g_t^{(n)}(z) - U_t^{(n)}|} \leq C\alpha + C|h(t)|.$$
By solving the simple differential equation, we see that
$$|h(t)| \leq \alpha [e^{CT} - 1] < \delta/4, \quad 0 \leq t \leq \sigma \wedge T.$$
In particular, $\sigma \geq T$, so we have
$$|g_t(z) - g_t^{(n)}(z)| \leq \delta/4, \quad z \in V_\delta, \ 0 \leq t \leq T.$$
□

PROPOSITION 4.48. *If $U : [0, \infty) \to \mathbb{R}$ is a continuous function, and g_t is the corresponding Loewner chain, then there exists a sequence of Loewner chains $g_t^{(n)}$ generated by simple curves $\gamma^{(n)}$ such that $g^{(n)} \xrightarrow{Cara} g$.*

PROOF. For each n let $U^{(n)}$ be the "interpolation by square root" using time intervals of length $1/n$, i.e., for $k/n \leq t \leq (k+1)/n$,
$$U_t^{(n)} = U_{k/n} + (U_{(k+1)/n} - U_{k/n}) \sqrt{n} \sqrt{t - (k/n)}.$$
Using (4.14), we see that each $g^{(n)}$ is generated by a simple curve, and it is easy to see that $U_t^{(n)} \to U_t$ uniformly on compact intervals. □

Suppose that $g^{(n)}$ is a sequence of Loewner chains generated by paths $\gamma^{(n)}$ and $g^{(n)} \xrightarrow{Cara} g$. Since we know there exist Loewner chains that are not generated by paths, the last proposition tells us that we cannot conclude that g is generated by a path. However, suppose we *know* that g is generated by a path γ. Can we conclude that $\gamma^{(n)}$ converges to γ uniformly on compact sets? Unfortunately, the answer is no; in fact, as the next example shows, there exist $g^{(n)}$ such that $U_t^{(n)} \to 0$

uniformly on compact intervals, but such that the corresponding curves have no subsequential limits.

EXAMPLE 4.49. Let $\gamma^{(n)}$ denote the polygonal path connecting the points
$$0, z_1, w_1, \hat{z}_1, \hat{w}_1, z_2, w_2, \hat{z}_2, \hat{w}_2, \ldots,$$
where
$$z_k = -\frac{1}{n} + \frac{ik}{n}, \quad w_k = \frac{ik}{2n}, \quad \hat{z}_k = \frac{1}{n} + \frac{ik}{n}, \quad \hat{w}_k = \frac{i}{n}\left[\frac{k}{2} + \frac{1}{4}\right].$$

This is a simple curve, which we assume is parametrized by capacity. Let $g_t^{(n)}$ denote the conformal transformation mapping $\mathbb{H} \setminus \gamma^{(n)}(0, t]$ onto $\mathbb{H}$ with $g_t^{(n)}(z) - z \to 0$ as $z \to \infty$. Let $U_t^{(n)} = g_t^{(n)}(\gamma^{(n)}(t))$ We claim that there is a c such that
$$|U_t^{(n)}| \leq \frac{c}{\sqrt{n}}, \quad 0 \leq t \leq n.$$

To see this, let $\tau = \tau_{\mathbb{H} \setminus \gamma^{(n)}(0,t]}$ be the first time that a Brownian motion hits $\mathbb{R} \cup \gamma^{(n)}(0,t]$. Then (see Proposition 3.43),
$$U_t^{(n)} = \lim_{y \to \infty} \pi y \left[\frac{1}{2} - \mathbf{P}^{iy}\{B_\tau \in l_+\}\right],$$
where l_+ denotes the union of $(0, \infty)$ and the part of $\gamma^{(n)}(0,t]$ to the "right" of $\gamma^{(n)}(t)$. Hence, we need to show that for all y sufficiently large
$$\left|\mathbf{P}^{iy}\{B_\tau \in l_+\} - \frac{1}{2}\right| \leq \frac{c}{y\sqrt{n}}.$$

Fix $t \leq n$, and choose k so that $\gamma^{(n)}(t)$ is on one of the line segments $[z_k, w_k]$, $[w_k, \hat{z}_k]$, $[\hat{z}_k, \hat{w}_k]$, $[\hat{w}_k, z_{k+1}]$. Since $\operatorname{hcap}(\gamma^{(n)}(0,t]) = t \leq n$, we know that $k \leq c' n^{3/2}$. Let $R = R_{k,n}$ be the rectangle
$$R = \left\{x + iy : -\frac{1}{n} \leq x \leq \frac{1}{n}, 0 < y \leq \frac{k+1}{n}\right\},$$
and let $\sigma = \tau_{\mathbb{H} \setminus R}$. Using the symmetry of R about the imaginary axis, we can see that
$$\left|\mathbf{P}^{iy}\{B_\tau \in l_+\} - \frac{1}{2}\right| \leq \mathbf{P}^{iy}\left\{B_\sigma \in \left[-\frac{1}{n} + \frac{i(k+1)}{n}, \frac{1}{n} + \frac{i(k+1)}{n}\right]\right\} +$$
$$\mathbf{P}^{iy}\left\{B_\sigma \in \left[-\frac{1}{n}, -\frac{1}{n} + \frac{i(k+1)}{n}\right]; B_\tau \in l_+\right\}$$
$$+ \mathbf{P}^{iy}\left\{B_\sigma \in \left[\frac{1}{n}, \frac{1}{n} + \frac{i(k+1)}{n}\right]; B_\tau \notin l_+\right\}.$$

The distribution of B_σ restricted to R can be bounded from above by the distribution of the first visit to
$$\left\{x + iy : |x| \leq \frac{1}{n}, \ y \leq \frac{k+1}{n}\right\}.$$
Given this and the Beurling estimate, we can see that
$$\mathbf{P}^{iy}\left\{B_\sigma \in \left[-\frac{1}{n} + \frac{i(k+1)}{n}, \frac{1}{n} + \frac{i(k+1)}{n}\right]\right\} \leq \frac{c}{y\sqrt{n}}.$$

Given $B_\sigma = -(1/n) + (((k+1)/n) - r)i$, the conditional probability that $B_\tau \in l_+$ is bounded above by the probability that a Brownian motion starting at $-(1/n) + (((k+1)/n) - r)i$ reaches $\{\text{Im}(z) \geq (k+1)/n\}$ before hitting $\{\text{Re}(z) = 1/n\}$. Using gambler's ruin we see that this probability is bounded above by a constant times $(nr)^{-1}$ (of course, it is also bounded by 1). Using Beurling again to estimate the density we get that

$$y\,\mathbf{P}^{iy}\left\{B_\sigma \in \left[-\frac{1}{n}, -\frac{1}{n} + \frac{i(k+1)}{n}\right]; B_\tau \in l_+\right\} \leq c\int_0^{c'\sqrt{n}} [\frac{1}{rn} \wedge 1][1 \vee r^{-1/2}]\,dr$$
$$\leq \frac{c}{\sqrt{n}}.$$

A similar argument gives

$$y\,\mathbf{P}^{iy}\left\{B_\sigma \in \left[\frac{1}{n}, \frac{1}{n} + \frac{i(k+1)}{n}\right]; B_\tau \notin l_+\right\} \leq \frac{c}{\sqrt{n}}.$$

We leave it to the reader to check that there is no subsequence of $\gamma^{(n)}$ that converges uniformly on compact sets.

The problem with the last example is that the sequence $\gamma^{(n)}$ is not precompact. If we know that our sequence of curves $\gamma^{(n)}$ has subsequential limits, then we can conclude that $\gamma^{(n)} \to \gamma$ uniformly on compact sets. We leave this as an exercise:

EXERCISE 4.50. *Suppose $g^{(n)}$ are Loewner chains generated by curves $\gamma^{(n)}$.*

- *If $\gamma^{(n)}$ converges uniformly to γ on compact intervals, then $g^{(n)} \xrightarrow{Cara} g$ where g is the Loewner chain generated by γ.*
- *If $g^{(n)} \xrightarrow{Cara} g$ and for every t_0, $\{\gamma_t^{(n)} : 0 \leq t \leq t_0\}$ is an equicontinuous family, then there exists a γ such that $\gamma^{(n)}$ converges uniformly to γ on compact intervals and g is generated by γ.*

CHAPTER 5

Brownian measures on paths

5.1. Measures on spaces of curves

In this chapter we will be considering a number of measures on spaces of curves in $\mathbb{C}$. Many of the measures will be obtained as Riemann integrals of continuous functions from subsets of $\mathbb{R}$ or $\mathbb{C}$ into the space of measures. To justify this construction we need to consider metric spaces of measures.

Let $C[0,1]$ denote the set of curves $\gamma : [0,1] \to \mathbb{C}$, and let $\mathcal{X}$ denote the complete metric space $C[0,1] \times [0,\infty)$ with the metric induced by the norm

$$\|(\gamma, r)\| = |r| + \|\gamma\|_\infty = |r| + \sup_{0 \leq t \leq 1} |\gamma(t)|.$$

Let $\mathcal{K}$ denote the set of curves $\gamma : [0, t_\gamma] \to \mathbb{C}$ where $t_\gamma \in (0, \infty)$. There is a natural one-to-one correspondence between $\mathcal{K}$ and $\mathcal{X}_+ := \{(\gamma, r) \in \mathcal{X} : r > 0\}$ given by $\gamma \leftrightarrow (\tilde\gamma, t_\gamma)$ where $\tilde\gamma(s) = \gamma(t_\gamma s), 0 \leq s \leq 1$. Let $d_\mathcal{X}$ denote the induced metric on $\mathcal{K}$,

$$d_\mathcal{X}(\gamma, \tilde\gamma) = |t_\gamma - t_{\tilde\gamma}| + \sup_{0 \leq s \leq 1} |\gamma(t_\gamma s) - \tilde\gamma(t_{\tilde\gamma} s)|.$$

It is sometimes convenient to consider another metric

$$d_\mathcal{K}(\gamma, \tilde\gamma) = \inf_\theta d(\gamma, \tilde\gamma; \theta),$$

where the infimum is over all increasing homeomorphisms $\theta : [0, t_\gamma] \to [0, t_{\tilde\gamma}]$, and

$$d(\gamma, \tilde\gamma; \theta) = \sup_{0 \leq s \leq t_\gamma} |\theta(s) - s| + \sup_{0 \leq s \leq t_\gamma} |\tilde\gamma(\theta(s)) - \gamma(s)|.$$

Note that $d_\mathcal{X}(\gamma, \tilde\gamma) = d(\gamma, \tilde\gamma; \theta)$ where $\theta(s) = (t_{\tilde\gamma}/t_\gamma)s$. The next lemma shows that $d_\mathcal{X}$ and $d_\mathcal{K}$ induce the same topology on $\mathcal{K}$.

LEMMA 5.1. *If $\gamma, \tilde\gamma \in \mathcal{K}$,*

(5.1) $$d_\mathcal{K}(\gamma, \tilde\gamma) \leq d_\mathcal{X}(\gamma, \tilde\gamma) \leq d_\mathcal{K}(\gamma, \tilde\gamma) + \mathrm{osc}(\tilde\gamma, 2\, d_\mathcal{K}(\gamma, \tilde\gamma)).$$

In particular, if $\gamma_n, \gamma \in \mathcal{K}$, then $d_\mathcal{K}(\gamma_n, \gamma) \to 0$ if and only if $d_\mathcal{X}(\gamma_n, \gamma) \to 0$.

PROOF. The first inequality is immediate. For the second inequality, note that if $s \in (0, 1)$ and θ is an increasing homeomorphism of $[0, t_\gamma]$ onto $[0, t_{\tilde\gamma}]$,

$$\begin{aligned}|\gamma(t_\gamma s) - \tilde\gamma(t_{\tilde\gamma} s)| &\leq |\gamma(t_\gamma s) - \tilde\gamma(\theta(t_\gamma s))| + |\tilde\gamma(\theta(t_\gamma s)) - \tilde\gamma(t_{\tilde\gamma} s)| \\ &\leq [\,d(\gamma, \tilde\gamma; \theta) - |t_\gamma - t_{\tilde\gamma}|\,] + \mathrm{osc}(\tilde\gamma, 2\, d(\gamma, \tilde\gamma; \theta)).\end{aligned}$$

The last inequality uses the estimate

$$|\theta(t_\gamma s) - t_{\tilde\gamma} s| \leq |\theta(t_\gamma s) - t_\gamma s| + |t_\gamma s - t_{\tilde\gamma} s| \leq d(\gamma, \tilde\gamma; \theta) + |t_\gamma - t_{\tilde\gamma}| \leq 2\, d(\gamma, \tilde\gamma; \theta).$$

Hence,

$$d_\mathcal{X}(\gamma, \tilde\gamma) \leq d(\gamma, \tilde\gamma; \theta) + \mathrm{osc}(\tilde\gamma, 2\, d(\gamma, \tilde\gamma; \theta)),$$

and we can take the infimum over all θ. □

When we speak of convergence or continuity in $\mathcal{K}$ we will mean with respect to $d_\mathcal{X}$ or $d_\mathcal{K}$. The previous lemma shows that these two notions of convergence are equivalent. If $a > 0$, let $\mathcal{K}^{(a)} = \{\gamma \in \mathcal{K} : t_\gamma \geq a\}$. Although the space $\mathcal{K}^{(a)}$ is complete under $d_\mathcal{X}$, it is not complete under $d_\mathcal{K}$. For example, the sequence of curves $\gamma_n(t) = t^n, 0 \leq t \leq 1$ is a Cauchy sequence in $(\mathcal{K}^{(1)}, d_\mathcal{K})$ with no limit. The previous lemma shows, however, that if γ_n is a Cauchy sequence in $(\mathcal{K}^{(a)}, d_\mathcal{K})$ that is equicontinuous, then it is a Cauchy sequence in $(\mathcal{K}^{(a)}, d_\mathcal{X})$ and hence has a limit.

By a measure on $\mathcal{K}$ will mean a σ-finite positive measure on $\mathcal{G} = \mathcal{G}_d$, the Borel σ-algebra induced by d. Since the topologies are equivalent, we can choose $d = d_\mathcal{X}$ or $d = d_\mathcal{K}$. Equivalently, a measure on $\mathcal{K}$ is a σ-finite measure on the complete metric space $X := C[0,1] \times (-\infty, \infty)$ that is carried on $C[0,1] \times (0, \infty)$. (We say μ is *carried on* V if $\mu(X \setminus V) = 0$.) Let $\mathcal{M}$ (resp., $\mathcal{M}_a$) denote the set of finite measures on $\mathcal{K}$ (resp., $\mathcal{K}^{(a)}$). If $\mu \in \mathcal{M}$ is a non-zero measure, we will write

$$\mu = |\mu|\, \mu^\#,$$

where $|\mu|$ denotes the total mass and $\mu^\#$ denotes the probability measure $\mu/|\mu|$. If $d = d_\mathcal{X}$ or $d = d_\mathcal{K}$, we will also use d for the *Prohorov metric* (see [**13**, Appendix III]) on $\mathcal{M}$ defined by: $d(\mu, \nu)$ is the infimum of all $\epsilon > 0$ such that for every $V \in \mathcal{G}$,

$$\mu(V) \leq \nu(V^\epsilon) + \epsilon, \quad \nu(V) \leq \mu(V^\epsilon) + \epsilon,$$

where $V^\epsilon = \{x : d(x, V^c) < \epsilon\}$. Note that

$$||\mu| - |\nu|| \leq d(\mu, \nu) \leq \max\{|\mu|, |\nu|\}, \quad d(\mu, (1+\epsilon)\mu) \leq \epsilon|\mu|,$$

$$d(\mu, \nu) \leq d((1+\epsilon)\mu, (1+\epsilon)\nu) \leq (1+\epsilon)\, d(\mu, \nu).$$

Hence we can see that $\mu_n \xrightarrow{d} \mu$ if and only if $|\mu_n| \to |\mu|$ and $\mu_n^\# \xrightarrow{d} \mu^\#$. (If μ is the zero measure, then the necessary and sufficient condition is $|\mu_n| \to 0$.) If μ, ν are probability measures on $\mathcal{K}$, then one way to show $d(\mu, \nu) \leq \epsilon$ is to find $\mathcal{K}$-valued random variables X, Y defined on the same probability space $(\Omega, \mathbf{P})$ such that X has distribution μ, Y has distribution ν, and

$$\mathbf{P}\{d(X, Y) \geq \epsilon\} \leq \epsilon.$$

Also, for probability measures μ, ν, $d(\mu, \nu)$ is the infimum of all ϵ such that $\mu(V) \leq \nu(V^\epsilon) + \epsilon$ for all Borel sets V. This is true since $[\mathcal{K} \setminus V^\epsilon]^\epsilon \subset \mathcal{K} \setminus V$ and hence $\mu(\mathcal{K} \setminus V^\epsilon) \leq \nu([\mathcal{K} \setminus V^\epsilon]^\epsilon) + \epsilon$ implies $\nu(V) \leq \mu(V^\epsilon) + \epsilon$.

Note that we have two different Prohorov metrics, $d_\mathcal{X}, d_\mathcal{K}$, since the definition of V^ϵ depends on the metric. However, the next lemma shows that they give the same topology on $\mathcal{M}$.

LEMMA 5.2. *if $\mu_n, \mu \in \mathcal{M}$, then $d_\mathcal{X}(\mu_n, \mu) \to 0$ if and only if $d_\mathcal{K}(\mu_n, \mu) \to 0$.*

PROOF. It suffices to prove this for probability measures μ_n, μ. Let us write $V_\mathcal{K}^\epsilon, V_\mathcal{X}^\epsilon$ for V^ϵ using the two different metrics. Since $d_\mathcal{K} \leq d_\mathcal{X}$ on $\mathcal{K}$, we have $V_\mathcal{K}^\epsilon \supset V_\mathcal{X}^\epsilon$ and hence $d_\mathcal{K} \leq d_\mathcal{X}$ on $\mathcal{M}$. For the other direction, first note that for every $u > 0$, there is a function $r_u(\delta)$ with $r_u(0+) = 0$ such that

$$\mu\{\gamma : \text{osc}(\gamma, \delta) > r_u(\delta) \text{ for some } \delta > 0\} \leq u.$$

This is easily seen by noting that for every fixed u, n, there is a $\delta = \delta(u,n) > 0$ such that
$$\mu\{\gamma : \operatorname{osc}(\gamma, \delta) > 2^{-n}\} \le u\, 2^{-n}.$$
Using (5.1), we see that if $d_\mathcal{K}(\gamma, \tilde\gamma) < \epsilon$, then
$$d_\mathcal{X}(\gamma, \tilde\gamma) \le \epsilon + 2\operatorname{osc}(\tilde\gamma, 2\epsilon).$$
Hence for every $u > 0$ we can find an $\epsilon = \epsilon(u) \in (0, u)$ such that for all V, $\mu[V_\mathcal{K}^\epsilon \setminus V_\mathcal{X}^u] \le u$. Hence, if n is sufficiently large so that $d_\mathcal{K}(\mu_n, \mu) \le \epsilon$, then
$$\mu_n(V) \le \mu(V_\mathcal{K}^\epsilon) + \epsilon \le \mu(V_\mathcal{X}^u) + \epsilon + u < \mu(V_\mathcal{X}^u) + 2u,$$
and hence $\operatorname{dist}(\mu, \mu_n) < 2u$. $\square$

Since $(\mathcal{K}^{(a)}, d_\mathcal{X})$ is a complete metric space, $\mathcal{M}_a$ is a complete metric space under the $d_\mathcal{X}$ (see [13, Appendix III]). It is not true that $(\mathcal{M}_a, d_\mathcal{K})$ is complete. For the remainder of this chapter, a continuous function from $\mathbb{R}$ or $\mathbb{C}$ to $\mathcal{M}$ will mean a function continuous with respect to $d_\mathcal{K}$, which is the same as continuous with respect to $d_\mathcal{X}$. . Suppose $t \mapsto \mu_t$ is a continuous function from $[a,b]$ into $\mathcal{M}$. We will need to consider the measure
$$\mu = \int_a^b \mu_t\, dt.$$
This is defined as the limit of the obvious Riemann sum approximations. It is not difficult to check that any such sequence of approximations forms a Cauchy sequence in $(\mathcal{M}, d_\mathcal{K})$ and in $(\mathcal{M}, d_\mathcal{X})$ Since $(\mathcal{M}, d_\mathcal{X})$ is complete, we know that a limit exists, and this limit is μ. Similarly if $z \mapsto \mu_z$ is a continuous map from a domain D into $\mathcal{M}$ we can define the integral
$$\int_D \mu_z\, dA(z) = \lim_{n\to\infty} \int_{K_n} \mu_z\, dA(z),$$
exists where A denotes area and K_n is a increasing sequence of compact sets whose union is D. Note that the measure on the left might not be a finite measure.

REMARK 5.3. We have introduced both $d_\mathcal{K}$ and $d_\mathcal{X}$ because it is often easier to prove continuity for $d_\mathcal{K}$, but $d_\mathcal{X}$ has the completeness property which allows the integral to be defined easily.

We let $\overline{\mathcal{K}}$ be the union of $\mathcal{K}$ and curves γ with $t_\gamma = 0$; the metric $d_\mathcal{K}$ extends naturally to $\overline{\mathcal{K}}$. If $\gamma, \eta \in \overline{\mathcal{K}}$, we define the *concatenation* $\gamma \oplus \eta$ by $t_{\gamma \oplus \eta} = t_\gamma + t_\eta$ and
$$\gamma \oplus \eta\,(t) = \begin{cases} \gamma(t), & 0 \le t \le t_\gamma \\ \gamma(t_\gamma) - \eta(0) + \eta(t - t_\gamma), & t_\gamma \le t \le t_\gamma + t_\eta. \end{cases}$$
Most of the time that we use concatenation, γ, η will satisfy $\gamma(t_\gamma) = \eta(0)$. It is easy to check that $(\gamma, \eta) \mapsto \gamma \oplus \eta$ is a continuous map from $\overline{\mathcal{K}} \times \overline{\mathcal{K}}$ to $\overline{\mathcal{K}}$. We define the *time reversal* $\gamma \mapsto \gamma^R$ by $t_{\gamma^R} = t_\gamma$ and $\gamma^R(t) = \gamma(t_\gamma - t), 0 \le t \le t_\gamma$. Clearly $\gamma \mapsto \gamma^R$ is continuous; in fact, $d_\mathcal{K}(\gamma, \eta) = d_\mathcal{K}(\gamma^R, \eta^R)$. If $\gamma \in \mathcal{K}$ and $0 \le s \le t \le t_\gamma$ define $\Gamma_s^t \gamma$ by $t_{\Gamma_s^t \gamma} = t - s$ and $\Gamma_s^t \gamma(r) = \gamma(s + r), 0 \le r \le t - s$. We write just Γ^t for Γ_0^t and Γ_s for $\Gamma_s^{t_\gamma}$. Note that
$$d_\mathcal{K}(\gamma, \Gamma_s^t \gamma) \le s + (t_\gamma - t) + \operatorname{diam}(\gamma[0, s]) + \operatorname{diam}(\gamma[t, t_\gamma]).$$
In particular, if $s_n \downarrow 0, t_n \uparrow t_\gamma$, then $\Gamma_{s_n}^{t_n} \gamma \to \gamma$.

If $z, w \in \mathbb{C}$ let $\mathcal{K}_z$ (resp., $\mathcal{K}^w$) be the set of $\gamma \in \mathcal{K}$ with $\gamma(0) = z$ (resp., $\gamma(t_\gamma) = w$). Let $\mathcal{K}_z^w = \mathcal{K}_z \cap \mathcal{K}_w$. If D is a domain, we let $\mathcal{K}(D)$ be the set of $\gamma \in \mathcal{K}$ with $t_\gamma > 0$ and $\gamma(0, t_\gamma) \subset D$. Note that we do not require $\gamma(0), \gamma(t_\gamma) \in D$. We also write $\mathcal{K}_z(D)$ for $\mathcal{K}_z \cap \mathcal{K}(D)$ and similarly for $\mathcal{K}^w(D), \mathcal{K}_z^w(D)$. These sets are empty unless $z, w \in \overline{D}$. The sets $\mathcal{K}_z, \mathcal{K}^w, \mathcal{K}_z^w$ are clearly closed. We can write $\mathcal{K}(D)$ as the countable intersection of open sets. We will write $\mathcal{K}(D)^o$ for the set of $\gamma \in \mathcal{K}(D)$ such that $\gamma(0), \gamma(t_\gamma) \in D$. If D_n is a sequence of domains increasing to D with $\overline{D}_n \subset D$, then $\mathcal{K}(D)^o = \cup_n \mathcal{K}(D_n)$. If $z, w \in D$ we can also write $\mathcal{K}_z(D)^o, \mathcal{K}^w(D)^o, \mathcal{K}_z^w(D)^o$.

If $z \in \mathbb{C} \setminus \{0\}$, the Brownian scaling map $\Upsilon_z : \mathcal{K} \to \mathcal{K}$ is defined by

$$t_{\Upsilon_z \gamma} = |z|^2 t_\gamma, \quad \Upsilon_z \gamma(t) = z\, \gamma(t/|z|^2), \quad 0 \le t \le |z|^2 t_\gamma.$$

This is a special case of a more general map. Suppose D is a domain and $f : D \to \mathbb{C}$ is a nonconstant analytic function. If $\gamma \in \mathcal{K}(D)$, let

$$S_{f,\gamma}(t) = \int_0^t |f'(\gamma(s))|^2 \, ds.$$

If $\gamma \in \{\gamma \in \mathcal{K}(D) : S_f(t_\gamma) < \infty\}$, we define $f \circ \gamma \in \mathcal{K}(f(D))$ by $t_{f \circ \gamma} = S_{f,\gamma}(t_\gamma)$ and $(f \circ \gamma)(t) = f(\gamma(S_{f,\gamma}^{-1}(t))), 0 \le t \le S_{f,\gamma}(t_\gamma)$. Note that S_{f_γ} is continuous and strictly increasing so $S_{f_\gamma}^{-1}$ is well defined. Note that $\Upsilon_z \gamma = f_z \circ \gamma$ where $f_z(w) = zw$. Suppose D is a bounded domain and f is a univalent function on a domain D' containing $\overline{D}$. Then $|f'|, |f''|$ and $1/|f'|$ are uniformly bounded on D and one can check that $\gamma \mapsto f \circ \gamma$ is continuous on $\mathcal{K}(D)$.

Let $C^*[0, 1] = \{\gamma \in C[0, 1] : \gamma(0) = \gamma(1)\}$ and let $\mathcal{X}^* = C^*[0, 1] \times (-\infty, \infty)$. Then $\mathcal{X}_+^* = \mathcal{X}_+ \cap \mathcal{X}^*$ is in one-to-one correspondence with $\tilde{\mathcal{C}} := \{\gamma \in \mathcal{K} : \gamma(0) = \gamma(t_\gamma)\}$. We call $\tilde{\mathcal{C}}$ the space of *(rooted) loops*. if $\gamma \in \tilde{\mathcal{C}}$ we can also consider γ as a function from $\mathbb{R}$ to $\mathbb{C}$ satisfying $\gamma(s) = \gamma(t_\gamma + s)$ for all s. An *unrooted loop* is an equivalence class $[\gamma]$ of curves in $\tilde{\mathcal{C}}$ under the equivalence relation $\theta_r \gamma \sim \gamma$ for all $r \in \mathbb{R}$ where $\theta_r \gamma(s) = \gamma(r + s)$ (here we consider $\theta_r \gamma, \gamma$ as periodic functions with domain $\mathbb{R}$). We let $\tilde{\mathcal{C}}_U$ denote the space of unrooted loops. This space is in one-to-one correspondence with $C_U^*[0, 1] \times (0, \infty)$, where $C_U^*[0, 1]$ denotes the corresponding collection of equivalence classes of curves in C^*.

LEMMA 5.4. *$C_U^*[0, 1]$ is a complete metric space under the metric*

$$d_{\mathcal{X}, U}(\gamma, \tilde{\gamma}) = \inf_r \|\gamma - \theta_r \tilde{\gamma}\|_\infty.$$

PROOF. Suppose $[\gamma_1], [\gamma_2], \ldots$ is a Cauchy sequence. Then

$$\sup_n \sup_{0 \le s \le 1} |\gamma_n(s)| < \infty.$$

Also, we claim that $\gamma_1, \gamma_2, \ldots$ is equicontinuous. Indeed, for every $\delta > 0$,

$$\begin{aligned} d_{\mathcal{X}, U}(\gamma_n, \gamma_m) &= \inf_r \|\gamma_n - \theta_r \gamma_m\|_\infty \\ &\ge \inf_r \frac{1}{2} |\mathrm{osc}(\gamma_n, \delta) - \mathrm{osc}(\theta_r \gamma_m, \delta)| \\ &= \frac{1}{2} |\mathrm{osc}(\gamma_n, \delta) - \mathrm{osc}(\gamma_m, \delta)|. \end{aligned}$$

Hence if $\epsilon > 0$, n is sufficiently large so that $d_{\mathcal{X}, U}(\gamma_n, \gamma_m) \le \epsilon/4$ for $m \ge n$, and δ is sufficiently small so that $\mathrm{osc}(\gamma_m, \delta) < \epsilon/2$ for $m \le n$, then $\mathrm{osc}(\gamma_m, \delta) < \epsilon$ for all

m. By the Arzela-Ascoli Theorem, there exists a γ and a subsequence n_j such that $\|\gamma_{n_j} - \gamma\|_\infty \to 0$. It is easy to see that $d_{\mathcal{X},U}(\gamma_n, \gamma) \to 0$. □

Measures on $\tilde{\mathcal{C}}$ can therefore be considered as measures on the complete metric space $C_U^*[0,1] \times (-\infty, \infty)$ supported on pairs $([\gamma], r)$ with $r > 0$.

5.2. Brownian measures on $\mathcal{K}$

We will now give a number of measures on $\mathcal{K}$ that are naturally derived from Brownian motion. Most of these are versions of what is generally called *Wiener measure*. If $z \in \mathbb{C}$ and $t > 0$, let $\mu(z, \cdot; t)$ denote the probability measure on $\mathcal{K}$ derived from $B_s, 0 \leq s \leq t$, where B_s is a standard Brownian motion with $B_0 = z$. Translation invariance and Brownian scaling can be used to see that $\mu(z, \cdot; t)$ is continuous in z, t. We can write

$$\mu(z, \cdot; t) = \int_\mathbb{C} \mu(z, w; t)\, dA(w),$$

where A denotes area and

$$\mu(z, w, t) = \lim_{\epsilon \to 0+} (\pi \epsilon^2)^{-1}\, \mu(z, \cdot; t)\, 1_{\{|\gamma(t) - w| \leq \epsilon\}}.$$

It is not difficult to show that the limit exists. In fact,

$$\mu(z, w; t) = p(z, w; t)\, \mu^\#(z, w, t),$$

where p is as defined in §2.4, and the probability measure

$$\mu^\#(z, w, t) := \mu(z, w; t)/|\mu(z, w; t)|$$

is that of the *Brownian bridge*, which has the distribution of

$$z + (w - z)\frac{s}{t} + [B_s - \frac{s}{t} B_t], \quad 0 \leq s \leq t.$$

From this one sees continuity in z, w, t, as well as the fact that $\mu(w, z; t)$ can be derived from $\mu(z, w; t)$ by the map $\gamma \mapsto \gamma^R$.

If D is a domain, let $\mu_D(z, \cdot; t)$ and $\mu_D(z, w; t)$ denote $\mu(z, \cdot; t)$ and $\mu(z, w; t)$ restricted to $\mathcal{K}(D)$. Note that $\mu_D(z, w; t)$ is supported on $\mathcal{K}(D)^o$ and

$$\mu_D(z, \cdot; t) = \int_D \mu_D(z, w; t)\, dA(w).$$

We can also define

$$\mu_D(\cdot, \cdot; t) = \int_D \int_D \mu_D(z, w; t)\, dA(w)\, dA(z).$$

If D is bounded this is a finite measure, but for unbounded D it can be infinite.

If m_1, m_2 are measures supported on $\mathcal{K}^w$ and $\mathcal{K}_w$, respectively, we write $m_1 \oplus m_2$ for the image of $m_1 \times m_2$ under the continuous map $(\gamma^1, \gamma^2) \mapsto \gamma^1 \oplus \gamma^2$. The Markovian property for Brownian motion can be expressed as a Chapman-Kolmogorov equation for path measures:

$$\mu_D(z, w; s + t) = \int_D [\mu_D(z, z'; s) \oplus \mu_D(z', w; t)]\, dA(z').$$

Define

(5.2) $$\mu_D(z, w) = \int_0^\infty \mu_D(z, w; t)\, dt.$$

If D is a domain with at least one regular boundary point and $z \neq w$, then this is a finite measure with total mass

$$|\mu_D(z,w)| = \int_0^\infty p(z,w;t)\, dt = \frac{1}{\pi} G_D(z,w),$$

where G_D is the Green's function as in §2.4. In this case, we can also write

(5.3) $$\mu_D(z,w) = \lim_{\epsilon \to 0+} -(\log \epsilon)\, \mu_D^*(z,w;\epsilon),$$

where $\mu_D^*(z,w,\epsilon)$ denotes the measure on paths obtained from starting a Brownian motion at z, stopping it at the first time it reaches $\partial D \cup \partial \mathcal{B}(w,\epsilon)$, and restricting the measure to paths that end at $\partial \mathcal{B}(w,\epsilon)$ (see (2.19)). If $D = \mathbb{C}$ or $z = w$, we can still define $\mu_D(z,w)$ as the limit as $\epsilon \to 0+$ of the integral in (5.2) from ϵ to $1/\epsilon$. In this case the measure is infinite, σ-finite (the integral blows up at infinity for all z, w and it blows up at zero if $z = w$). We can also define

$$\mu_D(z,\cdot) = \int_D \mu_D(z,w)\, dA(w) = \int_D \int_0^\infty \mu_D(z,w;t)\, dt\, dA(w),$$

$$\mu_D(\cdot,\cdot) = \int_D \int_D \mu_D(z,w)\, dA(w)\, dA(z)$$

$$= \int_D \int_D \int_0^\infty \mu_D(z,w;t)\, dt\, dA(w)\, dA(z).$$

If $f : D \to D'$ is a conformal transformation and m is a measure supported on $\mathcal{K}(D)$ we define $f \circ m$ to be the induced measure on $\mathcal{K}(D')$,

$$f \circ m(V) = m\{\gamma : f \circ \gamma \in V\}.$$

Here $f \circ \gamma$ is as defined in the previous section, and we define $f \circ m$ only if m is supported on curves γ such that $f \circ \gamma$ is well-defined (if $\gamma(0), \gamma(t_\gamma) \in D$ this is always true).

PROPOSITION 5.5. *If $f : D \to D'$ is a conformal transformation and $z, w \in D$, then $f \circ \mu_D(z,w) = \mu_{D'}(f(z),f(w))$.*

PROOF. In the case of bounded D and $z \neq w$, this follows from the conformal invariance of Brownian motion (Theorem 2.2) and (5.3). If D is unbounded or $z = w$, we can take limits. $\square$

If D is a regular domain and $z \in D$, we let $\mu_D(z,\partial D)$ denote the probability measure on $\mathcal{K}(D)$ derived from $B_t, 0 \leq t \leq \tau_D$ where $B_0 = z$. Suppose $z = 0, D = \mathbb{D}$. Then we can write

$$\mu_\mathbb{D}(0,\partial \mathbb{D}) = \int_0^{2\pi} \mu_\mathbb{D}(0,e^{i\theta})\, d\theta = \int_{\partial \mathbb{D}} \mu_\mathbb{D}(0,w)\, |dw|,$$

where $\mu_\mathbb{D}(0,w)$ denotes a measure supported on paths that leave $\mathbb{D}$ at w. In this case, these measures are very easy to define — their total mass is $1/2\pi$ and $\mu_\mathbb{D}^\#(0,w)$ is obtained from $\mu_\mathbb{D}(0,\partial \mathbb{D})$ by rotating each curve so that it ends at w. If D is a bounded Jordan domain and f is a conformal transformation of $\mathbb{D}$ onto D with $f(0) = z$, then conformal invariance of Brownian motion tells us that $f \circ \mu_\mathbb{D}(0,\partial \mathbb{D}) = \mu_D(z,\partial D)$. If D is a piecewise analytic Jordan domain (so that for all but a finite number of points $w \in \partial \mathbb{D}$, f is conformal in a neighborhood of w), we can write

$$\mu_D(z,\partial D) = \int_{\partial D} \mu_D(z,w)\, |dw| = \int_{\partial D} H_D(z,w)\, \mu_D^\#(z,w)\, |dw|.$$

for appropriate measures $\mu_D(z,w)$.[1] Here $|\mu_D(z,w)| = H_D(z,w)$, where H_D denotes the Poisson kernel. If $w = f(e^{i\theta})$, then

$$\mu_D^\#(z,w) = f \circ \mu_{\mathbb{D}}^\#(0, e^{i\theta}), \quad \mu_D(z,w) = |f'(e^{i\theta})|^{-1} f \circ \mu_{\mathbb{D}}(0, e^{i\theta}).$$

We can also obtain $\mu_D(z,w)$ by

$$\mu_D(z,w) = \frac{1}{2\epsilon} \lim_{\epsilon \to 0+} \mu_D(z, w + \epsilon \mathbf{n}_w),$$

where $\mathbf{n}_w$ denotes the unit inward normal at w. The factor $1/2$ comes from (2.20). This makes it natural to define for $z \in D, w \in \partial D$,

$$\mu_D(w,z) = \frac{1}{2\epsilon} \lim_{\epsilon \to 0+} \mu_D(w + \epsilon \mathbf{n}_w, z).$$

Under this definition, $\mu_D(w,z)$ can be obtained from $\mu_D(z,w)$ by time reversal $\gamma \mapsto \gamma^R$.

We will use the following fact which can be considered a case of the strong Markov property for Brownian motion. Suppose $D \subset D'$ where D is a piecewise analytic Jordan domain. Then if $z \in D, z' \in \partial D' \setminus \partial D$,

$$\mu_{D'}(z, z') = \int_{D' \cap \partial D} [\mu_D(z,w) \oplus \mu_{D'}(w, z')] \, |dw|$$
$$= \int_{D' \cap \partial D} [\mu_D^\#(z,w) \oplus \mu_{D'}^\#(w, z')] \, H_D(z,w) \, H_{D'}(w, z') \, |dw|.$$

Here we think of choosing a Brownian motion starting at z conditioned to leave D' at z' by first choosing the point w at which it first visits ∂D using the density $H_D(z,w) \, H_D(w, z')$, and then attaching paths from z to w and from w to z' using the appropriate conditional distributions. The measure $\mu_D^\#(z,w) \oplus \mu_{D'}^\#(w, z')$ can be obtained from $\mu_{\mathbb{D}}(0,1) \times \mu_{\mathbb{D}}(0,1)$ by conformal transformation and concatenation. In a similar fashion, if $z, z' \in D$ we can write

$$\mu_{D'}(z, z') = \mu_D(z, z') + \int_{D' \cap \partial D} [\mu_D(z,w) \oplus \mu_{D'}(w, z')] \, |dw|.$$

By using the same formula on $\mu_{D'}(z', w)$ and reversing time, we get

$$\mu_{D'}(w, z') = \int_{D' \cap \partial D} [\mu_{D'}(w, w') \oplus \mu_D(w', z')] \, |dw'|$$

(this is an example of a "last-exit" decomposition). Combining we get

$$\mu_{D'}(z, z') - \mu_D(z, z') =$$

(5.4)
$$\int_{D' \cap \partial D} \int_{D' \cap \partial D} [\mu_D(z,w) \oplus \mu_{D'}(w, w') \oplus \mu_D(w', z')] \, |dw'| \, |dw|.$$

Let $\overline{\mu}_{D'}(z, z'; D)$ denote the measure obtained from $\mu_{D'}(z, z')$ by considering the path only from its first to its last visit to $D' \cap \partial D$; the total mass of this measure is the $\mu_{D'}(z, z')$ measure of curves that intersect $D' \cap \partial D$. Then

$$\overline{\mu}_{D'}(z, z'; D) = \int_{D' \cap \partial D} \int_{D' \cap \partial D} H_D(z, w) \, H_D(z', w') \, \mu_{D'}(w, w') \, |dw'| \, |dw|.$$

---

[1] In this chapter we use $\mu_D(z,w)$ to mean different things depending on whether z, w are boundary or interior points.

There is a "last-exit" decomposition for $z \in D', z' \in \partial D'$

(5.5) $\quad \mu_{D'}(z, z') = \mu_D(z, z') + \int_{D' \cap \partial D} [\mu_{D'}(z, w) \oplus \mu_D(w, z')] \, |dw|.$

Using this we see that (5.4) and the equation following it hold if one or both of z, z' are in $\partial D' \setminus \partial D$.

If $x \in (0, 1)$, then $\mu_{\mathbb{D}}^{\#}(-x, 1) = f_x \circ \mu_{\mathbb{D}}^{\#}(0, 1)$ where $f_x(z) = (z - x)/(1 - xz)$ is the Möbius transformation with $f_x(0) = -x, f_x(1) = 1$. If $\gamma \in \mathcal{K}(\mathbb{D})$ with $\gamma(0) = 0, \gamma(t_\gamma) = 1$, then a straightforward argument using the explicit form of f_x shows that $\lim_{x \to 1} f_x \circ \gamma$ exists. (The limit is with respect to $d_\mathcal{K}$ or $d_\mathcal{X}$.) Hence we can define the probability measure $\mu_{\mathbb{D}}^{\#}(-1, 1) = \lim_{x \to -1} \mu_{\mathbb{D}}^{\#}(x, 1)$. By conformal invariance, we can define similarly $\mu_D^{\#}(z, w)$ for Jordan domains D at distinct analytic points z, w and we define

$$\mu_D(z, w) = H_D(z, w) \, \mu_D^{\#}(z, w).$$

Here $H_D(z, w)$ is the "boundary Poisson kernel" defined for $z, w \in \partial D$ by

$$H_D(z, w) = \lim_{\epsilon \to 0+} \epsilon^{-1} H_D(z + \epsilon \mathbf{n}_z, w) = \lim_{\epsilon \to 0+} \frac{1}{2\pi \epsilon^2} G_D(z + \epsilon \mathbf{n}_z, w + \epsilon \mathbf{n}_w),$$

where $\mathbf{n}_z, \mathbf{n}_w$ denote the inward normals at z, w. See (2.20). If $f : D \to D'$ is a conformal transformation that is extended in neighborhoods of distinct points $z, w \in \partial D$, then

$$H_D(z, w) = |f'(z)| \, |f'(w)| \, H_{D'}(f(z), f(w)).$$

In particular, if $f : \mathbb{D} \to D$ is a conformal transformation, Γ_1, Γ_2 disjoint arcs on $\partial \mathbb{D}$ and f has an extension to neighborhoods of Γ_1, Γ_2, then

$$\int_{f(\Gamma_1)} \int_{f(\Gamma_2)} H_{f(\mathbb{D})}(z', w') \, |dz'| \, |dw'| = \int_{\Gamma_1} \int_{\Gamma_2} H_{\mathbb{D}}(z, w) \, |dz| \, |dw|.$$

If $D \subset D', z \in \partial D \cap \partial D', w \in \partial D', \partial D, \partial D'$ are locally analytic with z, w not being exceptional points, then the strong Markov property gives

(5.6) $\quad \mu_{D'}(z, w) = \mu_D(z, w) + \int_{D' \cap \partial D} [\mu_D(z, z') \oplus \mu_{D'}(z', w)] \, |dz'|.$

The right-hand side splits $\mu_{D'}(z, w)$ into measures supported on $\mathcal{K}(D)$ and $\mathcal{K}(D') \setminus \mathcal{K}(D)$, respectively.

EXAMPLE 5.6.

- $$H_{\mathbb{D}}(e^{i\theta}, e^{i\theta'}) = H_{\mathbb{D}}(1, e^{i(\theta' - \theta)}) = \lim_{\epsilon \to 0+} \frac{1}{2\pi \epsilon} \frac{1 - (1 - \epsilon)^2}{|(1 - \epsilon) - e^{i(\theta' - \theta)}|^2}$$
  $$= \frac{1}{2\pi} \frac{1}{1 - \cos(\theta' - \theta)}$$

- (5.7) $\quad H_{\mathbb{H}}(x', x + x') = H_{\mathbb{H}}(0, x) = \lim_{\epsilon \to 0+} \epsilon^{-1} \frac{1}{\pi} \frac{\epsilon}{x^2 + \epsilon^2} = \frac{1}{\pi x^2}.$

- If $\mathbb{D}_+ = \{z \in \mathbb{H} : |z| < 1\}$, then

(5.8) $\quad H_{\mathbb{D}_+}(0, e^{i\theta}) = \frac{2}{\pi} \sin \theta.$

- Let $\mathcal{R} = \{z \in \mathbb{H} : \text{Im}(z) < \pi\}$. Then $f(z) = e^z$ maps $\mathcal{R}$ conformally onto $\mathbb{H}$ and hence if $x \in \mathbb{R}$,

$$H_\mathcal{R}(\pi i, x) = |f'(i\pi)| \, |f'(x)| \, H_\mathbb{H}(f(i\pi), f(x))$$
$$= \frac{e^x}{\pi (1+e^x)^2} = \frac{1}{4\pi \, \cosh^2(x/2)}.$$

- Let $\mathcal{R}_L = \{z \in \mathcal{R} : 0 < \text{Re}(z) < L\}$ and let ∂_1, ∂_2 denote the vertical boundaries as in §3.7. Then for $0 < x < L, 0 < y, y' < \pi$,

$$H_{\mathcal{R}_L}(x+iy, L+iy') = \sum_{n=1}^\infty \frac{2}{n\pi} (\sin ny)(\sin ny') \frac{\sinh nx}{\sinh nL},$$

$$H_{\mathcal{R}_L}(iy, L+iy') = \lim_{x \to 0+} x^{-1} H_{\mathcal{R}_L}(x+iy, L+iy')$$

(5.9)
$$= \sum_{n=1}^\infty \frac{2}{n\pi} \frac{(\sin ny)(\sin ny')}{\sinh nL}.$$

In particular, as $L \to \infty$,

$$H_{\mathcal{R}_L}(iy, L+iy') = \frac{4}{\pi} (\sin y)(\sin y') e^{-L} [1 + O(e^{-L})].$$

DEFINITION 5.7. If D is a piecewise analytic domain, then the *(Brownian) excursion measure* on D is defined by

$$\mu_D = \int_{\partial D} \int_{\partial D} \mu_D(z,w) \, |dz| \, |dw| = \int_{\partial D} \int_{\partial D} \mu_D^\#(z,w) \, H_D(z,w) \, |dz| \, |dw|.$$

If Γ_1, Γ_2 are arcs on ∂D, then $\mu_D(\Gamma_1, \Gamma_2)$ is μ_D restricted to paths going from Γ_1 to Γ_2, i.e.,

(5.10)
$$\mu_D(\Gamma_1, \Gamma_2) = \int_{\Gamma_2} \int_{\Gamma_1} \mu_D(z,w) \, |dz| \, |dw|.$$

The excursion measure is a σ-finite measure, but if Γ_1, Γ_2 are closed and disjoint, $\mu_D(\Gamma_1, \Gamma_2)$ is a finite measure with

$$|\mu_D(\Gamma_1, \Gamma_2)| = \int_{\Gamma_2} \int_{\Gamma_1} H_D(z,w) \, |dz| \, |dw| < \infty.$$

The discussion above proves the following proposition.

PROPOSITION 5.8. *If D, D' are piecewise analytic domains and $f : D \to D'$ is a conformal transformation, then $f \circ \mu_D = \mu_{D'}$.*

REMARK 5.9. We can define μ_D for any simply connected domain D (other than $\mathbb{C}$) by conformal invariance. Since $\mu_\mathbb{D}$ is invariant under Möbius transformations, μ_D is well defined. Moreover, the proposition holds for all simply connected domains.

EXAMPLE 5.10.
- Let $\Gamma_1 = [-1, 1], \Gamma_2 = (-\infty, -x] \cup [x, \infty)$ where $x > 1$. Then

$$|\mu_\mathbb{H}(\Gamma_1, \Gamma_2)| = \int_{|t|>x} \int_{-1}^1 \frac{ds \, dt}{\pi(t-s)^2} = \frac{2}{\pi} \log \frac{x+1}{x-1}.$$

- Let $F(L) = |\mu_{\mathcal{R}_L}(\partial_1, \partial_2)|$ with $\mathcal{R}_L, \partial_1, \partial_2$ as in §3.7. Then, $|\mu_D(\Gamma_1, \Gamma_2)| = F(L(D; \Gamma_1, \Gamma_2))$. From (5.9), we see that

$$(5.11) \qquad F(L) = \sum_{n \text{ odd}} \frac{8}{\pi n^3} \frac{1}{\sinh nL} = \frac{16}{\pi} e^{-L} + O(e^{-3L}), \quad L \to \infty.$$

We can also write

$$(5.12) \qquad \mu_D = \int_{\partial D} \mu_D(z) \, |dz|,$$

where $\mu_D(z)$ is defined by

$$\mu_D(z) = \int_{\partial D} \mu_D(z, w) \, |dw| = \lim_{\epsilon \to 0} \epsilon^{-1} \mu_D(z + \epsilon \mathbf{n}_z, \partial D).$$

Suppose $D_1 \subset D$ and ∂D_1 and ∂D agree in a neighborhood of $z \in \partial D_1 \cap \partial D$. Then $\mu_{D_1}(z)$ can be obtained from $\mu_D(z)$ by truncating paths at their first visit to ∂D_1.

The next proposition gives two decompositions for $\mu_{\mathbb{H}}(z, 0)$. The first comes from splitting the path at the point of the curve with maximal imaginary part, and the second from splitting at the point of maximal absolute value.

PROPOSITION 5.11. *Suppose* $\mathbb{H}_r = \{z \in \mathbb{H} : \text{Im}(z) < r\}$. *Then if* $z \in \mathbb{H}$,

$$\mu_{\mathbb{H}}(z, 0) = \int_{\text{Im}(z)}^{\infty} \int_{-\infty}^{\infty} [\mu_{\mathbb{H}_y}(z, x + iy) \oplus \mu_{\mathbb{H}_y}(x + iy, 0)] \, dx \, dy.$$

$$\mu_{\mathbb{H}}(z, 0) = \int_{|z|}^{\infty} r \int_0^{\pi} [\mu_{r\mathbb{D}_+}(z, re^{i\theta}) \oplus \mu_{r\mathbb{D}_+}(re^{i\theta}, 0)] \, d\theta \, dr.$$

PROOF. We will prove the second equality; the first is done similarly. Note that $\mu_{r\mathbb{D}_+}(z, 0)$ increases in $r > |z|$ and $\lim_{r \to |z|+} |\mu_{r\mathbb{D}_+}(z, 0)| = 0$. Hence it suffices to show that

$$\frac{d}{dr} \mu_{r\mathbb{D}_+}(z, 0) = r \int_0^{\pi} [\mu_{r\mathbb{D}_+}(z, re^{i\theta}) \oplus \mu_{r\mathbb{D}_+}(re^{i\theta}, 0)] \, d\theta.$$

By (5.5).

$$\mu_{(r+\epsilon)\mathbb{D}_+}(z, 0) - \mu_{r\mathbb{D}_+}(z, 0) = r \int_0^{\pi} [\mu_{r\mathbb{D}_+}(z, re^{i\theta}) \oplus \mu_{(r+\epsilon)\mathbb{D}_+}(re^{i\theta}, 0)] \, d\theta.$$

But

$$\lim_{\epsilon \to 0+} \epsilon^{-1} \mu_{(r+\epsilon)\mathbb{D}_+}(re^{i\theta}, 0) = \mu_{r\mathbb{D}_+}(re^{i\theta}, 0).$$

□

PROPOSITION 5.12 (Restriction Property). *Suppose* $D_1 \subset D$ *are Jordan domains and* Γ_1, Γ_2 *are disjoint closed arcs in both* ∂D_1 *and* ∂D. *Then* $\mu_{D_1}(\Gamma_1, \Gamma_2)$ *is* $\mu_D(\Gamma_1, \Gamma_2)$ *restricted to those curves that stay in* D_1.

5.2.1. Some lemmas about excursion measure.

We will prove here some lemmas that will be used in Chapter 8. Let $\mathcal{R}_L, \partial_1, \partial_2$ be as in §3.7 and let μ_L denote $\mu_{\mathcal{R}_L}$ restricted to excursions going from ∂_1 to ∂_2. As before, we let $\mu_L^\# = \mu_L/|\mu_L|$ be the probability measure obtained by normalization.

LEMMA 5.13. *There is a c such that if $L \geq 1$, $z_1, z_2, \ldots \in \partial \mathcal{R}_L \setminus (\partial_1 \cup \partial_2)$, and $\epsilon_1, \epsilon_2, \ldots$ are positive numbers, then the $\mu_L^\#$ measure of the set of excursions that intersect*

$$\bigcup_{j=1}^\infty \mathcal{B}(z_j, \epsilon_j),$$

is bounded above by

$$c \sum_{j=1}^\infty \epsilon_j^2.$$

PROOF. It suffices to prove the estimate for a single ball $\mathcal{B} = \mathcal{B}(z, \epsilon)$ with $z \in \partial \mathcal{R}_L \setminus (\partial_1 \cup \partial_2)$ and $0 < \epsilon < 1/8$. By symmetry, we may assume $r := \operatorname{Re}(z) \geq 1/2$. Let $w_y = r - (1/4) + iy$ where $0 < y < \pi$. It suffices to show that the probability that a Brownian motion starting at w_y hits $\mathcal{B}$ given that it exits $\mathcal{R}_L$ at ∂_2 is bounded above by $c \epsilon^2$. By symmetry we may assume $0 < y < \pi/2$.

Suppose $1/2 \leq r \leq L - (1/2)$. The probability that a Brownian motion starting at w_y leaves $\mathcal{R}_L$ at ∂_2 is bounded below by $c_1 y e^{-(L-r)}$. However, the probability that a Brownian motion starting at w_y visits $\mathcal{B}$ and leaves $\mathcal{R}_L$ at ∂_2 is bounded above by $c_2 y \epsilon^2 e^{-(L-r)}$. One can see this by separating the probability into: the probability that the path reaches $\{\operatorname{Re}(z') = r - (3/16)\}$, which is bounded by cy; the probability that the path then reaches $\mathcal{B}$ before leaving $\mathcal{R}_L$ which is bounded by $c\epsilon$; and the probability that the path then leaves $\mathcal{R}_L$ at ∂_2, which is bounded by $c \epsilon e^{-(L-r)}$ since points in $\mathcal{B}$ are within distance ϵ of $\partial \mathcal{R}_L$. A similar argument works for $\operatorname{Re}(z) \geq L - (1/2)$ (if $\operatorname{Re}(z) = L - \delta$ where $\epsilon < \delta < 1/2$, then the probabilities in the previous argument become $cy, c\delta\epsilon$ and $c(\epsilon/\delta)$, respectively). □

LEMMA 5.14. *There is a c such that the following true. Suppose $L \geq 1, 0 < \epsilon < 1/2$ and $\eta : [0,1] \to \mathbb{C}$ is a simple curve with $\operatorname{Re}(\eta(0)) = 0; \operatorname{Re}(\eta(1)) = L; 0 < \operatorname{Re}(\eta(t)) < L, 0 < t < 1$; and $\operatorname{Im}(\gamma(t)) > 0, 0 \leq t \leq 1$. Let D be the domain bounded by the vertical boundaries $\partial_1^* := [0, \eta(0)], \partial_2^* := [L, \eta(1)]$ and the "horizontal" boundaries $[0, L], \eta[0,1]$. If $\operatorname{Im}[\eta(1)] \geq 4\epsilon$ and*

$$\{x + iy : L - \epsilon < x < L, 0 < y < \operatorname{Im}[\eta(1)]/2\} \subset D,$$

then the $\mu_D^\#(\partial_1, \partial_2)$ measure of paths that do not intersect $\mathcal{B}(\eta(1), \epsilon)$ is at least c.

PROOF. Let $L' = L(D; \partial_1^*, \partial_2^*)$ and let $f : D \to \mathcal{R}_{L'}$ be the conformal transformation with $f(\partial_1^*) = \partial_1, f(\partial_2^*) = \partial_2$. Let $z = L - (\epsilon/2) + \operatorname{Im}[\eta(1)]i/4$. Let γ be the curve that starts at $L + \operatorname{Im}[\eta(1)]i/2$, proceeds along a straight line until $L - \epsilon + \operatorname{Im}[\eta(1)]i/2$, then goes upward in the imaginary direction until it hits η or reaches an imaginary value of $\operatorname{Im}[\eta(1)] + \epsilon$, and then if it has not hit η yet proceeds in the positive real direction until it hits η. Let $f(z) = \tilde{x} + i\tilde{y}$, and note that $c_1 < \tilde{y}$ (since the probability that a Brownian motion starting at z leaves D at $\eta(0,1)$ is at least a constant times the probability of leaving D at $[0, l]$). The Koebe 1/4 Theorem (Theorem 3.17) shows that, $L - \tilde{x} \leq c$. Note also that the probability that a Brownian motion starting at z leaves D without hitting γ given that it leaves

D at $\eta[0,1]$ is also bounded below by a constant. Hence the image of γ is contained in $\{x + iy : x \geq L' - c_3, y \geq c_3\}$ for some c_3. The estimate for the rectangle can be done in a straightforward manner, and the result then follows from conformal invariance. □

5.3. ℍ-excursions

Suppose $z \in D$, $w \in \partial D$ and ∂D is analytic at w and let $h(z) = H_D(z, w)$. Then the measure $\mu_D^\#(z, w)$ corresponds to "Brownian motion started at z conditioned to leave D at w". This measure is derived from a process $\tilde{B}$ that satisfies the stochastic differential equation

$$d\tilde{B}_t = \frac{\nabla h(\tilde{B}_t)}{h(\tilde{B}_t)} \, dt + dB_t$$

in D, where B denotes a standard Brownian motion (see (1.7)). To understand this process in simply connected domains D it suffices to study it in a particular domain and then to use conformal invariance. In this section we consider one choice, the upper half plane $\mathbb{H}$ with the boundary point ∞, even though this choice has the disadvantage of giving paths of infinite time duration. The advantage is that the real and imaginary parts of $\tilde{B}_t$ become independent.

In §1.9 we showed how a Bessel-3 process can be viewed as a one-dimensional Brownian motion "conditioned to always stay positive." For this reason we will also call such a process $\hat{Y}_t$ an $\mathbb{R}_+$-excursion. It can start at any $y \geq 0$. It can be realized as the absolute value of a standard three-dimensional Brownian motion. It satisfies the Bessel SDE:

$$d\hat{Y}_t = \frac{1}{\hat{Y}_t} \, dt + dY_t,$$

where Y_t denotes a standard one-dimensional Brownian motion. Let $\hat{Y}_t$ denote a Bessel-3 process and let $\hat{T}_y = \inf\{t : \hat{Y}_t = y\}$. Let Y_t denote a standard one-dimensional Brownian motion and let T_y denote the corresponding stopping times. We leave the following two facts to the reader.

- If $\hat{Y}_0 = 0$ and $0 < y < y'$, the random variables $\tilde{Y}_s := \hat{Y}_{s+\hat{T}_y}, 0 \leq s \leq \hat{T}_{y'} - \hat{T}_y$ are independent of $\hat{Y}_s, 0 \leq s \leq \hat{T}_y$ and have the same distribution as $Y_s, 0 \leq s \leq T_{y'}$, given $Y_0 = y$ and $T_{y'} < T_0$.
- Suppose $y > 0$, $Y_0 = 0$, and $\sigma_y = \sup\{t \leq T_y : Y_t = 0\}$. Then the distribution of $\tilde{Y}_t := Y_{t+\sigma_y}, 0 \leq t \leq T_y - \sigma_y$, is the same as $\hat{Y}_t, 0 \leq t \leq \hat{T}_y$, given $\hat{Y}_0 = 0$.

If X_t is a standard one-dimensional Brownian motion independent of $Y_t, \hat{Y}_t$, then $B_t = X_t + iY_t$ is a complex Brownian motion and $\hat{B}_t := X_t + i\hat{Y}_t$ is called an ℍ-*excursion*. An ℍ-excursion $\hat{B}$ can be started at any $z = x + iy$ with $y \geq 0$. Note that $T_y = \inf\{t : B_t \in \mathcal{I}_y\}$, $\hat{T}_y = \inf\{t : \hat{B}_t \in \mathcal{I}_y\}$ where $\mathcal{I}_y = \{z : \text{Im}(z) = y\}$. If $z = x + iy \in \mathbb{H}$ and $y' > y$, the distribution of $\hat{Y}_t, 0 \leq t \leq \hat{T}_{y'}$, given $\hat{Y}_0 = z$ is the same as the distribution of $Y_t, 0 \leq t \leq T_{y'}$, given $Y_0 = z$ and $T_{y'} < T_0$. The gambler's ruin estimate for one-dimensional Brownian motion tells us that $\mathbf{P}^{x+iy}\{T_{y'} < T_0\} = y/y'$.

Suppose $D \subset \mathbb{H}$ is a subdomain with $\sup\{\text{Im}(z) : z \in D\} < \infty$ and such that D has a piecewise analytic boundary. Let $H_D(z, w), z \in D, w \in \partial D$ be the Poisson kernel. Since $\text{Im}(\cdot)$ is a bounded harmonic function in D, $\text{Im}(z) = \mathbf{E}^z[\text{Im}(B_{\tau_D})] = \int_{\partial D} \text{Im}(w)\, H_D(z,w)\, |dw|$. Let $p_D(z,w)$ denote the density of the first visit to ∂D of an excursion $\hat{B}$. Then it is not difficult to check that

(5.13) $$p_D(z,w) = H_D(z,w)\, \frac{\text{Im}(w)}{\text{Im}(z)}.$$

In fact, (5.13) holds even if $\sup\{\text{Im}(z) : z \in D\} = \infty$; this can be seen by writing $p_D(z,w) = \lim_{r \to \infty} p_{D^r}(z,w)$ where $D^r = D \cap \{\text{Im}(z) < r\}$. In this case,

$$\int_{\partial D} p_D(z,w)\, |dw| = \mathbf{P}^z\{\hat{B}[0,\infty) \not\subset D\}.$$

Three important examples are the following:

- If $D_r = \{\text{Im}(z) > r\}$ and $s > r$, then (see Exercise 2.23),

(5.14) $$p_{D_r}(x+si, x'+ri) = \frac{r}{s}\, \frac{1}{\pi}\, \frac{s-r}{(x-x')^2 + (s-r)^2}.$$

In this case,

$$\int_{\partial D_r} p_{D_r}(x+si, z)\, |dz| = \frac{r}{s}.$$

- If $D = \mathbb{D}_+ = \{z \in \mathbb{H} : |z| < 1\}$, then (see (2.9))

(5.15) $$p_D(z, e^{i\theta}) = \frac{2}{\pi} \sin^2\theta\, [1 + O(|z|)], \quad z \to 0.$$

- If $D = \{z \in \mathbb{H} : |z| > 1\}$, then (see (2.12))

(5.16) $$p_D(z, e^{i\theta}) = \frac{2}{\pi}\, \frac{1}{|z|^2}\, \sin^2\theta\, [1 + O(|z|^{-1})], \quad z \to \infty.$$

PROPOSITION 5.15. *Suppose $A \in \mathcal{Q}$ and g_A is as defined in §3.4. If $z \in \mathbb{H} \setminus A$,*

$$\mathbf{P}^z\{\hat{B}[0,\infty) \cap A = \emptyset\} = \frac{\text{Im}[g_A(z)]}{\text{Im}(z)}.$$

If $x \in \mathbb{R} \setminus \overline{A}$,

(5.17) $$\mathbf{P}^x\{\hat{B}[0,\infty) \cap A = \emptyset\} = g_A'(x).$$

PROOF. Recall that $\text{hcap}(A) \leq \text{rad}(A)^2 < \infty$. From the expansion of g_A about infinity, we know that for all R sufficiently large

(5.18) $$g_A(\mathcal{I}_R) \subset \{R-1 \leq \text{Im}(z) \leq R\}.$$

Take $z \in \mathbb{H} \setminus A$, let σ_R be the first time that a Brownian motion B_t reaches $\mathcal{I}_R$, and let $\hat{\sigma}_R$ be the corresponding time for $\hat{B}_t$. Then

$$\mathbf{P}^z\{\hat{B}[0,\infty) \cap A = \emptyset\} = \lim_{R \to \infty} \mathbf{P}^z\{\hat{B}[0, \hat{\sigma}_R] \cap A = \emptyset\}$$
$$= \lim_{R \to \infty} \frac{\mathbf{P}^z\{B[0,\sigma_R] \cap (A \cup \mathbb{R}) = \emptyset\}}{\mathbf{P}^z\{B[0,\sigma_R] \cap \mathbb{R} = \emptyset\}}.$$

For $R > \text{Im}(z)$, the denominator equals $\text{Im}(z)/R$. By (5.18) and conformal invariance, for all R sufficiently large,

$$\mathbf{P}^{g_A(z)}\{B[0,\sigma_{R-1}] \cap \mathbb{R} = \emptyset\} \geq \mathbf{P}^z\{B[0,\sigma_R] \cap (A \cup \mathbb{R}) = \emptyset\}$$
$$\geq \mathbf{P}^{g_A(z)}\{B[0,\sigma_R] \cap \mathbb{R} = \emptyset\},$$

and hence the numerator is asymptotic to $\mathrm{Im}[g_A(z)]/R$ as $R \to \infty$. This gives the first equality. For the second, note that (5.15) implies that if $x \in \mathbb{R} \setminus \overline{A}$,

$$\mathbf{P}^x\{\hat{B}[0,\infty) \cap A = \emptyset\} = \lim_{\epsilon \to 0+} \mathbf{P}^{x+\epsilon i}\{\hat{B}[0,\infty) \cap A = \emptyset\}$$

$$= \lim_{\epsilon \to 0+} \frac{\mathrm{Im}[g_A(x+i\epsilon)]}{\epsilon} = g'_A(x).$$

The last inequality uses the fact that $g'_A(x) > 0$, which we know by construction of g_A. □

COROLLARY 5.16. *If $A \in \mathcal{Q}$ and $x \in \mathbb{R}$, then*

$$\lim_{y \to \infty} y^2 \, \mathbf{P}^{x+iy}\{\hat{B}[0,\infty) \cap A \neq \emptyset\} = \mathrm{hcap}(A).$$

If $y > \sup\{\mathrm{Im}(z) : z \in A\}$, then

$$y \int_{-\infty}^{\infty} \frac{1}{\pi} \mathbf{P}^{x+iy}\{\hat{B}[0,\infty) \cap A \neq \emptyset\} \, dx = \mathrm{hcap}(A).$$

If $r > \mathrm{rad}(A)$,

(5.19) $$r \int_0^\pi \mathbf{P}^{re^{i\theta}}\{\hat{B}[0,\infty) \cap A \neq \emptyset\} \left(\frac{2}{\pi} \sin^2 \theta\right) d\theta = \mathrm{hcap}(A).$$

PROOF. The first equality follows immediately from Proposition 5.15 and the expansion

$$\mathrm{Im}[g_A(x+iy)] = y - \frac{\mathrm{hcap}(A)}{y} + O(|y|^{-2}).$$

For the second equality, we use the first equality and (5.14) to get

$$\mathrm{hcap}(A) = \lim_{y_1 \to \infty} y_1^2 \, \mathbf{P}^{iy_1}\{\hat{B}[0,\infty) \cap A \neq \emptyset\}$$

$$= \lim_{y_1 \to \infty} y_1^2 \frac{y}{y_1} \int_{-\infty}^{\infty} \frac{1}{\pi} \frac{(y_1 - y)}{x^2 + (y_1 - y)^2} \mathbf{P}^{x+iy}\{\hat{B}[0,\infty) \cap A \neq \emptyset\} \, dx$$

$$= y \int_{-\infty}^{\infty} \frac{1}{\pi} \mathbf{P}^{x+iy}\{\hat{B}[0,\infty) \cap A \neq \emptyset\} \, dx.$$

The third equality is proved similarly using (5.16). □

EXERCISE 5.17. *Show that for every $a > 0$, there exist $c < \infty$ and $r > 0$ such that if $\gamma : [0,1] \to \mathbb{H}$ is a curve with $|\gamma(0)| = \epsilon < 1$, $|\gamma(1)| = 1$ and*

(5.20) $$\gamma[0,1] \subset \{x + iy \in \mathbb{H} : y \geq a \, |x|\},$$

then

$$\mathbf{P}^0\{\hat{B}[0,\infty) \cap \gamma[0,1] = \emptyset\} \leq c \, \epsilon^r.$$

Show that (5.20) is necessary by showing that for each $y > 0$,

$$\mathbf{P}^0\{\, \hat{B}[0,\infty) \cap \{x+iy : x \leq 0\} \,\} > 0$$

(this probability can easily be seen to be independent of y).

5.3. ℍ-EXCURSIONS

Define the ℍ-excursion Green's function $\hat{G}_\mathbb{H}$ by

$$\hat{G}_\mathbb{H}(z) = \lim_{\epsilon \to 0+} \frac{1}{\epsilon^2} \int_0^\infty \mathbf{P}\{|\hat{B}_s - z| \leq \epsilon\}\, ds.$$

In other words, the expected amount of the time that an ℍ-excursion spends in a set $V \subset \mathbb{H}$ is

$$\frac{1}{\pi} \int_V \hat{G}_\mathbb{H}(z)\, dA(z).$$

Using (5.13), (2.20), and (2.7), we see that

$$\hat{G}_\mathbb{H}(z) = \lim_{\delta \to 0+} \frac{\mathrm{Im}(z)}{\delta} G_\mathbb{H}(i\delta, z) = 2\pi\, \mathrm{Im}(z) H_\mathbb{H}(z, 0) = 2\pi \left(\frac{\mathrm{Im}(z)}{|z|}\right)^2.$$

In particular, $\hat{G}_\mathbb{H}(z) \leq 2\pi$, and the expected amount of time spent in the set V is bounded above by $2\,\mathrm{Area}(V)$. Suppose D is a Jordan domain and $f : \mathbb{H} \to D$ is a conformal transformation with $f(0) = z, f(\infty) = w$. Then $f \circ \hat{B}$ has the distribution $\mu_D^\#(z, w)$. Conformal invariance shows that the expected amount of time $f \circ \hat{B}$ spends in $f(V)$ is

$$\frac{1}{\pi} \int_V \hat{G}_\mathbb{H}(z)\, |f'(z)|^2\, dA(z).$$

In particular, the expected time spent in $f(V)$ is bounded above by $2\,\mathrm{Area}[f(V)]$, and expected time duration of a path distributed by $\mu^\#(z, w)$ is bounded above by $2\,\mathrm{Area}(D) < \infty$. Note that this estimate does not require any smoothness assumptions on ∂D.

REMARK 5.18. Suppose $r > 0$, $|z| < r$, and let $\hat{B}, \hat{B}^z$ denote ℍ-excursions starting at $0, z$, respectively. Let $\sigma_r = \inf\{t : |\hat{B}_t| = r\}$ and define σ_r^z similarly. It follows from (5.15) that we can define $(\hat{B}_t, \hat{B}_t^z)$ on the same probability space such that, except perhaps on an event of probability $O(|z|/r)$,

$$\hat{B}_{s+\tau_r} = \hat{B}^z_{s+\tau_r^z}, \quad 0 \leq t < \infty.$$

If $f : \mathbb{H} \to D$ is a conformal transformation, we get a similar coupling for processes defined under $\mu_D^\#(f(0), f(\infty))$ and $\mu_D^\#(f(z), f(\infty))$. From the previous paragraph, we have bounds on $\mathbf{E}[\tau_r]$ and the corresponding quantity obtained by conformal transformation. This allows us to give bounds on

$$d_\mathcal{K}(\mu_D^\#(f(0), f(\infty)), \mu_D^\#(f(z), f(\infty))),$$

and, in particular, to show that this goes to zero as $f(z)$ approaches $f(0)$.

We finish this section by stating some lemmas that tell us that we can approximate sets $A \in \mathcal{Q}$ by "nice" sets. Let $\mathcal{Q}_+$ denote the set of $A \in \mathcal{Q}$ such that $\overline{A} \cap \mathbb{R} \subset (0, \infty)$ and let $\mathcal{Q}_-$ denote the set of $A \in \mathcal{Q}$ such that $\overline{A} \cap \mathbb{R} \subset (-\infty, 0)$. Let $\mathcal{Q}_\pm$ denote the set of $A \in \mathcal{Q}$ with $0 \notin \overline{A}$; any such A can be written as $A_+ \cup A_-$ with $A_+ \in \mathcal{Q}_+, A_- \in \mathcal{Q}_-$. If $A \in \mathcal{Q}_\pm$, let $\Phi_A(z) = g_A(z) - g_A(0)$ be the conformal transformation of $\mathbb{H} \setminus A$ onto $\mathbb{H}$ with $\Phi_A(z) \sim z$ as $z \to \infty$ and $\Phi_A(0) = 0$. Then (5.17) tells us that the probability that an ℍ-excursion starting at the origin avoids A is $\Phi_A'(0)$. We call $A \in \mathcal{Q}_+$ a *smooth Jordan hull* if there is a smooth, simple curve $\gamma : (0, t) \to \mathbb{C}$ such that $\partial A \cap \mathbb{H} = \gamma(0, t)$. Smooth Jordan hulls in $\mathcal{Q}_-$ are defined similarly. The following simple lemma describes how to approximate hulls by smooth Jordan hulls.

LEMMA 5.19. *If $A \in \mathcal{Q}_+$, then there is a decreasing sequence of smooth Jordan hulls $A_1 \supset A_2 \supset \cdots$ such that $A = \cap_{n>0} A_n$. Moreover, $\Phi'_{A_n}(0) \uparrow \Phi'_A(0)$.*

PROOF. Let $0 < x_1 < x_2 < \infty$ be such that $\overline{A} \cap \mathbb{R} \subset [x_1, x_2]$ and let $x'_1 = \Phi_A(x_1), x'_2 = \Phi_A(x_2)$. Let $\psi : (-\infty, \infty) \to [0, \infty)$ be a smooth function that vanishes on $(\infty, x'_1]$ and $[x'_2, \infty)$ and is nonzero on (x'_1, x'_2). Let $\eta_n(t) = (x'_1 + t) + i\, n^{-1}\, \psi(t), 0 \le t \le x'_2 - x'_1,; \gamma_n(t) = \Phi_A^{-1} \circ \eta_n(t)$; and A_n the bounded hull bounded by γ. Then A_n decreases to A. Also
$$\Phi'_A(0) = \mathbf{P}^0\{\hat{B}[0,\infty) \cap A = \emptyset\} = \lim_{n \to \infty} \mathbf{P}^0\{\hat{B}[0,\infty) \cap A_n = \emptyset\}.$$
□

REMARK 5.20. Sometimes it is convenient to approximate hulls by smooth slits, i.e., by smooth simple curves $\gamma : (0, t] \to \mathbb{H}$ with $\gamma(0-) \in \mathbb{R}$. The last lemma shows that we can approximate hulls by smooth slits $\gamma_n(0, t - \epsilon]$ where γ_n is the boundary of a smooth Jordan hull. As an application, we can show that if $A \in \mathcal{Q}_+$ and $x_2 < x_1 \le 0$, then

(5.21) $$\Phi'_A(x_2) \ge \Phi'_A(x_1).$$

It suffices to prove this for $A = \gamma(0, t]$ where γ is a slit, and without loss of generality we may assume that γ is parametrized so that $\mathrm{hcap}[\gamma(0, t]] = 2t$. If we let $g_t = g_{\gamma(0,t]}$, then g_t satisfies the Loewner equation
$$\dot{g}_t(z) = \frac{2}{g_t(z) - U_t}, \quad g_0(z) = 0,$$
where $t \mapsto U_t$ is a continuous function with $g_t(x_2) < g_t(x_1) < U_t$. Differentiating this equation gives
$$g'_t(x_j) = \exp\left\{-\int_0^t \frac{2\, ds}{(U_s - g_s(x_j))^2}\right\},$$
from which (5.21) follows immediately. For later use, we note that if $b_1, b_2, k > 0$, then
$$g'_t(x_1)^{b_1} g'_t(x_2)^{b_2} \left[\frac{g_t(x_1) - g_t(x_2)}{x_1 - x_2}\right]^{-k} =$$
$$\exp\left\{2 \int_0^t -\frac{b_1\, ds}{(U_s - g_s(x_1))^2} + \frac{k\, ds}{(U_s - g_s(x_1))(U_s - g_s(x_2))} - \frac{b_2\, ds}{(U_s - g_s(x_2))^2}\right\}.$$
Since,
$$\frac{b_2}{(U_s - g_s(x_1))^2} - \frac{2\sqrt{b_2}}{(U_s - g_s(x_1))(U_s - g_s(x_2))} + \frac{b_2}{(U_s - g_s(x_2))^2} \ge 0,$$
we have
$$g'_t(x_1)^{b_1} g'_t(x_2)^{b_2} \left[\frac{g_t(x_1) - g_t(x_2)}{x_1 - x_2}\right]^{-k} \le \exp\left\{-2 \int_0^t \frac{\epsilon\, ds}{(U_s - g_s(x_1))^2}\right\},$$
where $\epsilon = b_1 - b_2 - k + 2\sqrt{b_2}$. In particular,

(5.22) $$\Phi'_A(x_1)^{b_1} \Phi'_A(x_2)^{b_2} \left[\frac{\Phi_A(x_1) - \Phi_A(x_2)}{x_1 - x_2}\right]^{-k} \le \Phi'_A(x_1)^{\epsilon} \le 1.$$

if $\epsilon = b_1 - b_2 - k + 2\sqrt{b_2} \ge 0$.

5.4. One-dimensional excursion measure

Although our primary study is of two-dimensional (complex) Brownian motions, it will be useful to discuss one-dimensional Brownian excursion measure. The Brownian bubble measure discussed in the next section can be written as the product of two one-dimensional loop measures with the imaginary part constrained to stay positive. Let $\mathcal{X}$ denote the set of curves $\gamma : [0, t_\gamma] \to \mathbb{R}$ and let $\mathcal{X}_+$ be the set of $\gamma \in \mathcal{X}$ with $\gamma(0, t_\gamma) \subset (0, \infty)$. Let $\mathcal{X}(x, y)$ denote the set of $\gamma \in \mathcal{X}$ with $\gamma(0) = x, \gamma(t_\gamma) = y$ and define $\mathcal{X}_+(x, y)$ similarly.

Let $\nu(x, \cdot; t)$ denote the probability measure on $\mathcal{X}$ induced by a one-dimensional Brownian motion $X_s, 0 \leq s \leq t$, with $X_0 = x$. We can write

$$\nu(x, \cdot; t) = \int_{-\infty}^{\infty} \nu(x, y; t) \, dy,$$

where $\nu(x, y; t)$ is a measure with total mass $(2\pi t)^{-1/2} e^{-(y-x)^2/2t}$ supported on $\mathcal{X}(x, y)$. It satisfies the Chapman-Kolmogorov relation

$$\nu(x_1, x_2; s + t) = \int_{-\infty}^{\infty} [\nu(x_1, y; s) \oplus \nu(y, x_2; t)] \, dy.$$

If $x, y > 0$, let $\nu_+(x, y; t), \nu_+(x, y)$ be $\nu(x, y; t), \nu(x, y)$ restricted to $\gamma \in \mathcal{X}_+(x, y)$. Since Brownian motion is time reversible, $\nu(y, x; t)$ and $\nu_+(y, x; t)$ can be obtained from $\nu(x, y; t)$ and $\nu_+(x, y; t)$ by time reversal of the paths.

We call $\nu(0, 0; t)$ the *Brownian loop measure (with time duration t)*. Note that $|\nu(0, 0; t)| = 1/\sqrt{2\pi t}$. Let $\theta_r^* : \mathcal{X}(0, 0) \to \mathcal{X}(0, 0)$ be the time and space translation map

$$t_{\theta_r^* \gamma} = t_\gamma, \quad \theta_r^* \gamma(t) = \gamma(t + r) - \gamma(r)$$

(here $\gamma, \theta_r^* \gamma$ are considered as functions on $\mathbb{R}$ with period t_γ). For each r and each t, $\nu(0, 0; t)$ is invariant under the map θ_r^*.

Suppose $X_0 = \epsilon > 0$, and let V_t be the event that $X_s > 0, 0 \leq s \leq t$. Let $q(y, t; \epsilon) = |\nu_+(\epsilon, y; t)|$ denote the density of X_t restricted to the event V_t so that

$$\mathbf{P}[V_t] = \int_0^\infty q(y, t; \epsilon) \, dy.$$

Brownian scaling implies that $q(y, t; \epsilon) = t^{-1/2} q(y/\sqrt{t}, 1; \epsilon/\sqrt{t})$. As $\epsilon \to 0+$, $\epsilon^{-1} q(y, t, \epsilon)$ approaches a positive solution of the heat equation with Dirichlet boundary conditions, i.e., $q(y, t) = c \bar{q}(y, t)$ where

$$\bar{q}(y, t) = t^{-3/2} \, y \, e^{-y^2/2t}.$$

Note that

(5.23) $$\int_0^\infty \bar{q}(y, t) \, dy = \sqrt{1/t}, \quad \int_0^\infty y \, \bar{q}(y, t) \, dy = \sqrt{\pi t/2}.$$

Let τ denote the first time that X reaches the origin. Since X_t is a martingale,

$$\epsilon = \mathbf{E}^\epsilon[X_{t \wedge \tau}] = \mathbf{P}^\epsilon[V_t] \, \mathbf{E}^\epsilon[X_t \mid V_t].$$

But (5.23) implies that as $\epsilon \to 0+$, $\mathbf{E}^\epsilon[X_t \mid V_t] \to \sqrt{\pi t/2}$. Hence,

$$\mathbf{P}^\epsilon[V_t] \sim \epsilon \sqrt{2/(t\pi)}, \quad \epsilon \to 0+,$$

and

(5.24) $$\lim_{\epsilon \to 0+} \epsilon^{-1} q(y, t; \epsilon) = \sqrt{2/\pi} \, t^{-3/2} \, y \, e^{-y^2/2t}.$$

For any $x > 0$, let $\nu_+(x, 0)$ denote the probability measure on paths given by $X_t, 0 \le t \le \tau$, where $X_0 = x$. It is supported on $\mathcal{X}_+(x, 0)$. We can write

$$\nu_+(x, 0) = \int_0^\infty \nu_+(x, 0; t) \, dt,$$

where $\nu_+(x, 0; t)$ is a measure supported on $\{\gamma \in \nu_+(x, 0) : t_\gamma = t\}$. In fact,

(5.25) $$|\nu_+(x, 0; t)| = \sqrt{\frac{1}{2\pi}} \, t^{-3/2} \, x \, e^{-x^2/2t}.$$

(This can be derived from the reflection principle for one dimensional Brownian motion,

$$\mathbf{P}^x\{\tau \le t\} = \mathbf{P}^0\{\max_{0 \le s \le t} X_s \ge x\} = 2\mathbf{P}^0\{X_t \ge x\} = 2\mathbf{P}^0\{X_1 \ge x/\sqrt{t}\}.)$$

If $x > 0$, the measure $\nu_+(0, x; t)$ is defined by

$$\nu_+(0, x; t) = \lim_{\epsilon \to 0+} \epsilon^{-1} \nu_+(\epsilon, x; t).$$

Under this definition $\nu_+(0, x; t)$ is 2 times the time reversal of $\nu_+(x, 0; t)$ (see (5.24) and (5.25)). We also define

$$\nu_+(0, x) = \lim_{\epsilon \to 0+} \epsilon^{-1} \nu_+(\epsilon, x) = \int_0^\infty \nu_+(0, x; t) \, dt.$$

The *one-dimensional Brownian excursion measure* $\nu_+(0, 0)$ is defined by

$$\nu_+(0, 0) = \lim_{\epsilon \to 0+} \epsilon^{-1} \nu_+(\epsilon, 0).$$

The measure $\nu_+(0, 0)$ is an infinite measure supported on $\mathcal{X}_+(0, 0)$. The gambler's ruin estimate tells us that $\nu_+(0, 0)$ gives measure $1/x$ to the set of curves of diameter at least x. The restriction of $\nu_+(0, 0)$ to paths γ with $t_\gamma \ge t$ is

$$\int_0^\infty [\nu_+(0, x; t) \oplus \nu_+(x, 0)] \, dx,$$

which has total mass

(5.26) $$\int_0^\infty |\nu_+(0, x; t)| \, dx = \int_0^\infty \sqrt{\frac{2}{\pi}} \, t^{-3/2} \, x \, e^{-x^2/2t} \, dx = \sqrt{\frac{2}{\pi t}}.$$

If we let $\bar{\nu}_{+,t}(0, 0)$ denote the measure derived from $\nu_+(0, 0)$ by considering the path only from time t to t_γ, then

$$\bar{\nu}_{+,t}(0, 0) = \int_0^\infty \sqrt{\frac{2}{\pi}} \, t^{-3/2} \, x \, e^{-x^2/2t} \, \nu_+(x, 0) \, dx.$$

We can write

$$\nu_+(0, 0) = \int_0^\infty \nu_+(0, 0; t) \, dt,$$

where the measures $\nu_+(0, 0; t)$ can be obtained by appropriate limits or by

$$\nu_+(0, 0; t) = \int_0^\infty \nu(0, x; s) \oplus \nu(x, 0; t - s) \, dx.$$

Here we can choose any $s \in (0,t)$. Note that

$$|\nu_+(0,0;t)| = \int_0^\infty [\frac{2}{\sqrt{\pi}} s^{-3/2} x\, e^{-x^2/2s}] [\frac{1}{\sqrt{2\pi}} (t-s)^{-3/2} x\, e^{-x^2/2(t-s)}]\, dx$$
(5.27)
$$= (1/\sqrt{2\pi})\, t^{-3/2}.$$

Note that this is consistent with (5.26) and the fact that the $\nu_+(0,0)$ measure of $\{\gamma \in \mathcal{X}_+(0,0) : t_\gamma \geq t\}$ can also be written as

$$\int_t^\infty |\nu_+(0,0;s)|\, ds.$$

Let $\mathcal{X}^*(0,0)$ be the set of $\gamma \in \mathcal{X}(0,0)$ such that there is a unique $t_\gamma^- \in [0, t_\gamma)$ such that $X_t > X_{t_\gamma^-}$ for $t \in [0, t_\gamma) \setminus \{t_\gamma^-\}$. The measure $\nu(0,0)$ is supported on $\mathcal{X}^*(0,0)$. For each $\epsilon > 0$, define $T_\epsilon : \mathcal{X}^*(0,0) \to [0,\infty)$ by: $T_\epsilon(\gamma) = 1/t_\gamma$ if $t_\gamma < \epsilon$ and, if $t_\gamma > \epsilon$, $T_\epsilon(\gamma) = (1/\epsilon)\, 1\{t_\gamma^- < \epsilon\}$. Note that for all $\gamma \in \mathcal{X}^*(0,0)$,

(5.28) $$\int_0^{t_\gamma} T_\epsilon(\theta_r^* \gamma)\, dr = 1.$$

Let $\nu^\epsilon(0,0;t)$ be the measure whose Radon-Nikodym derivative with respect to $\nu(0,0;t)$ is T_ϵ. Note that (5.28) and the invariance of $\nu(0,0;t)$ under θ_r^* imply that

$$|\nu^\epsilon(0,0;t)| = t^{-1} |\nu(0,0;t)| = \frac{1}{\sqrt{2\pi}} t^{-3/2}.$$

Then it is not difficult to show that

(5.29) $$\nu_+(0,0;t) = \lim_{\epsilon \to 0+} \nu^\epsilon(0,0;t).$$

We will now show how to describe the excursion measure $\mu_\mathbb{H}(0)$ in terms of these measures on one-dimensional paths. We write

$$\mu(0) = \int_{-\infty}^\infty \mu_\mathbb{H}(0,x)\, dx = \int_{-\infty}^\infty \int_0^\infty \mu_\mathbb{H}(0,x;t)\, dt\, dx.$$

The measure $\mu_\mathbb{H}(0,x;t)$ is supported on $\mathcal{K}_0^x(\mathbb{H}) \cap \{t_\gamma = t\}$ and can be given by

(5.30) $$\mu_\mathbb{H}(0,x;t) = \nu(0,x;t) \times \nu_+(0,0;t).$$

Note that

$$|\mu_\mathbb{H}(0,x;t)| = e^{-x^2/2t}/(2\pi t^2),$$

so that

$$H_\mathbb{H}(0,x) = \int_0^\infty |\mu_\mathbb{H}(0,x;t)|\, dt = \frac{1}{\pi x^2},$$

which agrees with (5.7).

5.5. Boundary bubbles

The *(Brownian) boundary bubble measure (in $\mathbb{H}$ rooted at 0)* is a σ-finite measure on curves $\gamma \in \mathcal{K}(\mathbb{H})$ with $\gamma(0) = 0, \gamma(t_\gamma) = 0$. It can be defined in a number of equivalent ways, for example

$$\mu_\mathbb{H}^{\text{bub}}(0) := \lim_{x \to 0+} \pi \mu_\mathbb{H}(0,x),$$

provided this limit is interpreted correctly. To be precise, for each $r > 0$, let $\mu(0, x; r)$ denote $\mu(0, x)$ restricted to curves γ that intersect $\mathbb{H} \setminus r\mathbb{D}$ and define

$$\mu_{\mathbb{H}}^{\text{bub}}(0; r) = \lim_{x \to 0+} \pi \, \mu_{\mathbb{H}}(0, x; r).$$

Since $|\mu_{\mathbb{H}}(0, x; r)|$ is uniformly bounded as $x \to 0+$, we can consider this limit as in §5.1. We then set $\mu_{\mathbb{H}}^{\text{bub}}(0) = \lim_{r \to 0+} \mu_{\mathbb{H}}^{\text{bub}}(0; r)$. Note that (5.6) and (5.8) imply

$$
\begin{aligned}
\mu^{\text{bub}}(0; r) &:= \pi \int_{r\mathbb{D}_+} [\mu_{r\mathbb{D}_+}(0, z) \oplus \mu_{\mathbb{H}}(z, 0)] \, |dz| \\
&= r^{-2} \int_0^\pi \frac{2}{\pi} \sin^2 \theta \, [\mu_{r\mathbb{D}_+}^{\#}(0, re^{i\theta}) \oplus \mu_{\mathbb{H}}^{\#}(re^{i\theta}, 0)] \, d\theta.
\end{aligned}
$$
(5.31)

In particular, $\mu_{\mathbb{H}}^{\text{bub}}(0)$ is normalized so that $|\mu_{\mathbb{H}}^{\text{bub}}(0; r)| = 1/r^2$. Alternatively, we can write

$$\mu_{\mathbb{H}}^{\text{bub}}(0) = \lim_{\epsilon \to 0+} \frac{\pi}{\epsilon} \mu_{\mathbb{H}}(\epsilon i, 0) = \lim_{\epsilon \to 0+} \frac{\pi}{2\epsilon^2} \mu_{\mathbb{H}}(\epsilon i, \epsilon i).$$

To check this, we first note that for each $r > 0$, the conditional measures converge,

$$[\mu_{\mathbb{H}}^{\text{bub}}(0; r)]^{\#} = \lim_{\epsilon \to 0+} [\mu_{\mathbb{H}}(\epsilon i, 0; r)]^{\#} = \lim_{\epsilon \to 0+} [\mu_{\mathbb{H}}(\epsilon i, \epsilon i; r)]^{\#},$$

where in each case the $(\cdot; r)$ indicates the measure restricted to curves that intersect $\partial(r\mathbb{D})$. To check that the constants are chosen correctly, for ease let $r = 1$. Then as $\epsilon \to 0+$,

$$
\begin{aligned}
|\mu_{\mathbb{H}}(\epsilon i, 0; 1)| &= \int_0^\pi H_{\mathbb{D}_+}(\epsilon i, e^{i\theta}) \, H_{\mathbb{H}}(e^{i\theta}, 0) \, d\theta \\
&\sim \epsilon \int_0^\pi [\frac{2}{\pi} \sin \theta] \, [\frac{1}{\pi} \sin \theta] \, d\theta = \frac{\epsilon}{\pi}.
\end{aligned}
$$

and, using (2.20),

$$|\mu_{\mathbb{H}}(\epsilon i, \epsilon i; 1)| \sim 2\epsilon \, |\mu_{\mathbb{H}}(\epsilon i, 0; 1)|.$$

By (5.30), we see that

$$\mu_{\mathbb{H}}^{\text{bub}}(0) = \pi \int_0^\infty [\nu(0, 0; t) \times \nu_+(0, 0; t)] \, dt,$$

where $\nu(0, 0; t), \nu_+(0, 0; t)$ denote one-dimensional loop measures as in the previous section. In particular, the $\mu_{\mathbb{H}}^{\text{bub}}(0)$ measure of the set of γ with $t_\gamma \geq t$ is

$$\pi \int_t^\infty |\mu(0, 0; s) \oplus \mu_+(0, 0; s)| \, ds = \pi \int_t^\infty (2\pi s)^{-2} \, ds = (2t)^{-1}.$$

DEFINITION 5.21. The *Schwarzian derivative* of a C^3 function f is defined by

$$Sf(z) = \left[\frac{f''(z)}{f'(z)}\right]' - \frac{1}{2}\left[\frac{f''(z)}{f'(z)}\right]^2 = \frac{f'''(z)}{f'(z)} - \frac{3}{2}\left[\frac{f''(z)}{f'(z)}\right]^2.$$

Suppose D is a simply connected subdomain of $\mathbb{H}$ that contains $\epsilon\mathbb{D}_+$ for some $\epsilon > 0$. For any $r < \epsilon$,

(5.32) $\qquad \mu_{\mathbb{H}}^{\text{bub}}(0)\{\gamma : \gamma(0, t_\gamma) \not\subset D\} = \mu_{\mathbb{H}}^{\text{bub}}(0; r)\{\gamma : \gamma(0, t_\gamma) \not\subset D\} < \infty.$

PROPOSITION 5.22. *Suppose D is a simply connected subdomain of $\mathbb{H}$ containing $\epsilon \mathbb{D}_+$ for some $\epsilon > 0$. Suppose $f : D \to \mathbb{H}$ is a conformal transformation with $f(0) = 0$, and let A be the image of $\mathbb{H} \setminus D$ under the map $z \mapsto -1/z$. Then*

$$\mu_{\mathbb{H}}^{\text{bub}}(0)\{\gamma : \gamma(0, t_\gamma) \not\subset D\} = \text{hcap}(A) = -\frac{1}{6} S f(0).$$

PROOF. We may assume that $f'(0) = 1$ for otherwise we consider $\tilde{f}(z) = f(z)/f'(0)$. Note that $\phi(z) := -1/f(-1/z)$ is a conformal transformation of $\mathbb{H} \setminus A$ onto $\mathbb{H}$ with $\phi(\infty) = \infty$, $\phi'(\infty) = 1$. Hence there is an $x \in \mathbb{R}$ such that

$$f(z) = -\frac{1}{g_A(-1/z) + x}.$$

However, (5.19) and (5.32) show that the $\mu_{\mathbb{H}}^{\text{bub}}(0)$ measure of the set of curves that intersect $\mathbb{H} \setminus D$ is $\text{hcap}(A)$. Since $g_A(w) = w + \text{hcap}(A)/w + O(|w|^{-2})$ as $w \to \infty$,

$$f(z) = z + x z^2 + [x^2 - \text{hcap}(A)] z^3 + \cdots,$$

i.e., $f''(0) = 2x$, $f'''(0) = 6[x^2 - \text{hcap}(A)]$, $Sf(0) = -6\,\text{hcap}(A)$. □

If D is a domain (not necessarily simply connected), $z \in \partial D$, and ∂D is analytic near z we can define $\mu_D^{\text{bub}}(z)$ in a similar way:

$$\mu_D^{\text{bub}}(z) = \pi \lim_{w \to z} \mu_D(w, z).$$

Suppose $f : \mathbb{H} \to D$ is a conformal transformation with $f(0) = z$ and such that f extends analytically to $\mathcal{B}(0, \delta)$ for some $\delta > 0$. Then for $x \in (-\delta, \delta)$,

$$f \circ \mu_{\mathbb{H}}(x, 0) = |f'(x)| \, |f'(0)| \, \mu_D(f(x), f(0)).$$

This gives the scaling rule

(5.33) $$f \circ \mu_{\mathbb{H}}^{\text{bub}}(0) = |f'(0)|^2 \, \mu_D^{\text{bub}}(z).$$

The boundary bubble measure also satisfies the following *restriction property*: if D is a subdomain of $\mathbb{H}$ as in Proposition 5.22, then $\mu_D^{\text{bub}}(0)$ is $\mu_{\mathbb{H}}^{\text{bub}}(0)$ restricted to curves with $\gamma(0, t_\gamma) \subset D$. Proposition 5.22 can be restated as

$$|\mu_{\mathbb{H}}^{\text{bub}}(0) - \mu_D^{\text{bub}}(0)| = -\frac{1}{6} S f(0).$$

REMARK 5.23. Suppose V_n is a sequence of sets in $\mathcal{Q}$ with $\text{rad}(V_n) \to 0$ and let $h_n = \text{hcap}(V_n)$, $r_n = \text{rad}(V_n)$. By Proposition 3.46, there is a constant c such that for all $|z| \geq 1/2$ and all n with $r_n \leq 1/4$,

$$|g_{V_n}(z) - (z + h_n z^{-1})| \leq c h_n r_n.$$

By Schwarz reflection, g_{V_n} can be extended to $\{z : |z| > r_n\}$. Using the Cauchy integral formula on the function $g_{V_n}(z) - z - (h_n/z)$, we see that

$$g'_{V_n}(1) = 1 - h_n + O(h_n r_n), \quad g''_{V_n}(1) = 2h_n + O(h_n r_n)$$

$$g'''_{V_n}(1) = -6h_n + O(h_n r_n),$$

and hence

$$|\mu_{\mathbb{H}}^{\text{bub}}(1; V_n)| = -\frac{1}{6} S g_{V_n}(1) = h_n + O(h_n r_n),$$

where $\mu_{\mathbb{H}}^{\text{bub}}(1; V_n)$ denotes $\mu_{\mathbb{H}}^{\text{bub}}(1)$ restricted to curves that intersect V_n. Hence,

$$\lim_{n \to \infty} h_n^{-1} \mu_{\mathbb{H}}^{\text{bub}}(1; V_n) = \mu_{\mathbb{H}}^{\#}(1, 0) \oplus \mu_{\mathbb{H}}^{\#}(0, 1) = \pi^2 [\mu_{\mathbb{H}}(1, 0) \oplus \mu_{\mathbb{H}}(0, 1)].$$

Suppose $f : \mathbb{H} \to D$ is a conformal transformation with $f(0) = 0, f(1) = w$, and suppose f can be extended analytically in neighborhoods of 0 and 1. If $W_n = f(V_n)$, then $\mathrm{hcap}(W_n) \sim |f'(0)|^2 \, h_n$ as $n \to \infty$. Also,

$$f \circ \mu_{\mathbb{H}}^{\mathrm{bub}}(1; V_n) = |f'(1)|^2 \, \mu_D^{\mathrm{bub}}(z; W_n),$$

$$f \circ \mu_{\mathbb{H}}(1, 0) = |f'(0)| \, |f'(1)| \, \mu_D(z, 0),$$

$$f \circ \mu_{\mathbb{H}}(0, 1) = |f'(0)| \, |f'(1)| \, \mu_D(0, z).$$

Therefore,

$$\lim_{n \to \infty} [\mathrm{hcap}(W_n)]^{-1} \mu_D^{\mathrm{bub}}(z; W_n)$$

(5.34)
$$= |f'(0)|^{-2} |f'(1)|^{-2} \, \pi^2 \, f \circ [\mu_{\mathbb{H}}(1,0) \oplus \mu_{\mathbb{H}}(0,1)]$$
$$= \pi^2 \, [\mu_D(z,0) \oplus \mu_D(0,z)].$$

The boundary bubble measure can also be given in terms of the excursion measures. Let

$$\mathbb{H}_r = \mathbb{H} \cap \{\mathrm{Im}(z) < r\}, \quad \mathbb{H}^s = \mathbb{H} \cap \{\mathrm{Im}(z) > s\}, \quad \mathbb{H}_r^s = \mathbb{H}_r \cap \mathbb{H}^s.$$

It is easy to check that $\mu_{\mathbb{H}}^{\mathrm{bub}}(0)$ is supported on curves γ such that there is a unique time t_γ^+ such that $\mathrm{Im}[\gamma(t)] < \mathrm{Im}[\gamma(t_\gamma^+)]$ for $t \in [0, \gamma_t] \setminus \{t_\gamma^+\}$. (This is equivalent to the statement that a one-dimensional Brownian bridge has a unique maximum.) By splitting $\gamma[0, t_\gamma]$ into $\gamma[0, t_\gamma^+]$ and $\gamma[t_\gamma^+, t_\gamma]$ we get the following.

PROPOSITION 5.24.

$$\mu_{\mathbb{H}}^{\mathrm{bub}}(0) = \pi \int_0^\infty \int_{-\infty}^\infty [\mu_{\mathbb{H}_y}(0, x + iy) \oplus \mu_{\mathbb{H}_y}(x + iy, 0)] \, dx \, dy.$$

PROOF. Let m_y denote $(1/\pi) \, \mu_{\mathbb{H}}^{\mathrm{bub}}(0)$ restricted to curves that intersect $\mathcal{I}_y$. Then, using Proposition 5.11,

$$m_y = \int_{-\infty}^\infty [\mu_{\mathbb{H}_y}(0, x + iy) \oplus \mu_{\mathbb{H}_r}(x + iy, 0)] \, dx$$

$$= \int_{-\infty}^\infty \int_y^\infty \int_{-\infty}^\infty [\mu_{\mathbb{H}_y}(0, x + iy) \oplus \mu_{\mathbb{H}_r}(x + iy, x' + ir) \oplus \mu(x' + ir, 0)] dz' \, dr \, dx$$

But, (5.6) gives

$$\int_{-\infty}^\infty [\mu_{\mathbb{H}_y}(0, x + iy) \oplus \mu_{\mathbb{H}_r}(x + iy, x' + ir)] \, dx = \mu_{\mathbb{H}_r}(0, x' + ir).$$

$\square$

Proposition 5.24 can be considered as a way to obtain $\mu_{\mathbb{H}}^{\mathrm{bub}}(0)$ by integrating over all possible points in $\mathbb{H}$ that can be the maximal point of the loop in the sense of the largest imaginary part. There is a similar expression obtained by focusing on the maximal point where maximal is defined in terms of largest absolute value. The proof is essentially the same so we just state the result.

PROPOSITION 5.25.

$$\mu_{\mathbb{H}}^{\mathrm{bub}}(0) = \pi \int_0^\infty \int_{\partial(r\mathbb{D}_+)} [\mu_{r\mathbb{D}_+}(0, z) \oplus \mu_{r\mathbb{D}_+}(z, 0)] \, |dz|.$$

We end this section by describing another way to construct $\mu_{\mathbb{H}}^{\text{bub}}(0)$. Let $\mathcal{X}$ be the set of curves $\gamma \in \tilde{\mathcal{C}}$ such that there is a unique $t_\gamma^- \in [0, t_\gamma)$ with $\text{Im}[\gamma(t)] > \text{Im}[\gamma(t_\gamma^-)]$ for $t \in [0, t_\gamma) \setminus \{t_\gamma^-\}$. It is straightforward to check that the measure $\mu(0,0)$ is supported on $\mathcal{X}$. For every $\epsilon > 0$, define $T_\epsilon : \tilde{\mathcal{C}} \to [0, \infty)$ as follows:

$$T_\epsilon(\gamma) = 1/t_\gamma \text{ if } t_\gamma < \epsilon \text{ or } \gamma \notin \mathcal{X},$$

and if $\gamma \in \mathcal{X}$ with $t_\gamma \geq \epsilon$,

$$T_\epsilon(\gamma) = \epsilon^{-1} \text{ if } t_\gamma^- \leq \epsilon$$

and

$$T_\epsilon(\gamma) = 0 \text{ if } t_\gamma^- > \epsilon.$$

Note that for each $\gamma \in \tilde{\mathcal{C}}$,

(5.35) $$\int_0^{t_\gamma} T(\theta_r \gamma) \, dr = 1.$$

Let ν_ϵ be the measure whose Radon-Nikodym derivative with respect to $\mu(0,0)$ is T_ϵ. Recall that we can write

$$\mu(0,0) = \int_0^\infty \mu(0,0;t) \, dt,$$

where $\mu(0,0;t)$ is supported on curves γ with $t_\gamma = t$ and $|\mu(0,0;t)| = 1/(2\pi t)$. Hence, if $s > \epsilon$, then we can write ν_ϵ, restricted to curves with $t_\gamma \geq s$ as $\int_s^\infty \nu_{\epsilon, t} \, dt$, where $|\nu_{\epsilon, tt}| = (1/t)|\mu(0,0;t)| = 1/(2\pi t^2)$. Then we can write

(5.36) $$\mu_{\mathbb{H}}^{\text{bub}}(0) = \lim_{\epsilon \to 0+} \pi \nu_\epsilon.$$

To be more precise, if we let $\nu_{\epsilon, \delta}$ be ν_ϵ restricted to curves γ with $t_\gamma \geq \delta$, then for fixed $\delta > 0$, as $\epsilon \to 0+$, $\nu_{\epsilon, \delta}$ approaches $\mu_{\mathbb{H}}^{\text{bub}}(0)$ restricted to curves with $t_\gamma \geq \delta$. This is derived in a similar way as (5.29). Note that the normalization of ν_ϵ is such that the limit gives measure $1/(2t)$ to curves of time duration at least t which we have already seen is true for $\mu_{\mathbb{H}}^{\text{bub}}(0)$.

5.6. Loop measure

The *Brownian loop measure* $\mu^{\text{loop}}(D)$ in a domain D is a σ-finite conformally invariant measure on *unrooted* loops in $\tilde{\mathcal{C}}_U(D)$ that satisfies the restriction property, i.e., if $D' \subset D$, then $\mu^{\text{loop}}(D')$ is the same as $\mu^{\text{loop}}(D)$ restricted to loops that stay in D'. Since it satisfies the restriction property, it suffices to define $\mu^{\text{loop}} = \mu^{\text{loop}}(\mathbb{C})$ and then $\mu^{\text{loop}}(D)$ is μ^{loop} restricted to curves in $\tilde{\mathcal{C}}_U(D')$. It is easiest to define μ^{loop} by first considering measures on rooted loops and then forgetting about the loops. We call a Borel measurable function $T : \tilde{\mathcal{C}} \to [0, \infty)$ a *unit weight* if for each γ,

$$\int_0^{t_\gamma} T(\theta_r \gamma) \, dr = 1.$$

Here, as before, θ_r is defined by $\theta_r \gamma(t) = \gamma(t+r)$ where curves γ in $\tilde{\mathcal{C}}$ are considered as functions $\gamma : (-\infty, \infty) \to \mathbb{C}$ with $\gamma(s + t_\gamma) = \gamma(s)$.

EXAMPLE 5.26.   • $T(\gamma) = 1/t_\gamma$.

- Suppose $f : D \to D'$ is a conformal transformation. Define T by $T(\gamma) = 1/t_\gamma$ if $\gamma \in \tilde{\mathcal{C}} \setminus \tilde{\mathcal{C}}(D)$ and $T(\gamma) = |f'(\gamma(0))|^2/t_{f \circ \gamma}$ if $\gamma \in \tilde{\mathcal{C}}(D)$. Then T is a unit weight. To see this, suppose $\gamma \in \tilde{\mathcal{C}}(D)$. Note that $t_{f \circ \gamma} = t_{f \circ \theta_r \gamma}$. Hence,

$$\int_0^{t_\gamma} T(\theta_r \gamma)\, dr = (1/t_{f \circ \gamma}) \int_0^{t_\gamma} |f'(\theta_r \gamma(0))|^2\, dr = 1.$$

- Let $\mathcal{Y}$ be the set of curves in $\tilde{\mathcal{C}}$ such that there is a unique $t_\gamma^- \in [0, t_\gamma)$ such that $\mathrm{Im}[\gamma(t)] > \mathrm{Im}[\gamma(t_\gamma^-)]$ for $t \in [0, t_\gamma) \setminus \{t_\gamma^-\}$. For every $\epsilon > 0$, define T_ϵ as follows:

$$T_\epsilon(\gamma) = 1/t_\gamma \text{ if } t_\gamma < \epsilon \text{ or } \gamma \notin \mathcal{Y},$$

and if $\gamma \in \mathcal{Y}$ with $t_\gamma \geq \epsilon$,

$$T_\epsilon(\gamma) = \epsilon^{-1} \text{ if } 0 \leq t_\gamma^- \leq \epsilon$$

and

$$T_\epsilon(\gamma) = 0 \text{ if } \epsilon < t_\gamma^- < t_\gamma.$$

Then T_ϵ is a unit weight.

- Informally, we can define T_- by $T_- = \lim_{\epsilon \to 0+} T_\epsilon$. Then if $\gamma \in \mathcal{Y}$, $T_-(\gamma)$ is the "δ-function" of the set $\{\gamma \in \mathcal{Y} : t_\gamma = 0\}$. Appropriately interpreted, this is a unit weight.

If T is any Borel measurable function from $\tilde{\mathcal{C}}$ to $[0, \infty)$ and D is a domain in $\mathbb{C}$, we define the rooted loop measure with respect to T by

$$\mu^{\mathrm{rooted}}(D; T) = \int_D T(\gamma)\, \mu_D(z, z)\, dA(z).$$

This measure induces a measure $\mu^{\mathrm{loop}}(D; T)$ on unrooted loops in D in an obvious way; note that if

$$\int_0^{t_\gamma} T(\theta_r \gamma)\, dr = \int_0^{t_\gamma} T_1(\theta_r \gamma)\, dr,$$

for every $\gamma \in \tilde{\mathcal{C}}(D)$, then $\mu^{\mathrm{loop}}(D; T) = \mu^{\mathrm{loop}}(D; T_1)$. We define $\mu^{\mathrm{loop}}(D)$ to be $\mu^{\mathrm{loop}}(D; T)$ where T is a unit weight, and $\mu^{\mathrm{loop}} = \mu^{\mathrm{loop}}(\mathbb{C})$. It is immediate that this satisfies the restriction property. The next proposition shows that it satisfies conformal invariance.

PROPOSITION 5.27 (Conformal Invariance). *If $f : D \to D'$ is a conformal transformation, then*

$$f \circ \mu^{\mathrm{loop}}(D) = \mu^{\mathrm{loop}}(D').$$

PROOF. Let $T^*(\gamma) = 1/t_\gamma$ and define T by $T(\gamma) = T^*(\gamma)$ if $\gamma \notin \tilde{\mathcal{C}}(D)$, and if $\gamma \in \tilde{\mathcal{C}}(D)$, $T(\gamma) = |f'(\gamma(0))|^2/t_{f \circ \gamma}$. Then T, T^* are unit weights and $\mu^{\mathrm{loop}}(D)$ is induced by

$$\int_D T(\gamma)\, \mu_D(z, z)\, dA(z).$$

which is a measure on $\tilde{\mathcal{C}}(D)$. But,

$$\begin{aligned}
f \circ \int_D T(\gamma)\, \mu_D(z,z)\, |dz| &= \int_D |f'(\gamma(0))|^2/t_{f\circ\gamma}\, [f \circ \mu_D(z,z)]\, dA(z) \\
&= \int_D t_{f\circ\gamma}^{-1} |f'(\gamma(0))|^2\, \mu_{D'}(f(z), f(z))\, dA(z) \\
&= \int_{D'} T^*(\gamma)\, \mu_{D'}(w,w)\, dA(w),
\end{aligned}$$

Since T^* is a unit weight, the last measure induces $\mu^{\mathrm{loop}}(D')$ on $\tilde{\mathcal{C}}_{\mathrm{U}}(D')$. □

REMARK 5.28. A rooted loop $\gamma \in \tilde{\mathcal{C}}$ can be considered as a triple (z, t, η) where $z \in \mathbb{C}, t \in (0, \infty)$ and $\eta \in \tilde{\mathcal{C}}$ with $t_\eta = 1, \eta(0) = 0$. The correspondence is given by $z = \gamma(0)$, $t = t_\gamma$, $\eta(s) = t_\gamma^{-1/2} [\gamma(st_\gamma) - z]$. Let $\mu^\# = \mu^\#(0,0;1)$ denote the probability measure associated to the two-dimensional Brownian bridge (loop) with time duration 1 at the origin. If $T(\gamma) = 1/t_\gamma$, then the measure $\mu^{\mathrm{rooted}}(\mathbb{C}; T)$ is the same as the measure

$$\mathrm{area} \times \frac{dt}{2\pi t^2} \times \mu^\#$$

on $\mathbb{C} \times (0, \infty) \times \tilde{\mathcal{C}}$. By considering this as a measure on unrooted loops, we get μ^{loop}.

Using the unit weight T_ϵ and letting ϵ go to zero, we get another expression for the loop measure. Let $\mu_\mathbb{H}^{\mathrm{bub}}(0)$ be the boundary bubble measure and let $\mu_{\mathbb{H}+z}^{\mathrm{bub}}(z)$ be the corresponding translation by z. Then,

$$\mu^{\mathrm{loop}} = \int_\mathbb{C} T_-\, \mu(z,z)\, dA(z) = \frac{1}{\pi} \int_\mathbb{C} \mu_{\mathbb{H}+z}^{\mathrm{bub}}(z)\, dA(z).$$

Here we are considering $\mu_{\mathbb{H}+z}^{\mathrm{bub}}(z)$ as a measure on unrooted loops. This representation associates with each unrooted loop the rooted loop rooted at the point with minimal imaginary part. See (5.36) to see why there is a factor of $1/\pi$. Using Proposition 5.24, we get the next representation that we write as another proposition.

PROPOSITION 5.29. *The measure*

$$\int_{-\infty}^\infty \int_{-\infty}^\infty \int_{-\infty}^\infty \int_y^{-\infty} [\mu_{\mathbb{H}_y^s}(x+iy, r+is) \oplus \mu_{\mathbb{H}_y^s}(r+is, x+iy)]\, ds\, dy\, dr\, dx,$$

considered as a measure on unrooted loops, is the Brownian loop measure μ^{loop}.

We will now like to show that the "derivative" of the loop measure is given by the boundary bubble measure. If V, V_1 are subsets of D, let $\mu^{\mathrm{loop}}(D; V)$ denote $\mu^{\mathrm{loop}}(D)$ restricted to loops that intersect both V and V_2. Let $\mu_D^{\mathrm{bub}}(z; V)$ denote $\mu_D^{\mathrm{bub}}(z)$ restricted to loops that intersect V, considered as a measure on unrooted loops.

PROPOSITION 5.30. *Let V_n be a sequence of hulls in $\mathcal{Q}$ with $h_n = \mathrm{hcap}(V_n) > 0$ and $r_n := \mathrm{rad}(V_n) \to 0$. Then for every $r > 0$,*

$$\lim_{n \to \infty} h_n^{-1}\, \mu^{\mathrm{loop}}(\mathbb{H}; \mathcal{I}_r, V_n) = \mu_\mathbb{H}^{\mathrm{bub}}(0; \mathcal{I}_r).$$

PROOF. Let $f_r(z) = rz$. Then
$$f_r \circ \mu^{\text{loop}}(\mathbb{H}, \mathcal{I}_1, V_n) = \mu^{\text{loop}}(\mathbb{H}, \mathcal{I}_r; rV_n),$$
$$f_r \circ \mu^{\text{bub}}_{\mathbb{H}}(0; \mathcal{I}_1) = r^2 \mu^{\text{bub}}_{\mathbb{H}}(0, \mathcal{I}_r),$$
and $\text{hcap}[rV_n] = r^2 h_n$. Hence it suffices to prove the result for $r = 1$. Note that
$$\mu^{\text{loop}}(\mathbb{H}; \mathcal{I}_1, V_n) = \frac{1}{\pi} \int_1^\infty \int_{-\infty}^\infty \mu^{\text{bub}}_{\mathbb{H}_y}(x + iy; V_n) \, dx \, dy.$$
$$\mu^{\text{bub}}_{\mathbb{H}}(0; \mathcal{I}_1) = \pi \int_1^\infty \int_{-\infty}^\infty [\mu_{\mathbb{H}_y}(0, x + iy) \oplus \mu_{\mathbb{H}_y}(x + iy, 0)] \, dx \, dy.$$
As a measure on unrooted loops, $\mu_{\mathbb{H}_y}(0, x+iy) \oplus \mu_{\mathbb{H}_y}(x+iy, 0) = \mu_{\mathbb{H}_y}(x+iy, 0) \oplus \mu_{\mathbb{H}_y}(0, x+iy)$. Also, (5.34) gives
$$\lim_{n \to \infty} h_n^{-1} \mu^{\text{bub}}_{\mathbb{H}_y}(x + iy, V_n) = \pi^2 \left[\mu_{\mathbb{H}_y}(x+iy, 0) \oplus \mu_{\mathbb{H}_y}(0, x+iy) \right].$$
□

5.7. Brownian loop soup

If m is a nonatomic Borel σ-finite measure on a metric space $\mathcal{M}$, a *Poisson point process* with measure m is collection of random variables $I(E)$ Borel sets E with $m(E) < \infty$, satisfying the following:

- W.p.1, $E \mapsto I(E)$ is a measure, i.e., if $E_1, E_2, \ldots$ are mutually disjoint, then
$$I(E_1 \cup E_2 \cup \cdots) = I(E_1) + I(E_2) + \cdots \ .$$
- For fixed E, $I(E)$ is a Poisson with mean $m(E)$.
- If $E_1, \ldots, E_n$ are mutually disjoint, $I(E_1), \ldots, I(E_n)$ are independent.

The random set of points
$$\mathcal{P} = \{x \in \mathcal{M} : I(\{x\}) = 1\}$$
is called a *Poissonian realization* of the measure m. Note that $I(E) = \#[E \cap \mathcal{P}]$. The nonatomic nature of m implies that with probability one $I(\{x\}) \in \{0, 1\}$ for all $x \in \mathcal{M}$, i.e., no point appears twice in a Poissonian realization. We will write $I(E; \lambda), \lambda \geq 0$ for a Poisson point process on $\mathcal{M} \times [0, \infty)$ with measure $m \times [\text{length}]$; Here $I(E; \lambda)$ is shorthand for $I(E \times [0, \lambda])$. For fixed λ, $I(\cdot; \lambda)$ is a Poisson point process with measure λm For fixed E, $I(E; \lambda)$ is a Poisson process with parameter $\lambda m(E)$. We can write a realization from this process as an increasing family of sets
$$\mathcal{P}_\lambda = \{x \in \mathcal{M} : I(\{x\}; t) = 1\}.$$
The nonatomic nature of m implies that with probability one for all t, $\mathcal{P}_\lambda \setminus \cup_{s < \lambda} \mathcal{P}_s$ contains at most one point.

DEFINITION 5.31. The *(Brownian) bubble soup (in $\mathbb{H}$ rooted at 0)* is a Poisson point process on $\tilde{\mathcal{C}} \times [0, \infty)$ with measure $\mu^{\text{bub}}_{\mathbb{H}}(0) \times$ length. We write $\mathcal{O}_\lambda$ for a realization of this process.

5.7. BROWNIAN LOOP SOUP

DEFINITION 5.32. The *(Brownian) loop soup* is a Poisson point process on $\tilde{\mathcal{C}}_U \times [0, \infty)$ with measure $\mu^{\text{loop}} \times \text{length}/2$. We write $\mathcal{L}_\lambda$ for a realization of this process. If D is a domain, then we let

$$\mathcal{L}_\lambda(D) = \{\gamma \in \mathcal{L}_\lambda : \gamma[0, t_\gamma) \subset D\}.$$

Note that $\mathcal{L}_\lambda(D)$ is a realization of the point process with measure $\mu^{\text{loop}}(D) \times \text{length}/2$. Proposition 5.27 can be restated as follows.

PROPOSITION 5.33. *[Conformal invariance] Suppose $f : D \to D'$ is a conformal transformation and $\mathcal{L}_\lambda$ is the realization of the loop soup. Then the distribution of*

$$f \circ \mathcal{L}_\lambda(D) := \{f \circ \gamma : \gamma \in \mathcal{L}_\lambda(D)\}$$

is the same as that of $\mathcal{L}_\lambda(D')$.

Now suppose that $\eta : [0, \infty) \to \mathbb{C}$ is a simple curve with $\eta(0) = 0$, $\eta(0, \infty) \subset \mathbb{C}$, and such that $\text{hcap}(\gamma(0, t]) = 2t$. Let $H_t = \mathbb{H} \setminus \gamma[0, t]$, and let $g_t = g_{\eta(0,t]}, f_t = g_t^{-1}$. as in §3.4,. Let $\mathcal{L}$ be a realization of the loop soup. Then $\mathcal{L}_\lambda \setminus \mathcal{L}_\lambda(H_t)$ is the set of loops in $\mathcal{L}_\lambda$ that intersect $\eta[0, t]$. Let $\mathcal{O}_s$ denote a realization of the bubble soup. For each $\gamma \in \mathcal{O}_\infty$ let $s(\gamma)$ denote the "time at which γ was created", i.e., the s such that $\gamma \in \mathcal{O}_s$ but $\gamma \notin \mathcal{O}_{s'}, s' < s$. Then Propositions 5.30 and 5.33 give the following.

PROPOSITION 5.34. *Under the assumptions of the previous paragraph, the distribution of*

$$\{f_{s(\eta)} \circ \eta : \eta \in \mathcal{O}_{t\lambda}\}$$

is the same as that of $\mathcal{L}_\lambda \setminus \mathcal{L}_\lambda(D_t)$.

CHAPTER 6

Schramm-Loewner evolution

6.1. Chordal *SLE*

DEFINITION 6.1. The *chordal Schramm-Loewner evolution with parameter* $\kappa \geq 0$ *(with the standard parametrization)*, or *chordal* SLE_κ, is the random collection of conformal maps g_t obtained from solving the initial value problem

$$(6.1) \qquad \dot{g}_t(z) = \frac{2}{g_t(z) - \sqrt{\kappa}\, B_t}, \quad g_0(z) = z \quad (z \in \mathbb{H}),$$

where B_t is a standard one-dimensional Brownian motion.

In other words, chordal SLE_κ is the (random) Loewner chain derived from solving (4.10) with driving function $U_t = \sqrt{\kappa}\, B_t$. Unless we state otherwise, we will assume that the Brownian motion B_t starts at the origin. Let H_t be the domain of g_t and let $K_t = \mathbb{H} \setminus H_t$ be the random continuously growing hulls.

REMARK 6.2. Schramm invented chordal SLE_κ while considering possible scaling limits for the loop-erased random walk [76]. The nature of the model led to the conjecture that the scaling limit would be a random simple curve $\gamma : [0, \infty) \to \bar{\mathbb{H}}$ with $\gamma(0) = 0$, $\gamma(0, \infty) \subset \mathbb{H}$, and such that the following is true. If we parametrize the curve by capacity, let $g_t = g_{\gamma(0,t]}$, and define

$$\bar{\gamma}^s(t) = g_s(\gamma(s+t)) - g_s(\gamma(s)),$$

then the conditional distribution of $\bar{\gamma}^s$ given $\gamma[0, s]$ is the same as the distribution of γ. If we let $\bar{g}_t(z) = g_t(z) - U_t$, and define $\bar{g}_{s,t}$ for $s < t$ by $\bar{g}_t = \bar{g}_{s,t} \circ \bar{g}_s$, then these assumptions become:

- *Identically distributed increments*: the distribution of $\bar{g}_{s,t}$ depends only on $t - s$;
- *Markovian property*: $\bar{g}_{s,t}$ is independent of g_r, $0 \leq r \leq s$.

If we convert these assumptions to assumptions about the driving function U_t of g_t, we see that U_t is a continuous process satisfying:

- the distribution of $U_t - U_s$ depends only on $t - s$;
- $U_t - U_s$ is independent of $U_r, 0 \leq r \leq s$.

It is well known that these assumptions imply that U_t must be a Brownian motion, i.e., of the form $U_t = \mu t + \sqrt{\kappa} B_t$ for some standard Brownian motion B_t. If we make the further assumption that the distribution of K_t is symmetric about the imaginary axis (the loop-erased walk has this property), then this implies that $\mu = 0$.[1] Hence we are left with one parameter to choose, κ. One can also define SLE_κ with drift μ by choosing driving function $U_{t,\mu} = \mu t + \sqrt{\kappa} B_t$. For fixed $t < \infty$, the distribution

---

[1] Alternatively, we could postulate that the distribution of K_t is invariant under the scaling, i.e., that $r^{-1} K_{r^2 t}$ has the same distribution as K_t for all $r > 0$. This also implies $\mu = 0$.

of $U_{s,\mu}, 0 \le s \le t$, is absolutely continuous with the distribution of $\sqrt{\kappa}\,B_t$. Hence the distributions of $g_s, 0 \le s \le t$, are absolutely continuous. As we will soon see, when we vary κ we get the behavior of g_t changes significantly.

The existence of the random maps g_t and corresponding continuously growing hulls K_t follows from the results in §4.1 since w.p.1 B_t is a continuous function of t. However, it is not so easy to see whether the chain is generated by a path. Recall from §4.4 that the borderline for being able to prove existence of a generating curve is Hölder continuity with $\alpha = 1/2$. If the Hölder-$(1/2)$ norm is small then the chain is generated by a simple path. If the norm is large, then it is possible that the chain is not generated by a path. A Brownian path is not quite Hölder-$(1/2)$ continuous (although, it can be thought of loosely as a Hölder-$(1/2)$ path with a randomly varying norm). Hence we cannot use the deterministic result to conclude that the maps g_t come from a random curve. However, the following theorem shows that this is the case. For convenience, we will use this theorem in this chapter, although most of the results could be stated without assuming it.

THEOREM 6.3. *W.p.1, SLE_κ is generated by a path.*

PROOF. For $\kappa \ne 8$, see Theorem 7.4. We do not include a proof for $\kappa = 8$ in this book. However, for $\kappa = 8$, it follows from results in [**57**] where SLE_8 is given as a limit of discrete measures on paths (see Exercise 4.50). □

DEFINITION 6.4. *A (chordal) SLE_κ path or curve in $\mathbb{H}$ is the random curve $\gamma(t)$ that generates chordal SLE_κ.*

In particular, $g_t(\gamma(t)) = \sqrt{\kappa}\,B_t$ in the sense $\lim_{z \to \gamma(t)} g_t(z) = \sqrt{\kappa}\,B_t$. The term *trace* is also used by some for the path, but this term is ambiguous since trace usually refers to $\gamma[0,t]$ or $\gamma[0,\infty)$ as a set rather than the function $\gamma : [0,\infty) \to \mathbb{C}$.

PROPOSITION 6.5 (*SLE scaling*). *Suppose g_t is chordal SLE_κ and $r > 0$. Then $\hat{g}_t(z) := r^{-1}\,g_{r^2 t}(r\,z)$ has the distribution of SLE_κ. Equivalently, if γ is an SLE_κ path, and $\hat{\gamma}(t) := r^{-1}\gamma(r^2 t)$, then $\hat{\gamma}$ has the distribution of an SLE_κ path.*

PROOF. Clearly, $\hat{g}_0(z) = z$. Suppose $U_t = \sqrt{\kappa}\,B_t$ is the driving function for g_t. Let $\hat{B}_t := r^{-1} B_{r^2 t}$ which is a standard Brownian motion. then,

$$\dot{\hat{g}}_t(z) = r\,\dot{g}_{r^2 t}(r\,z) = \frac{2\,r}{g_{r^2 t}(r\,z) - \sqrt{\kappa}\,B_{r^2 t}} = \frac{2}{\hat{g}_t(z) - \sqrt{\kappa}\,\hat{B}_t}.$$

□

REMARK 6.6. The equation (6.1) is also valid for $x \in \mathbb{R} \setminus \{0\}$ and is valid up to time $T_x = \inf\{t : x \in \overline{K}_t\}$. If $T_x < \infty$, then $\lim_{t \to T_x^-} g_t(x) - \sqrt{\kappa}\,B_t = 0$.

REMARK 6.7. If $h_t(z) = g_{r^2 t}(z)$, then h_t is the solution of the initial value problem

$$\dot{h}_t(z) = \frac{2\,r^2}{h_t(z) - r\,\sqrt{\kappa}\,\hat{B}_t}, \quad h_0(z) = z,$$

where $\hat{B}_t$ is a standard Brownian motion. In particular the solution to

$$\dot{g}_t(z) = \frac{(2/\kappa)}{g_t(z) - B_t}, \quad g_0(z) = z,$$

6.1. CHORDAL SLE

is SLE_κ parametrized so that $\text{hcap}(K_t) = (2/\kappa)\,t$. If $g_t(z)$ is SLE_κ, i.e., satisfies (6.1), we will define

$$\hat{g}_t(z) = \frac{g_t(z) - \sqrt{\kappa}\,B_t}{\sqrt{\kappa}}. \tag{6.2}$$

Note that $\hat{g}_t(\gamma(t)) = 0$ and $\hat{g}_t(z)$ satisfies the SDE

$$d[\hat{g}_t(z)] = \frac{2/\kappa}{\hat{g}_t(z)}\,dt + dW_t, \quad \hat{g}_0(z) = z, \tag{6.3}$$

where $W_t = -B_t$ is a standard one-dimensional Brownian motion. In fact, it may be more convenient to *define* SLE_κ as the solution to

$$\dot{\bar{g}}_t(z) = \frac{a}{\bar{g}_t(z) + B_t}, \quad h_0(z) = z,$$

where B_t is a standard Brownian motion (which is the negative of the Brownian motion in the previous expressions) and $a = 2/\kappa$. If $Z_t = Z_t(z) = \bar{g}_t(z) + B_t$, then this equation becomes the stochastic differential equation

$$dZ_t = \frac{a}{Z_t}\,dt + dB_t.$$

We take this approach in Chapter 7.

More generally, suppose $t \mapsto \beta(t)$ is a continuous function, possibly random, but such that $\beta(t)$ is measurable with respect to $\{B_s : 0 \le s \le t\}$. Let

$$Y_t = \int_0^t \beta(s)\,dB_s,$$

and let h_t be the solution of the Loewner equation

$$\dot{h}_t(z) = \frac{2\,\beta(s)^2}{h_t(z) - \sqrt{\kappa}\,Y_t}, \quad h_0(z) = z.$$

Define $a(t)$ by

$$\int_0^{a(t)} \beta(s)^2\,ds = t.$$

Then $\tilde{B}_t = Y_{a(t)}$ is a standard Brownian motion. Let $g_t(z) = h_{a(t)}(z)$. Then

$$\dot{g}_t(z) = \frac{2}{g_t(z) - \sqrt{\kappa}\,\tilde{B}_t}.$$

In other words, h_t is a random time change of SLE_κ.

Let D be a simply connected domain and z, w distinct points on ∂D. Let $F : D \to \mathbb{H}$ be a conformal transformation with $F(z) = 0, F(w) = \infty$. The map F is not unique; however, any other such transformation $\hat{F}$ can be written as rF for some $r > 0$. We *define* chordal SLE_κ in D to be the collection of maps $h_t(z) = F^{-1}[g_t(F(z))]$, where g_t is chordal SLE_κ in $\mathbb{H}$ from 0 to ∞. If we had chosen $\hat{F}$ instead of F we would have $\hat{h}_t(z) = \hat{F}^{-1}[g_t(\hat{F}(z))] = F^{-1}[r^{-1}\,g_t(rF(z))] = F^{-1}[\hat{g}_{t/r^2}(F(z))]$. Hence the definition is independent of the choice of map *up to a time change*. We consider chordal SLE_κ in D connecting z and w as being defined only modulo time change. If γ is the SLE_κ path, then $F^{-1} \circ \gamma$ gives the SLE_κ path in D. We consider this as a measure on unparametrized paths.

In the next sections we will analyze SLE_κ. The arguments are not very difficult, but they require a different way of thinking, at least for probabilists. The main object of interest is the evolution of the random curve $\gamma(t)$ (or perhaps the random hulls K_t), but the dynamics are defined instead for the maps g_t. In order to understand what is happening on the curve, one studies what happens off of the curve!

6.2. Phases

The proof that SLE_κ is generated by a curve γ is difficult. However, if we assume this fact, we can derive some facts about this curve. Recall that the time T_z satisfies $\lim_{t \to T_z^-} \hat{g}_t(z) = 0$ and that $\hat{g}_t(z)$ satisfies the Bessel equation (6.3). For every $s \geq 0$, let γ^s denote the curve $\gamma^s(t) = g_s(\gamma(t+s)) - \sqrt{\kappa} B_s$. Note that the distribution of γ^s is the same as that of γ.

PROPOSITION 6.8.
- If $\kappa \leq 4$, then w.p.1 $T_x = \infty$ for all $x > 0$.
- If $\kappa > 4$, then w.p.1 $T_x < \infty$ for all $x > 0$.
- If $\kappa \geq 8$, then w.p.1 $T_x < T_y$ for all $0 < x < y$.
- If $4 < \kappa < 8$ and $0 < x < y$, then $\mathbf{P}\{T_x = T_y\} > 0$.

PROOF. This is a restatement of Proposition 1.21 with $a = 2/\kappa$. □

PROPOSITION 6.9. If $\kappa \leq 4$, then γ is a simple curve with $\gamma(0, \infty) \subset \mathbb{H}$.

PROOF. The previous proposition and symmetry show that w.p.1, $\gamma(0, \infty) \cap \mathbb{R} = \emptyset$. Hence w.p.1, $\gamma^q(0, \infty) \cap \mathbb{R} = \emptyset$ for each rational q. Suppose there exist $t_1 < t_2$ with $\gamma(t_1) = \gamma(t_2)$. Then for each $q \in (t_1, t_2)$, $\gamma^q(0, \infty) \cap \mathbb{R} \neq \emptyset$. □

PROPOSITION 6.10. If $4 < \kappa < 8$, then with probability one,
$$\bigcup_{t>0} \overline{K_t} = \overline{\mathbb{H}},$$
but $\gamma[0, \infty) \cap \mathbb{H} \neq \mathbb{H}$. Also, $\mathrm{dist}(0, \mathbb{H} \setminus K_t) \to \infty$. In particular, $|\gamma(t)| \to \infty$ as $t \to \infty$.

PROOF. Call a point z *swallowed* if $T_z < \infty$ but z is not contained in the closure of $\cup_{t<T_z} K_t$. Note that if z is swallowed then there is an open disk $\mathcal{B}$ about z such that for $w \in \mathcal{B}$, $T_w = T_z$ and each w is swallowed. Proposition 6.8 shows that with positive probability there is an $x > 1$ with $T_x = T_1$; in fact, using scaling we can see that the probability that some such $x > 1$ exists is one. Also, $\gamma(T_1)$ is the largest real x with $T_x = T_1$. Let $\epsilon = \mathrm{dist}(1, \gamma[0, T_1]) > 0$. Then all z in $\mathbb{H} \cap \mathcal{B}(1, \epsilon)$ are swallowed. This shows that $\gamma[0, \infty) \cap \mathbb{H} \neq \mathbb{H}$. Let T be the first time that both 1 and -1 are swallowed. Then topological considerations show that there is a disk $\mathcal{B}'$ about the origin such that $\mathcal{B}' \cap \mathbb{H} \subset K_T$. In particular, for each $u > 0$ there is an $\epsilon > 0$ such that $\mathbf{P}\{\mathcal{B}(0, \epsilon) \cap \mathbb{H} \subset K_T\} \geq 1 - 2u$, and hence there is a $t = t_{\epsilon, u}$ such that $\mathbf{P}\{\mathcal{B}(0, \epsilon) \cap \mathbb{H} \subset K_t\} \geq 1 - u$. By scaling this inequality holds for all ϵ (for a t depending on ϵ), and this gives the first assertion. □

PROPOSITION 6.11. If $\kappa \geq 8$, then γ is a space-filling curve, i.e., $\gamma[0, \infty) = \overline{\mathbb{H}}$.

PROOF. Note that Proposition 6.8 implies that $x \in \gamma[0, \infty)$ for all $x > 0$ (and hence by symmetry for $x \in \mathbb{R}$). The proof for $z \in \mathbb{H}$ follows from a similar argument, which we delay until Theorem 7.9. □

6.2. PHASES

PROPOSITION 6.12. *W.p.1, for $0 < \kappa \leq 4$,*
$$\lim_{t \to \infty} |\gamma(t)| = \infty.$$

PROOF. Let $b = \liminf_{t \to \infty} |\gamma(t)|$. By scaling, $\mathbf{P}\{b \geq r\}$ is the same for all $r > 0$, so it suffices to prove that $\mathbf{P}\{b > 0\} = 1$. For $r \in (0,1)$, let σ_r be the smallest t such that $|\gamma(t) - 1| \leq r$. It suffices to show that $\inf\{r : \sigma_r < \infty\} > 0$ w.p.1, for then a similar argument shows that w.p.1 the distance between $\gamma^1[0,\infty)$ and the two images of the origin under g_1 is strictly positive. Hence, $\mathbf{P}\{b > 0\} = 1$.

First assume $\kappa < 4$, Then an easy harmonic measure estimate shows that if $\sigma_r < \infty$, then $g_{\sigma_r}(1) - U_{\sigma_r} < cr$. But by Proposition 1.21, we know that $\inf_t g_t(1) - U_t > 0$ for $\kappa < 4$. Hence w.p.1 $\sigma_r = \infty$ for some $r > 0$.

Now assume $\kappa = 4$, and let $X_t = \hat{g}_t(1/4), Y_t = \hat{g}_t(1/2), Z_t = \hat{g}_t(1)$. If it were the case that $\sigma_r < \infty$ for all r one could see (using a Beurling estimate, e.g.) that $[Z_t - X_t]/[Y_t - X_t] \to \infty$. We will now show that w.p.1
$$\lim_{t \to \infty} \frac{Z_t - X_t}{Y_t - X_t} < \infty$$
which will establish the proposition. If $\kappa = 4$, then $a = 2/\kappa = 1/2$ and (see (1.9))
$$d\left[\log \frac{Z_t - X_t}{Z_t}\right] = -\frac{1}{2} \frac{1}{X_t Z_t} dt - \frac{1}{Z_t} dB_t.$$

$$d\left[\log \frac{Z_t - X_t}{Y_t - X_t}\right] = \frac{1}{2} \frac{Z_t - Y_t}{X_t Y_t Z_t} dt.$$

Hence,
$$\lim_{t \to \infty} \log \frac{Z_t - X_t}{Y_t - X_t} = \log \frac{Z_0 - X_0}{Y_0 - X_0} + \frac{1}{2} \int_0^\infty \frac{Z_t - Y_t}{X_t Y_t Z_t} dt.$$

We claim that w.p.1 the last integral is finite. First note that $(Z_t - Y_t)/Z_t \leq (Z_t - X_t)/Z_t \leq R_t$ where R_t is defined by $R_0 = (Z_0 - X_0)/Z_0 = 3/4$ and
$$d[\log R_t] = -\frac{1}{2 Z_t^2} dt - \frac{1}{Z_t} dB_t.$$

If we define the time change $r(t)$ by $\int_0^{r(t)} Z_s^{-2} ds = t$, then $\log R_{r(t)} = \log(3/4) - (t/2) - B_t$. In particular, w.p.1 there is a (random) C_1 such that $R_{r(t)} \leq C_1 e^{-t/4}$ for all t. This implies for all t sufficiently large, $R_t \leq 1/2$, i.e., $X_t \geq Z_t/2$. Therefore it suffices to prove that w.p.1
$$\int_0^\infty R_t Z_t^{-2} dt < \infty.$$

But,
$$\int_0^\infty R_t Z_t^{-2} dt = \sum_{k=0}^\infty \int_{r(k)}^{r(k+1)} R_t Z_t^{-2} dt \leq \sum_{k=0}^\infty C_1 e^{-k/4} < \infty.$$

□

6.3. The locality property for $\kappa = 6$

Suppose $\gamma(t)$ is an SLE_κ curve starting at the origin with corresponding maps g_t, and let $\mathcal{N}$ be an $\mathbb{H}$-neighborhood of the origin as in §4.6. Let $\hat{\mathcal{N}}$ denote the union of $\mathcal{N}$ and all $x \in \mathbb{R}$ such that $\mathcal{N}$ is an $\mathbb{H}$-neighborhood of x. Let $t_0 = \inf\{t : \gamma(t) \notin \hat{\mathcal{N}}\}$. Let Φ be a locally real conformal transformation of $\mathcal{N}$ into $\mathbb{H}$; note that Φ extends by Schwarz reflection to $\{z : z \text{ or } \bar{z} \in \hat{\mathcal{N}}\}$. For $t < t_0$, let $\gamma^*(t) = \Phi \circ \gamma(t)$. In this section, we study the distribution of $\gamma^*(t)$; in particular, we ask if it locally has the same distribution as (a time change of) $\gamma(t)$. The method of analysis is to write a Loewner differential equation for $\gamma^*(t)$, which defines a driving function U_t^*, and then to combine the results of §4.6 with Itô's formula to derive a stochastic differential equation for U_t^*.

Let H_t^* be the unbounded component of $\mathbb{H} \setminus \gamma^*[0,t]$ and let $g_t^* : H_t^* \to \mathbb{H}$ be the unique conformal transformation such that

$$g_t^*(z) = z + \frac{b^*(t)}{z} + \cdots, \quad z \to \infty.$$

Here $b^*(t) = \operatorname{hcap}(\gamma^*(0,t])$. Let $\Phi_t = g_t^* \circ \Phi \circ g_t^{-1}$ as in §4.6.1. The maps g_t^* satisfy the Loewner equation

$$\dot{g}_t^*(z) = \frac{\dot{b}^*(t)}{g_t^*(z) - U_t^*} = \frac{2\,\Phi_t'(U_t)^2}{g_t^*(z) - U_t^*}, \quad g_t^*(z) = z,$$

where $U_t = \sqrt{\kappa}\, B_t$ and $U_t^* = g_t^*(\gamma^*(t)) = \Phi_t(U_t)$. Note that the map Φ_t is random, depending on $B_s, 0 \leq s \leq t$. Proposition 4.40 shows that Φ_t is C^1 in t with $\dot{\Phi}_t(U_t) = -3\,\Phi_t''(U_t)$. Hence Itô's formula (Proposition 1.9) shows that U_t^* satisfies the stochastic differential equation

$$\begin{aligned}
dU_t^* &= \left[\dot{\Phi}_t(U_t) + \frac{\kappa}{2}\,\Phi_t''(U_t)\right] dt + \sqrt{\kappa}\,\Phi_t'(U_t)\,dB_t \\
&= \left[\frac{\kappa}{2} - 3\right] \Phi_t''(U_t)\,dt + \sqrt{\kappa}\,\Phi_t'(U_t)\,dB_t.
\end{aligned}$$

The last equation is easier to interpret if we reparametrize the curve γ^* so that $\operatorname{hcap}[\gamma^*[0,t]] = 2t$. Define the change of time $r(t)$ by

$$t = \int_0^{r(t)} \Phi_s'(U_s)^2\,ds.$$

Then $\tilde{\gamma}^*(t) := \gamma^*(r(t))$ is parametrized by capacity and $\tilde{U}_t^* := U_{r(t)}^*$ satisfies

$$d\tilde{U}_t^* = \left[\frac{\kappa - 6}{2}\right] \frac{\Phi_{r(t)}''(U_{r(t)})}{\Phi_{r(t)}'(U_{r(t)})^2}\,dt + \sqrt{\kappa}\,d\tilde{B}_t,$$

where $\tilde{B}_t := \int_0^{r(t)} \Phi_s'(U_s)\,dB_s$ is a standard Brownian motion. If $\tilde{h}_t(z) = h_{r(t)}(z)$, then

$$\dot{\tilde{h}}_t(z) = \frac{2}{\tilde{h}_t(z) - \tilde{U}_t^*}.$$

In particular, if $\kappa = 6$, $\tilde{U}_t^* = \sqrt{\kappa}\,\tilde{B}_t$ and $\tilde{\gamma}^*$ has the distribution of an SLE_6 curve. In other words, γ^* is a time change of an SLE_6 curve.

THEOREM 6.13 (Locality). *If $\kappa = 6$, then $\gamma^*(s)$ is a time change of SLE_6 stopped at the first time it leaves $\Phi(\hat{\mathcal{N}})$.*

For $\kappa \neq 6$, we see that the image of $\gamma(t)$ is absolutely continuous with respect to a time change of SLE_κ. This follows from the Girsanov transformation, that implies roughly that Brownian motion with drift is absolutely continuous with driftless Brownian motion (assuming the same variance for both).

Recall that chordal SLE_κ is defined in other simply connected domains only up to a time parametrization. For the remainder of this section, we will say that $\gamma(t)$ is an SLE_κ connecting boundary points z and w in a domain D if it is the image of a time change of chordal SLE_κ in $\mathbb{H}$ connecting 0 and ∞.

Suppose $\gamma^*(t)$ is SLE_6 in $\mathbb{H}$ connecting 0 and $x > 0$. We can construct this by $\Phi \circ \gamma(t)$ where $\Phi(z) = zx/(z+x)$ is a Möbius transformation of $\mathbb{H}$ with $\Phi(0) = 0$, $\Phi(\infty) = x$. Let t_0 be the first time t with $\gamma(t) \in [x, \infty)$; since $4 < \kappa < 8$, w.p.1, $t_0 < \infty$ and $x < \gamma(t_0) < \infty$. For $t < t_0$, the curve $\gamma[0,t]$ is contained in a neighborhood $\hat{\mathcal{N}}$ as above ($\hat{\mathcal{N}}$ may depend on the realization), but for $t \geq t_0$ this is not true since any choice of $\hat{\mathcal{N}}$ would have to include x and $\Phi^{-1}(x) = \infty$. Therefore, for $t < t_0$, $\gamma^*(t)$ is a time change of SLE_6. After time t_0 the two curves $\gamma, \tilde{\gamma}$ evolve differently; γ heads toward infinity while γ^* heads toward x. By conformal invariance, we get a similar result for other domains.

PROPOSITION 6.14. *Suppose D is a Jordan domain and z, w, w' are distinct boundary points. Consider chordal SLE_6 from z to w and chordal SLE_6 from z to w', both in D. Then (modulo change of time) the two processes have the same distribution up to the first time they reach the arc of ∂D between w and w' not containing z.*

6.4. The restriction property for $\kappa = 8/3$

In this section we consider SLE_κ for $\kappa \leq 4$ so that the curve $\gamma(t)$ is simple with $\gamma(0, \infty) \subset \mathbb{H}$. Suppose $A \in \mathcal{Q}_\pm$ (see the end of §5.3 for definitions), and let $\Phi_A(z) = g_A(z) - g_A(0)$ be the conformal transformation of $\mathbb{H} \setminus A$ onto $\mathbb{H}$ with $\Phi_A(0) = 0, \Phi_A(z) \sim z$ as $z \to \infty$. It follows from Proposition 6.12 and scaling that $0 < \mathbf{P}\{\gamma[0, \infty) \cap A = \emptyset\} < 1$. On the event $V_A := \{\gamma[0, \infty) \cap A = \emptyset\}$ we can consider the path $\Phi_A \circ \gamma(t)$. We say that SLE_κ satisfies the *restriction property* if the distribution of $\Phi_A \circ \gamma$ conditioned on the event V_A is the same as (a time change of) SLE_κ. We will show that the only $\kappa \leq 4$ for which SLE_κ satisfies the restriction property is $\kappa = 8/3$.

When one considers a *simple* curve from 0 to ∞ modulo time change, then one can specify where the curve visits by specifying those $A \in \mathcal{Q}_\pm$ for which $\gamma[0, \infty) \cap A = \emptyset$. Hence the distribution of SLE_κ up to time change for $\kappa \leq 4$ is given by specifying $\mathbf{P}(V_A)$ for each $A \in \mathcal{Q}_\pm$. By symmetry it suffices to give $\mathbf{P}(V_A)$ for $A \in \mathcal{Q}_+$, and by Lemma 5.19 it suffices to give the value for smooth Jordan hulls in $\mathcal{Q}_+$.

LEMMA 6.15. *Suppose $\kappa \leq 4$ and there is an $\alpha > 0$ such that $\mathbf{P}(V_A) = \Phi'_A(0)^\alpha$ for all $A \in \mathcal{Q}_\pm$. Then SLE_κ satisfies the restriction property.*

PROOF. Suppose $\mathbf{P}(V_A) = \Phi'_A(0)^\alpha$ for all $A \in \mathcal{Q}_\pm$ and let $A, A_1 \in \mathcal{Q}_\pm$. Then

$$\mathbf{P}\{\Phi_A \circ \gamma[0, \infty) \cap A_1 = \emptyset; \gamma[0, \infty) \cap A = \emptyset\} = \mathbf{P}(V_{A \cup \Phi_A^{-1}(A_1)}).$$

But $\Phi_{A \cup \Phi_A^{-1}(A_1)} = \Phi_{A_1} \circ \Phi_A$, so the right hand side is $\Phi'_{A_1}(0)^\alpha \Phi'_A(0)^\alpha$. □

REMARK 6.16. Recall from (5.17) that $\Phi'_A(0)$ is the probability that an $\mathbb{H}$-excursion avoids A. One can also prove a converse of this lemma. See Proposition 9.4.

Fix $\kappa \leq 4$ and let $\gamma(t)$ be an SLE_κ curve. Let $A \in \mathcal{Q}_\pm$, let $\mathcal{F}_t$ denote the filtration of the Brownian motion B_t, and let $\tilde{M}_t = \tilde{M}_{t,A} = \mathbf{E}[1_{V_A} \mid \mathcal{F}_t]$. Then $\tilde{M}_t$ is a bounded martingale with $\tilde{M}_0 = \mathbf{P}(V_A)$ and $\lim_{t \to \infty} \tilde{M}_t = 1_{V_A}$ w.p.1. Let $t_A = \inf\{t : \gamma(t) \in A\}$. The assumptions for SLE_κ (identical increments and Markovian property) imply that

$$(6.4) \qquad \tilde{M}_t = 1\{t < t_A\} \, \mathbf{P}(V_{g_t(A) - g_t(0)}).$$

We claim that if M_t is another bounded martingale with respect to $\mathcal{F}_t$ such that $\lim_{t \to \infty} M_t = 1_{V_A}$ w.p.1, then $\tilde{M}_t = M_t$ w.p.1. This can be seen in a number of ways; for example, $\mathbf{E}[(M_t - \tilde{M}_t)^2]$ increases with t but converges to 0. Hence, if we find a bounded martingale M_t with $\lim_{t \to \infty} M_t = 1_{V_A}$, then $\mathbf{P}(V_A) = M_0$. In view of the previous lemma and (6.4), it is natural to try

$$M_t = 1\{t < t_A\} \, \Phi'_{g_t(A) - g_t(0)}(0)^\alpha = 1\{t < t_A\} \, \Phi'_t(U_t)^\alpha,$$

where $U_t = \sqrt{\kappa} B_t$ and Φ_t is as in the previous section with $\Phi_0 = \Phi = \Phi_A$. From (4.37), we know that

$$\dot{\Phi}'_t(U_t) = \frac{\Phi''_t(U_t)^2}{2 \Phi'_t(U_t)} - \frac{4 \Phi'''_t(U_t)}{3}.$$

Using this we can apply Itô's formula to see that for $t < t_A$,

$$(6.5) \qquad \frac{dM_t}{\alpha M_t} = \left[\frac{(\alpha - 1)\kappa + 1}{2} \frac{\Phi''_t(U_t)^2}{\Phi'_t(U_t)^2} + \left(\frac{\kappa}{2} - \frac{4}{3}\right) \frac{\Phi'''_t(U_t)}{\Phi'_t(U_t)} \right] dt + \frac{\Phi''_t(U_t)}{\Phi'_t(U_t)} \sqrt{\kappa} \, dB_t.$$

THEOREM 6.17. $SLE_{8/3}$ satisfies the restriction property. In fact, if γ is a chordal $SLE_{8/3}$ curve in $\mathbb{H}$ and $A \in \mathcal{Q}_\pm$, then

$$(6.6) \qquad \mathbf{P}\{\gamma[0, \infty) \cap A = \emptyset\} = \Phi'_A(0)^{5/8}.$$

PROOF. Using Lemma 5.19 we can see that it suffices to prove (6.6) for smooth Jordan hulls $A \in \mathcal{Q}_+$. Fix such an A generated by the simple curve $\beta : (0, 1) \to \mathbb{H}$, and let $M_t = 1\{t < t_A\} \Phi'_t(U_t)^{5/8}$. Then (6.5) shows that M_t is a martingale for $0 < t < \tau_A$ with $0 \leq M_t \leq 1$; to show that $M_{t \wedge t_A}$ is a continuous martingale we need to show that $M_{t_A-} = 0$ if $t_A < \infty$. The martingale convergence theorem then tells us that $M_{t \wedge t_A} \to M_\infty$ with probability one for some random variable M_∞ with $0 \leq M_\infty \leq 1$. We need to show that $M_\infty = 1_{V_A}$. Hence to prove the theorem, it suffices to show that, except perhaps on an event of probability zero, $\limsup_{t \to \infty} M_t^{8/5} = 1$ on V_A and $\liminf_{t \to t_A-} M_t^{8/5} = 0$ on $(V_A)^c$. Recall (see (5.17)) that $M_t^{8/5}$ is the probability that a Brownian excursion in $\mathbb{H} \setminus \gamma(0, t]$ from $\gamma(t)$ to infinity avoids A.

By scaling we may assume that $\sup\{\text{Im}(w) : w \in A\} = 1$. Suppose $r > 2$, and let σ_r be the first time that $\text{Im}[\gamma(t)] = r$. Note that w.p.1 $\sigma_r < \infty$ for each r (since, e.g., scaling implies that $\mathbf{P}\{\sigma_r < \infty\}$ is the same for all $r > 0$). Consider a

Brownian excursion in $\mathbb{H} \setminus \gamma(0, \sigma_r]$ going to infinity started at $z = z_\epsilon := \gamma(\sigma_r) + \epsilon i$. Then the probability that the excursion hits A is given by

$$(6.7) \qquad \lim_{R \to \infty} \frac{\mathbf{P}^z\{B[0, \tau_R] \subset \mathbb{H} \setminus \gamma(0, \sigma_r]; B[0, \tau_R] \cap A \neq \emptyset\}}{\mathbf{P}^z\{B[0, \tau_R] \subset \mathbb{H} \setminus \gamma(0, \sigma_r]\}},$$

where B denotes a Brownian motion and τ_R is the first time it reaches $\{\text{Im}(w) = R\}$. Let $\eta = \eta_r$ be the first time that B reaches the square of side length two centered at $\gamma(\sigma_r)$, with sides parallel to the axes. Let l denote the side of this square with imaginary part $r + 1$. Note that $\mathbf{P}^z\{B(\eta) \in l\} > 1/4$ but for all w in the region bounded by the square with $\text{Im}(w) \leq r$, $\mathbf{P}^w\{B(\eta) \in l\} \leq 1/4$. Therefore, by the strong Markov property, $\mathbf{P}^z\{B_\eta \in l \mid B[0, \eta] \cap \gamma(0, \sigma_r] = \emptyset\} \geq 1/4$, and hence, using gambler's ruin, the denominator above is bounded below by $(1/4)\mathbf{P}\{B[0, \eta] \cap \gamma(0, \sigma_r] = \emptyset\}/(R - r)$. The strong Markov property and the Beurling estimates imply that the probability that the Brownian motion starting at z reaches A without hitting $\mathbb{R} \cup \gamma(0, \sigma_r]$ is bounded above by $c\,r^{-1/2}\,\mathbf{P}^z\{B[0, \eta] \cap \gamma(0, \sigma_r] = \emptyset\}$. Starting on A, the probability that a Brownian motion reaches $\{\text{Im}(w) = R\}$ without hitting $\mathbb{R}$ is no more than $1/R$. Therefore, the numerator in (6.7) is bounded above by $c\,r^{-1/2}\,R^{-1}\,\mathbf{P}^z\{B[0, \eta] \cap \gamma(0, \sigma_r] = \emptyset\}$, and hence the limit is bounded by $c\,r^{-1/2}$. By letting $\epsilon \to 0+$, we see that the probability that an $(\mathbb{H} \setminus \gamma(0, \sigma_r])$-excursion starting at $\gamma(\sigma_r)$ hits A is bounded above by $c\,r^{-1/2}$. In particular, $\limsup_{t \to \infty} M_t^{8/5} = 1$ on the event V_A.

We now consider what happens on the event $(V_A)^c$. In this case, except for an event of probability zero, $\gamma(t_A) = \beta(s)$ for some $s \in (0, 1)$. We need to show that $\liminf_{t \to t_A^-} \Phi_t'(U_t) = 0$. We will consider the sequence of times $t^m = \inf\{t : |\gamma(t) - \beta(s)| = 1/m\}$. Since β is smooth at s, there is a $\delta > 0$ such that the line segment $l = [s, s + \delta\mathbf{n}] \subset A$ where $\mathbf{n}$ is the unit normal at $\beta(s)$ pointing into A. Let L_t denote the image of this line segment under Φ_t. Consider a Brownian motion starting on the line l; then there is a positive probability that the first time it visits $\mathbb{R} \cup \gamma[0, t_m]$ will be on the "right" side of $\gamma(t^m)$ (and similarly for "left" side). From this one can see that L_{t^m} lies in a wedge $\{\text{Im}(w) \geq a\,|\text{Re}(w)|\}$. From this we see that the probability that a Brownian excursion hits L_{t^m} goes to one as $m \to \infty$ (see Exercise 5.17). $\square$

REMARK 6.18. The theorem tells us that for bounded A such that $\mathbb{H} \setminus \overline{A}$ is simply connected, then the probability that $SLE_{8/3}$ avoids A is $p(A)^{5/8}$ where $p(A)$ denotes the probability that an $\mathbb{H}$-excursion avoids A. In fact, the assumption of boundedness is not necessary. If $A \subset \mathbb{H}$ with $0 \notin \overline{A}$ and such that $\mathbb{H} \setminus A$ is simply connected, then

$$(6.8) \qquad \mathbf{P}\{\gamma[0, \infty) \cap A = \emptyset\} = p(A)^{5/8}.$$

This follows easily from the theorem by approximation by compact A_n. Note that if such an A satisfies $\sup\{\text{Im}(z) : z \in A\} < \infty$, then $p(A) > 0$ and we can write $p(A) = \Phi_A'(0)$, where

$$\Phi_A(z) = \lim_{n \to \infty} \Phi_{A_n}(z).$$

This limit is well defined "away from infinity"; although $\Phi_{A_n}(z) \sim z$, it is not necessarily true that $\Phi_A(z) \sim z$.

Although the restriction property does not hold for $\kappa \neq 8/3$, we do get an interesting martingale by considering $\Phi_t'(U_t)^\alpha$ with a weighting of the paths. This will be used in Chapter 9 to construct "restriction measures". Recall that the Schwarzian derivative Sf is defined by

$$Sf(z) = \frac{f'''(z)}{f'(z)} - \frac{3}{2}\left[\frac{f''(z)}{f'(z)}\right]^2.$$

The "dt" term in (6.5) can be rewritten as

$$(\frac{\kappa}{2} - \frac{4}{3})\, S\Phi_t(U_t) + \frac{\kappa(2\alpha+1) - 6}{4} \frac{\Phi_t''(U_t)^2}{\Phi_t'(U_t)^2}.$$

Let

(6.9) $\qquad \alpha = \alpha(\kappa) = \dfrac{6-\kappa}{2\kappa}, \quad \lambda = \lambda(\kappa) = (8-3\kappa)\,\alpha = \dfrac{(8-3\kappa)(6-\kappa)}{2\kappa}.$

Then for these values of α, λ, (6.5) becomes

(6.10) $\qquad dM_t = M_t\left[-\dfrac{\lambda}{6} S\Phi_t(U_t)\, dt + \alpha\sqrt{\kappa}\, \dfrac{\Phi_t''(U_t)}{\Phi_t'(U_t)}\, dB_t\right].$

PROPOSITION 6.19. *Suppose $\kappa \leq 4$ and α, λ are defined as in (6.9). Let*

$$Y_t = \Phi_t'(U_t)^\alpha \, \exp\left\{\frac{\lambda}{6} \int_0^t S\Phi_s(U_s)\, ds\right\}.$$

Then Y_t is a local martingale for $t < t_A$. If $\kappa \leq 8/3$, then Y_t is a bounded martingale.

PROOF. Itô's formula and (6.10) show that Y_t is a local martingale. Also $S\Phi_s(U_s)$ is negative (see Proposition 5.22 where $-(1/6)S\Phi_s(U_s)$ gives the measure of a set) and $\lambda \geq 0$ for $\kappa \leq 8/3$. Hence $Y_t \leq 1$ for these κ.

6.5. Radial SLE

DEFINITION 6.20. The *radial Schramm-Loewner evolution with parameter κ (with standard parametrization)*, or *radial SLE_κ*, is the random collection of maps g_t obtained from solving the initial value problem

(6.11) $\qquad \dot{g}_t(z) = g_t(z)\, \dfrac{e^{i\sqrt{\kappa}B_t} + g_t(z)}{e^{i\sqrt{\kappa}B_t} - g_t(z)}, \quad g_0(z) = z \quad (z \in \mathbb{D}),$

where B_t is a standard one-dimensional Brownian motion.

In other words, radial SLE_κ is the solution of the radial Loewner equation as in §4.2 with driving function $U_t = \sqrt{\kappa}\, B_t$. We can give any initial distribution on the Brownian motion. Usually we will assume that $B_0 = x \in \mathbb{R}$ or that U_0 has a uniform distribution on $[0, 2\pi)$. For each $z \in \overline{\mathbb{D}}$, the solution of (6.11) exists up to a time $T_z \in [0, \infty]$ ($T_z = 0$ if $B_0 = x, z = e^{i\sqrt{\kappa}x}$). If $D_t = \{z \in \mathbb{D} : T_z > t\}$, then g_t is the conformal transformation of D_t onto $\mathbb{D}$ with $g_t(0) = 0$ and $g_t'(0) > 0$. In fact, $g_t'(0) = e^t$. Note that if $|z| = 1$, then $|g_t(z)| = 1$ for all $t < T_z$.

Theorem 6.3 holds for radial SLE, i.e., there is a random curve $\gamma : [0, \infty) \to \overline{\mathbb{D}}$ such that D_t is the connected component of $\mathbb{D} \setminus \gamma[0, t]$ containing the origin. In fact, as we will see below, locally (for small times t) the distribution of the curve is absolutely continuous with respect to that obtained by taking the image of chordal SLE_κ under the map $z \mapsto e^{iz}$.

When analyzing radial SLE_κ it is often convenient to think of the maps g_t as mapping a subdomain of $\mathbb{H}$ into $\mathbb{H}$. The map $z \mapsto e^{iz}$ takes $\mathbb{H}$ into $\mathbb{D}$ and is locally

6.5. RADIAL SLE

conformal although it is not one-to-one. If $z \in \mathbb{H}$ with $e^{iz} \in D_t \setminus \{0\}$, we can define $h_t(z) = -i \log g_t(e^{iz})$ at least in a neighborhood of z (or we can consider log as a multi-valued function). Then (6.11) implies that

$$\dot h_t(z) = \cot\left[\frac{h_t(z) - \sqrt{\kappa}\, B_t}{2}\right], \quad h_0(z) = z. \tag{6.12}$$

If $z = x \in \mathbb{R}$, this is a real differential equation for $h_t(x) = \arg g_t(e^{ix})$. If we let $Y_t = Y_t(z) = h_t(z) - \sqrt{\kappa}\, B_t$ and $W_t = -B_t$, then Y_t satisfies the stochastic differential equation

$$dY_t = \cot(Y_t/2)\, dt + \sqrt{\kappa}\, dW_t.$$

If we let $\tilde Y_t = Y_{t/\kappa}$, then $\tilde Y_t$ satisfies

$$d\tilde Y_t = \frac{a}{2} \cot\left[\frac{\tilde Y_t}{2}\right] + d\tilde B_t, \tag{6.13}$$

where $a = 2/\kappa$ and $\tilde B_t := \sqrt{\kappa}\, W_{t/\kappa}$ is a standard Brownian motion. Equivalently, if $\tilde g_t = g_{t/\kappa}$, then $\tilde g_t$ satisfies

$$\dot{\tilde g}_t(z) = (a/2)\, \tilde g_t(z)\, \frac{e^{-i\tilde B_t} - \tilde g_t(z)}{e^{-i\tilde B_t} + \tilde g_t(z)}.$$

This is the equation for radial $SLE_{2/a}$ parametrized so that $\tilde g_t'(0) = e^{at/2}$. The equation (6.12) is valid up to the first time t that $h_t(z) \equiv \sqrt{\kappa} B_t \pmod{2\pi}$; if $z = x \in (\sqrt{\kappa}\, B_0, \sqrt{\kappa}\, B_0 + 2\pi)$, the equation is valid until $Y_t \in \{0, 2\pi\}$. As $u \to 0$, $\cot u = (1/u) - (u/3) + O(u^3)$. Hence near zero, (6.13) looks like (6.3), the Bessel SDE for chordal SLE_κ. From this we can see that $T_x = \infty$ if $\kappa \le 4$ (the $\kappa = 4$ case needs a little more argument which we leave to the reader) and $T_x < \infty$ if $\kappa > 4$. In fact, this indicates that there is a close relationship between chordal SLE_κ and radial SLE_κ for the same value of κ. We now investigate this relationship by starting with chordal SLE_κ, exponentiating, using the radial Loewner equation, and then taking a logarithm to return to $\mathbb{H}$.

It will be convenient to think in terms of curves. Let γ denote a chordal SLE_κ curve with $\gamma(0) = 0$ with corresponding maps g_t and driving function $U_t = \sqrt{\kappa}\, B_t$. Let $f_t = g_t^{-1}$. Let $\mathcal{N}$ be a simply connected $\mathbb{H}$-neighborhood of 0 such that $z \mapsto e^{iz}$ is one-to-one on $\mathcal{N}$ and let t_* be the first time t that $\gamma(t) \notin \hat{\mathcal{N}}$. We will consider $t < t_*$. Let $\eta(t) = \exp\{i\gamma(t)\}$ and let D_t be the connected component of $\mathbb{D} \setminus \eta[0, t]$ containing the origin. Let $\tilde g_t$ be the conformal transformation of D_t onto $\mathbb{D}$ with $\tilde g_t(0) = 0, \tilde g_t'(0) > 0$, and let $h_t(z) = -i \log \tilde g_t(e^{iz})$. Let $\Phi_t = h_t \circ f_t$ which is a conformal transformation of an $\mathbb{H}$-neighborhood of U_t onto an $\mathbb{H}$-neighborhood of $U_t^* := \Phi_t(U_t)$. By scaling (see the remark after Proposition 3.58) we can see that

$$\frac{d}{dt} \log \tilde g_t'(0) = \Phi_t'(U_t)^2, \tag{6.14}$$

and hence h_t satisfies the equation

$$\dot h_t(z) = \Phi_t'(U_t)^2\, \cot\left[\frac{h_t(z) - U_t}{2}\right].$$

Itô's formula gives

$$d[U_t^*] = [\dot\Phi_t(U_t) + (\kappa/2)\, \Phi_t''(U_t)]\, dt + \sqrt{\kappa}\, \Phi_t'(U_t)\, dB_t,$$

This use of Itô's formula uses the fact that Φ_t is C^1 in t at U_t. In fact, from (4.43) we know that $\dot{\Phi}_t(U_t) = -3\Phi''_t(U_t)$. Hence,

$$d[U^*_t] = [(\kappa/2) - 3]\,\Phi''_t(U_t)\,dt + \sqrt{\kappa}\,\Phi'_t(U_t)\,dB_t.$$

As in §6.3, it is useful to reparametrize so that γ^* is parametrized by capacity. Define $r(t)$ by $\int_0^{r(t)} \Phi'_t(U_s)^2\,ds = t$ and let $\tilde{U}^*_t = U^*_{r(t)}$. Then $\tilde{U}^*_t$ satisfies

$$d\tilde{U}^*_t = \left[\frac{\kappa - 6}{2}\right] \frac{\Phi''_{r(t)}(U_{r(t)})}{\Phi'_{r(t)}(U_{r(t)})^2}\,dt + \sqrt{\kappa}\,d\tilde{B}_t,$$

where $\tilde{B}_t := \int_0^{r(t)} \Phi'_s(U_s)\,dB_s$ is a standard Brownian motion. If $\tilde{h}_t = h_{r(t)}$, then

(6.15) $$\dot{\tilde{h}}_t(z) = \cot\left[\frac{\tilde{h}_t(z) - \tilde{U}^*_t}{2}\right].$$

REMARK 6.21. Suppose U_t satisfies $dU_t = R_t\,dt + \sqrt{\kappa}\,dB_t$ where R_t is a continuous, adapted process. The same argument gives

$$dU^*_t = (\frac{\kappa - 6}{2})\,\Phi''_t(U_t)\,dt + \Phi'_t(U_t)\,dU_t$$

(6.16) $$= \left[(\frac{\kappa - 6}{2})\,\Phi''_t(U_t) + R_t\,\Phi'_t(U_t)\right]\,dt + \sqrt{\kappa}\,\Phi'_t(U_t)\,dB_t.$$

If we let $r(t), \tilde{h}(t)$ be defined as in the previous paragraph, then $\tilde{h}_t$ satisfies (6.15) with

$$d\tilde{U}^*_t = \left[(\frac{\kappa - 6}{2})\,\frac{\Phi''_{r(t)}(U_{r(t)})}{\Phi'_{r(t)}(U_{r(t)})^2} + \frac{R_{r(t)}}{\Phi'_{r(t)}(U_t)}\right]\,dt + \sqrt{\kappa}\,dB_t.$$

In particular, if

(6.17) $$dU_t = (\frac{6 - \kappa}{2})\,\frac{\Phi''_t(U_t)}{\Phi'_t(U_t)}\,dt + \sqrt{\kappa}\,dB_t,$$

then $\tilde{U}^*_t$ is a Brownian motion and γ^* is radial SLE_κ.

For $\kappa = 6$, we see that $U_t = \sqrt{6}\,B_t$ implies that $\tilde{U}^*_t = \sqrt{6}\,\tilde{B}_t$. Hence in this case, we get that radial SLE_6 can be obtained from chordal SLE_6 by mapping $\mathbb{H}$ to $\mathbb{D}$ as long as we restrict to a simply connected neighborhood in which the exponential map is one-to-one. We summarize this as a proposition.

PROPOSITION 6.22. *Let γ be a chordal SLE_6 path and let $\gamma^*(t) = \exp\{i\gamma(t)\}$. Let t_* be the first time that $\gamma^*[0, t]$ disconnects 0 from $\partial \mathbb{D}$. Then $\gamma^*(t), 0 \leq t \leq t_*$, has the same distribution as (a time change of) radial SLE_6 up to the first time the radial path disconnects the origin from the unit circle.*

PROPOSITION 6.23. *Suppose D is a simply connected subdomain of $\mathbb{D}$ whose boundary includes the origin and an open arc l of the unit circle containing 1. Let γ be a chordal SLE_6 path from 1 to 0 in D and let $\tilde{\gamma}$ be a radial SLE_6 path from 1 to 0 in $\mathbb{D}$. Then γ and $\tilde{\gamma}$ have the same distribution (modulo time change) up to the first time the paths leave $D \cup l$.*

For $\kappa \neq 6$, the equivalence is not as strong. However, we can see that the chordal SLE_κ paths and radial SLE paths have absolutely continuous distributions provided that we look at the paths in neighborhoods in which the exponential is one-to-one. In particular, "local" almost sure facts about chordal SLE_κ paths also hold for radial SLE_κ paths. Global behavior can be different, however. For example, if $\kappa \leq 4$, chordal SLE_κ paths and radial SLE_κ paths have no double points. However, the image of a chordal SLE_κ path under the exponential map *does* have double points since the curve wraps around the origin and disconnects 0 from $\partial \mathbb{D}$.

Radial SLE_κ can be defined on any simply connected domain D from a boundary point z to an interior point w by conformal transformation. Note that the conformal transformation is unique so we do not need a scaling rule to show this is well defined.

Proposition 6.23 is a kind of locality result relating chordal SLE_6 and radial SLE_6. There is also a locality result for radial SLE_6. Suppose D is a simply connected subdomain of $\mathbb{D}$ containing the origin and such that ∂D contains an open arc l about 1. Let $\Psi_D : D \to \mathbb{D}$ be the unique conformal transformation with $\Psi_D(0) = 0, \Psi'_D(0) > 0$. Suppose γ is a radial SLE_κ path starting at 1 with driving function $U_t = \sqrt{\kappa}\, B_t$ and corresponding maps g_t. Let t_* be the first time t that $\gamma(t) \notin D \cup l$; we will consider $t < t_*$. Define $\Phi = \Phi_D$ by $\Phi(z) = -i \log \Psi_D(e^{iz})$. Let $\gamma^*(t) = \Psi \circ \gamma(t)$ and let g_t^* be the unique conformal transformation of the component of $\mathbb{D} \setminus \gamma^*[0,t]$ containing the origin onto $\mathbb{D}$ with $g_t^*(0) = 0$, $(g_t^*)'(0) = 0$ and let $\Phi_t(z) = -i \log[g_t^* \circ \Psi \circ g_t^{-1}](e^{iz})$.

Let $U_t^* = \Phi_t(U_t)$. Then Itô's formula and (4.40) give

(6.18) $$dU_t^* = [(\kappa/2) - 3]\, \Phi_t''(U_t)\, dt + \sqrt{\kappa}\, \Phi_t'(U_t)\, dB_t.$$

This is a local martingale if and only if $\kappa = 6$.

PROPOSITION 6.24. *Suppose D is a simply connected subdomain of $\mathbb{D}$ containing the origin such that ∂D contains an open arc l containing 1. Let γ be an SLE_6 path in $\mathbb{D}$ from 1 to 0 and γ' an SLE_6 path in D from 1 to 0. Then (modulo time change) γ and γ' have the same distribution up to the first time the paths leave $D \cup l$.*

6.5.1. Radial $SLE_{8/3}$.

Let $\gamma(t)$ be a radial $SLE_{8/3}$ path started at 1 with driving function $U_t = \sqrt{8/3}\, B_t$. This is a simple curve $\gamma : [0, \infty) \to \overline{\mathbb{D}}$ with $\gamma(0) = 1, \gamma(0, \infty) \subset \mathbb{D}, \gamma(t) \to 0$ as $t \to \infty$. We will investigate some properties of this path that are particular to the value $\kappa = 8/3$. The first is a radial version of the restriction property. Let $A \subset \mathbb{D} \setminus \{0\}$ be a set as in §3.5, i.e., such that $A = \overline{A} \cap \mathbb{D}$ and $D_A := \mathbb{D} \setminus A$ is simply connected. Let V_A be the event $\{\gamma[0, \infty) \cap A = \emptyset\}$, and let $\Psi_A = \Psi_{D_A}$ be the conformal transformation of D_A onto $\mathbb{D}$ with $\Psi_A(0) = 0, \Psi'_A(0) > 0$.

REMARK 6.25. Suppose $1 \notin \overline{A}$. Then (see Exercise 2.22)

$$H_{D_A}(0,1) = |\Psi'_A(1)|\, H_\mathbb{D}(0, \Psi_A(1)) = |\Psi'_A(1)|/2\pi,$$

where H denotes the Poisson kernel. From this we can see that $|\Psi'_A(1)|$ is the probability that a Brownian excursion in $\mathbb{D}$ from 0 to 1 avoids A.

THEOREM 6.26. *If γ is a radial $SLE_{8/3}$ path from 1 to 0 in $\mathbb{D}$, and A, Ψ_A are as above with $1 \notin \overline{A}$,*

$$\mathbf{P}\{\gamma(0,\infty) \cap A = \emptyset\} = \Psi'_A(0)^{5/48}\, |\Psi'_A(1)|^{5/8}.$$

LEMMA 6.27. *There is a constant c such that if A, Ψ_A are as in Theorem 6.26 and $d = \text{dist}(0, A)$,*

$$\Psi'_A(0)^{5/48} |\Psi'_A(1)|^{5/8} \leq c \, d^{5/24}.$$

PROOF. Corollary 3.19 implies that $\Psi'_A(0) \leq 4/d$. The strong Markov property gives

$$H_{D_A}(0,1) \leq q \sup_{|z|=1/2} H_{D_A}(z,1) \leq q \sup_{|z|=1/2} H_{\mathbb{D}}(z,1) \leq c\,q.$$

where $q = q_A$ is the probability that a Brownian motion starting at the origin reaches the circle of radius $1/2$ before leaving D_A. The Beurling estimate (Theorem 3.76) shows that $q \leq c\, d^{1/2}$. Since $|\Psi'_A(1)| = 2\pi H_{D_A}(0,1)$, the lemma follows. □

Proof of Theorem 6.26. As in the proof of Theorem 6.17, we let V_A be the event $\{\gamma(0,\infty) \cap A = \emptyset\}$ and let $\tilde{M}_t = \mathbf{P}[V_A \mid \mathcal{F}_t]$. Here $\mathcal{F}_t$ is the filtration of the Brownian motion B_t. Then $\tilde{M}_t$ is a bounded martingale with $\tilde{M}_\infty = \lim_{t \to \infty} \tilde{M}_t = 1_{V_A}$. Hence, if we find another bounded martingale M_t with respect to $\mathcal{F}_t$ such that $\lim_{t \to \infty} M_t = 1_{V_A}$, then $M_t = \tilde{M}_t$, and, in particular, $M_0 = \mathbf{P}(V_A)$.

Let $t_A = \inf\{t > 0 : \gamma(t) \in A\}$. If $t < \tau_A$, let $A_t = g_t(A) \cap \mathbb{D}$ and $\Psi_t = \Psi_{A_t}$. Let

$$M_t = 1\{t < t_A\} \, \Psi'_t(0)^{5/48} \, |\Psi'_t(e^{iU_t})|^{5/8}.$$

By Lemma 6.27, we know that M_t is uniformly bounded.

We will first show that M_t is a martingale for $t < t_A$. Assume $t < t_A$, let $\gamma^*(t) = \Phi_A \circ \gamma(t)$, and let D_t^* be the component of $\mathbb{D} \setminus \gamma^*[0,t]$ containing the origin. Let g_t^* be the conformal transformation of D_t^* onto $\mathbb{D}$ with $g_t^*(0) = 0, (g_t^*)'(0) > 0$. Note that $\Psi_t = g_t^* \circ \Psi \circ g_t^{-1}$ and g_t^* satisfies the Loewner equation

$$\dot{g}_t^*(z) = |\Psi'_t(e^{iU_t})|^2 \, (g_t^*)'(z) \, \frac{g_t^*(z) + e^{iU_t}}{g_t^*(z) - e^{iU_t}}.$$

This uses the fact

$$\frac{d}{dt}[\log \Psi'_{A \cup \gamma[0,t]}(0)] = |\Psi'_t(e^{iU_t})|^2,$$

which can be deduced from the remark after Proposition 3.58. Since $\Psi_{A \cup \gamma[0,t]} = \Psi_t \circ g_t$, $\Psi'_t(0) = e^{-t} \Psi'_{A \cup \gamma[0,t]}(0)$. Using this, we can see that

$$\dot{\Psi}'_t(0) = (|\Psi'_t(e^{iU_t})|^2 - 1) \, \Psi'_t(0).$$

Let $Y_t = |\Psi'_t(e^{iU_t})|$. The product rule gives

$$dM_t = M_t \left[\frac{5}{48} (Y_t^2 - 1)\, dt + Y_t^{-5/8} d(Y_t^{5/8}) \right].$$

Let $\Phi_t(z) = -i \log \Psi_t(e^{iz})$. Note that $|\Phi'_t(U_t)| = |\Psi'_t(e^{iU_t})|$ and since $\Phi'_t(U_t) > 0$, $Y_t = \Phi'_t(U_t)$. Itô's formula gives

$$\begin{aligned} dY_t &= [\dot{\Phi}'_t(U_t) + (1/2)(8/3)\Phi'''_t(U_t)]\, dt + \sqrt{8/3}\, \Phi''_t(U_t)\, dB_t \\ &= \left[\frac{\Phi''_t(U_t)^2}{2Y_t} + \frac{Y_t - Y_t^3}{6} \right] dt + \sqrt{8/3}\, \Phi''_t(U_t)\, dB_t. \end{aligned}$$

The second inequality uses (4.42). Therefore,

$$\frac{d[Y_t^{5/8}]}{Y_t^{5/8}} = \frac{5(1-Y_t^2)}{48} dt + \frac{5}{48} \sqrt{8/3} \, \frac{\Phi_t''(U_t)}{Y_t} \, dB_t.$$

$$d[M_t] = M_t \, \frac{5}{48} \sqrt{8/3} \, \frac{\Phi_t''(U_t)}{Y_t} \, dB_t.$$

Therefore, M_t is a martingale.

The proof that $M_\infty = 1$ on V_A and $M_{t_A-} = 0$ on $(V_A)^c$ is very similar to that of Proposition 6.17 so we omit it. In proving the second fact, we use Corollary 3.19 to see that $\lim_{t \to t_A-} \Psi_t'(0) \leq 4/\mathrm{dist}(0, A \cup \gamma[0,t_A])$. Hence it suffices to show that $\lim_{t \to t_A-} \Psi_t'(U_t) = 0$. □

In the remainder of this subsection, we want to show that radial $SLE_{8/3}$ can be obtained from chordal $SLE_{8/3}$ by "conditioning to avoid a disconnection". Let $\gamma : [0, \infty) \to \overline{\mathbb{H}}$ be a chordal $SLE_{8/3}$ curve, and let $\gamma^*(t) = \exp\{i\gamma(t)\}$. For each $t \geq 0$, let

$$J_t = \{z : z = \gamma(s) + 2\pi i k \text{ for some } 0 \leq s \leq t, \, k \in \mathbb{Z} \setminus \{0\}\}.$$

In other words, J_t is the union of all the $2\pi i k$ translates of $\gamma[0,t]$ other than the trivial ($k=0$) translation. Let t_* be the first time t that $\gamma[0,t] \cap J_t \neq \emptyset$, or equivalently, the first time t that $\gamma^*(t) \in \gamma^*[0,t)$. We call t_* the "disconnection time" since it is the first time at which 0 is not connected to $\partial \mathbb{D}$ in $\mathbb{D} \setminus \gamma^*[0,t]$. Let $S_t = \mathbf{P}\{\gamma[0,\infty) \cap J_t = \emptyset \mid \mathcal{F}_t\}$. The remark after the proof of Theorem 6.17 shows that

$$S_t = 1\{t_* > t\} \, \Phi_t'(U_t)^{5/8},$$

where $\Phi_t = \Phi_{g_t(J_t)}$. Although $g_t(J_t)$ is not a bounded set, Φ_t can be defined by an appropriate limit. In fact, Φ_t is the same as the Φ_t in §4.6.3. From (4.45), we know that

$$\dot{\Phi}_t'(U_t) = \frac{\Phi_t''(U_t)^2}{2\,\Phi_t'(U_t)} - \frac{4\,\Phi_t'''(U_t)}{3} - \frac{\Phi_t'(U_t)^3}{6}.$$

Itô's formula then gives for $t < t_*$,

$$d[\Phi_t'(U_t)] = \left[\frac{\Phi_t''(U_t)^2}{2\,\Phi_t'(U_t)} - \frac{\Phi_t'(U_t)^3}{6}\right] dt + \sqrt{8/3}\, \Phi_t''(U_t)\, dB_t,$$

(6.19) $\quad d[\Phi_t'(U_t)^{5/8}] = \Phi_t'(U_t)^{5/8} \left[-\frac{5\,\Phi_t'(U_t)^2}{48} dt + \frac{5}{8}\sqrt{8/3}\,\frac{\Phi_t''(U_t)}{\Phi_t'(U_t)} dB_t\right].$

In particular, if

$$M_t = 1\{t < t_*\} \exp\left\{\frac{5}{48} \int_0^t \Phi_s'(U_s)^2 \, ds\right\} \Phi_t'(U_t)^{5/8},$$

then M_t is a martingale with

$$dM_t = M_t \, \frac{5}{8} \sqrt{8/3} \, \frac{\Phi_t''(U_t)}{\Phi_t'(U_t)} \, dB_t.$$

(Itô's formula shows it is a local martingale and the bound $M_t \leq e^{5t/48}$ implies that it is, in fact, a martingale.)

Let $\mathbf{Q}_t$ be the measure whose Radon-Nikodym derivative with respect to $\mathbf{P}$ is M_t and let $\mathbf{Q}$ be as in §1.9. Proposition 1.15 shows that with respect to $\mathbf{Q}$, B_t satisfies

$$dB_t = \frac{5}{8} \sqrt{8/3} \frac{\Phi_t''(U_t)}{\Phi_t'(U_t)} \, dt + d\tilde{B}_t,$$

where $\tilde{B}_t$ is a standard Brownian motion. Hence,

$$dU_t = \frac{5}{3} \frac{\Phi_t''(U_t)}{\Phi_t'(U_t)} \, dt + \sqrt{8/3} \, d\tilde{B}_t.$$

Using (6.17), we see that with respect to $\mathbf{Q}$, $\tilde{U}_t^*$ is a time change of Brownian motion, i.e., γ^* is a time change of radial $SLE_{8/3}$.

6.6. Whole-plane SLE_κ

Let $\kappa > 0$ and $B_t, -\infty < t < \infty$ be a two-sided Brownian motion such that $\sqrt{\kappa} B_0$ is uniformly distributed on $[0, 2\pi)$. This can be constructed by taking independent Brownian motions B_t^1, B_t^2 starting at the origin and an independent Y, uniform on $[0, 2\pi/\sqrt{\kappa})$ and setting $B_t = Y + B_t^1, t \geq 0$ and $B_t = Y + B_{-t}^2, t \leq 0$.

DEFINITION 6.28. *Whole-plane SLE_κ (from 0 to infinity)* is the family of conformal maps g_t satisfying

(6.20) $$\dot{g}_t(z) = g_t(z) \frac{e^{-iU_t} + g_t(z)}{e^{-iU_t} - g_t(z)}$$

with $U_t = \sqrt{\kappa} B_t$ and the initial condition,

$$\lim_{t \to -\infty} e^t g_t(z) = z, \quad z \in \mathbb{C} \setminus \{0\}.$$

It was shown in §4.3 that such a family of maps is well defined. As was noted there, the whole-plane Loewner equation is really the radial Loewner equation started at time $t = -\infty$. There is a curve $\gamma : (-\infty, \infty) \to \mathbb{C}$ with $\lim_{t \to -\infty} \gamma(t) = 0, \lim_{t \to \infty} \gamma(t) = \infty$ such that for each t, g_t is the conformal transformation of H_t onto $\mathbb{C} \setminus \overline{\mathbb{D}}$ with $g_t(z) \sim e^{-t} z$ as $z \to \infty$. Here H_t is the unbounded component of $\mathbb{C} \setminus \gamma[-\infty, t]$ and we write $\gamma(-\infty) = 0$. This curve is also called a whole-plane SLE_κ path. We let $K_t = \mathbb{C} \setminus H_t$; then $K_t \in \mathcal{H}$, where $\mathcal{H}$ is defined as in §3.2.

LEMMA 6.29 (Scaling). *Suppose g_t is whole-plane SLE_κ, $r \in \mathbb{R}$, and $\tilde{g}_t(z) = g_{t+r}(e^r z)$. Then $\tilde{g}_t$ has the distribution of whole-plane SLE_κ. Equivalently, if γ is a whole-plane SLE_κ path and $\tilde{\gamma}(t) = e^{-r}\gamma(t+r)$, then $\tilde{\gamma}$ has the distribution of a whole-plane SLE_κ path.*

PROOF. This is an easy consequence of (6.20) and the fact that e^{-iU_t} is a stationary process. $\square$

If z, w are distinct points in $\mathbb{C}$ we can define whole-plane SLE_κ connecting z and w by conformal transformation using a linear fractional transformation sending 0 to z and ∞ to w. The scaling rule shows that the distribution, at least up to time change, is independent of the choice of transformation. As in the chordal case, we consider this distribution as being defined modulo a change of time.

The locality property for chordal and radial SLE_6 immediately translates into a locality property for whole-plane SLE_6. We state one version here. Remember that whole-plane SLE_6 from 0 to w is defined only up to change of time.

PROPOSITION 6.30 (Locality). *Suppose γ is a whole-plane SLE_6 curve connecting 0 and ∞ and suppose $w \in \mathbb{C} \setminus \{0\}$. Let t_* be the first time t that $\gamma[0,t]$ disconnects w from ∞. Then $\gamma(t), 0 \le t \le t_*$ has the distribution of a whole-plane SLE_6 path from 0 to w stopped at the first time that it disconnects w from ∞.*

PROPOSITION 6.31. *Let D be a simply connected domain other than $\mathbb{C}$ containing the origin. Let γ be a whole-plane SLE_6 path (from 0 to ∞) and let*

$$\sigma_D = \inf\{t : \gamma(t) \in \partial D\}.$$

Let W_t denote a complex Brownian motion starting at the origin and let

$$\tau_D = \inf\{W_t \in \partial D\}.$$

Then $\gamma(\sigma_D)$ and $W(\tau_D)$ have the same distribution, i.e., the measure on ∂D induced by $\gamma(\sigma_D)$ is harmonic measure in D started at 0.

PROOF. Harmonic measure in D started at 0 can be characterized as the family of probability distributions that are conformally invariant in the sense: if $f : D \to D'$ is a conformal transformation with $f(0) = 0$, then $\mathbf{P}\{W(\tau_D) \in V\} = \mathbf{P}\{W(\tau_{D'}) \in f(V)\}$. Let $D \ne \mathbb{C}$ be a simply connected domain containing the origin, and let $w \in \partial D$. By the locality property, the distribution of $\gamma(\sigma_D)$ is not changed if we let γ be an SLE_6 path from 0 to w. From this we can see that that the distribution of $\gamma(\sigma_D)$ is a conformal invariant, i.e., if $f : D \to D'$ is a conformal transformation with $f(0) = 0$, then $\mathbf{P}\{\gamma(\sigma_D) \in V\} = \mathbf{P}\{\gamma(\sigma_{D'}) \in f(V)\}$. □

Let K_t denote the hull generated by whole-plane SLE_6. If W_t is a complex Brownian motion starting at the origin, let $\hat{K}_t$ denote the hull, i.e., the complement of the unbounded component of $\mathbb{C} \setminus W[0,t]$. The next proposition shows that the hulls of SLE_6 are very closely related to the hulls of Brownian motion (although the curves γ and W are significantly different).

PROPOSITION 6.32. *The distributions of $K_{\sigma_\mathbb{D}}$ and $\hat{K}_{\tau_\mathbb{D}}$ are the same, where $\sigma_\mathbb{D}, \tau_\mathbb{D}$ are as in Proposition 6.31.*

PROOF. To give the distribution of a hull $K \in \mathcal{H}$ containing 0 and such that $K \cap \{|w| \ge 1\}$ is a single point, one needs to give the probability that K lies in D for every simply connected $D \subset \mathbb{D}$. But these are exactly the probability that $\gamma(\sigma_D) \in \partial \mathbb{D}$ and the probability that $\gamma(\tau_D) \in \partial \mathbb{D}$, respectively. In both cases this is the harmonic measure (in D from 0) of $\partial \mathbb{D}$. □

6.7. Cardy's formula

In this section we compute some crossing probabilities for SLE_κ for $\kappa > 4$. These are generalizations of a formula that Cardy derived [20, 21] for percolation using (nonrigorous) conformal field theory techniques. Let $\gamma(t)$ be a chordal SLE_κ ($\kappa > 4$) path and $x, y > 0$. We will first compute the probability that $\gamma[0, T_x]$ has "swallowed" $-y$, i.e., $\mathbf{P}\{T_{-y} < T_x\}$. Geometrically, the event $\{T_{-y} > T_x\}$ corresponds to the event that $[-y, 0]$ is connected to $(0, \infty)$ by a path that avoids $\gamma[0, T_x]$. By scaling and symmetry, $\mathbf{P}\{T_{-y} > T_x\} = \mathbf{P}\{T_y > T_{-x}\} = \mathbf{P}\{T_{-y/x} > T_1\}$. The probability is given in terms of a hypergeometric function $F(\alpha, \beta, \gamma; z)$; see Appendix B for definitions.

PROPOSITION 6.33. *If $y > 0$ and γ is an SLE_κ curve with $4 < \kappa$, then*
$$\mathbf{P}\{T_{-y} > T_1\}$$
$$= \frac{\Gamma(2-4a)}{\Gamma(2-2a)\,\Gamma(1-2a)} \left(\frac{y}{y+1}\right)^{1-2a} F\left(2a, 1-2a, 2-2a; \frac{y}{y+1}\right)$$
$$= \frac{\Gamma(2-4a)}{\Gamma(1-2a)^2} \int_0^{\frac{y}{y+1}} \frac{du}{u^{2a}\,(1-u)^{2a}},$$
where $a = 2/\kappa$.

PROOF. Fix $y > 0$ and let
$$X_t = \hat{g}_t(1), \quad Y_t = \hat{g}_t(-y), \quad Z_t = \frac{X_t}{X_t - Y_t},$$
where $\hat{g}_t$ is as defined in (6.2). Let σ be the first time t that $Z_t \in \{0, 1\}$. Then $\mathbf{P}\{T_{-y} > T_1\} = \mathbf{P}\{Z_\sigma = 1\} = \psi(y/(y+1))$ where
$$\psi(r) = \mathbf{P}\{Z_\sigma = 1 \mid Z_0 = r\} = \mathbf{P}\{Z_\sigma = 0 \mid Z_0 = 1 - r\}.$$
The last equality uses symmetry about the imaginary axis. Recall that
$$dX_t = \frac{a}{X_t}\,dt + dB_t, \quad d[X_t - Y_t] = \left[\frac{a}{X_t} - \frac{a}{Y_t}\right] dt = -\frac{a\,(X_t - Y_t)}{X_t Y_t}\,dt,$$
where B_t is a standard Brownian motion. The product rule gives
$$dZ_t = \left[\frac{a}{X_t\,(X_t - Y_t)} + \frac{a}{Y_t\,(X_t - Y_t)}\right] dt + \frac{1}{X_t - Y_t}\,dB_t$$
$$= \frac{a}{(X_t - Y_t)^2} \left[\frac{1}{Z_t} + \frac{1}{Z_t - 1}\right] dt + \frac{1}{X_t - Y_t}\,dB_t.$$
Let $r(t)$ denote the time change given by
$$\int_0^{r(t)} \frac{ds}{(X_s - Y_s)^2} = t,$$
and let $\tilde{Z}_t = Z_{r(t)}$. Then $\tilde{Z}_t$ satisfies
$$d\tilde{Z}_t = \left[\frac{a}{\tilde{Z}_t} - \frac{a}{1 - \tilde{Z}_t}\right] dt + d\tilde{B}_t,$$
where $\tilde{B}_t = \int_0^{r(t)} (X_s - Y_s)^{-1}\,dB_s$ is a standard Brownian motion. Note that $\psi(\tilde{Z}_t) = \mathbf{E}[\tilde{Z}_\infty \mid \tilde{\mathcal{F}}_t]$ is a martingale, where $\tilde{\mathcal{F}}_t$ denotes the filtration of $\tilde{Z}_t$. Hence, by Itô's formula, ψ must satisfy[2]

(6.21) $$\psi''(u) + 2a \left[\frac{1}{u} - \frac{1}{1-u}\right] \psi'(u) = 0,$$

$$u\,(1-u)\,\psi''(u) + [2a - 4au]\,\psi'(u) = 0.$$

Clearly, constants satisfy this equation, and by solving the first-order equation for ψ', we get a second solution,
$$\psi'(u) = \frac{1}{u^{2a}\,(1-u)^{2a}}.$$

---

[2] This use of Itô's formula assumes that ψ is C^2. However, once we have a solution to (6.21), we can dispense with this assumption; see the first paragraph of the proof of Proposition 1.21. We will leave out this detail in a number of the arguments in this chapter.

We can also view (6.21) as the hypergeometric equation (B.3) with $\alpha = 4a - 1, \beta = 0, \gamma = 2a \in (0,1)$. A nonconstant solution to this equation (see (B.4)) is $u^{1-2a} F(2a, 1-2a, 2-2a; u)$. Since $\psi(0) = 0$ and $\psi(1) = 1$, (B.6) gives the first equality. The second equality follows from

$$\int_0^1 \frac{du}{u^{2a}(1-u)^{2a}} = \frac{\Gamma(1-2a)^2}{\Gamma(2-4a)},$$

see (B.2). □

Let
$$t_* = \inf\{t : \gamma(t) \in [1, \infty)\}.$$

If $\kappa \leq 4$, then $t_* = \infty$ w.p.1, and if $\kappa \geq 8$, $\gamma(t_*) = 1$ w.p.1. If $4 < \kappa < 8$, $\gamma(t_*)$ has a nontrivial distribution which we now compute.

PROPOSITION 6.34. Suppose γ is a chordal SLE_κ curve with $4 < \kappa < 8$, and let $t_* = \inf\{t : \gamma(t) \in [1, \infty)\}$. Then

$$(6.22) \quad \mathbf{P}\{\gamma(t_*) < 1+x\} = \frac{\Gamma(2a)}{\Gamma(4a-1)\Gamma(1-2a)} \int_0^{\frac{x}{1+x}} \frac{du}{u^{2-4a}(1-u)^{2a}},$$

where $a = 2/\kappa$.

PROOF. Let $x > 0$, and let $Y_t = \hat{g}_t(1+x), X_t = \hat{g}_t(1)$. The event $\{\gamma(t_*) < 1+x\}$ is the same as the event $\{T_{1+x} > T_1\}$ which is the same as the event that $R_t := [Y_t - X_t]/X_t$ goes to infinity as $t \to T_1-$, see (1.8). Since

$$dY_t = \frac{a}{Y_t} dt + dB_t, \quad dX_t = \frac{a}{X_t} dt + dB_t,$$

an Itô's formula calculation gives

$$dR_t = \left[\frac{R_t}{X_t^2} - \frac{a R_t}{(R_t+1) X_t^2} - \frac{a R_t}{X_t^2}\right] dt - \frac{R_t}{X_t} dB_t.$$

This equation is easier to understand if we change time so that the quadratic variation grows linearly, i.e., if we define $r(t), \tilde{B}_t$ by

$$\int_0^{r(t)} \left(\frac{R_s}{X_s}\right)^2 ds = t, \quad \tilde{B}_t = -\int_0^{r(t)} \frac{R_s}{X_s} dB_s.$$

Under this time change, $\tilde{R}_t := R_{r(t)}$ satisfies

$$d\tilde{R}_t = \left[\frac{1-2a}{\tilde{R}_t} + \frac{a}{\tilde{R}_t+1}\right] dt + d\tilde{B}_t,$$

and $\tilde{B}_t$ is a standard Brownian motion. If $\psi(x) = \mathbf{P}\{T_{1+x} > T_1\}$, then scaling shows that $\psi(\tilde{R}_t)$ is a martingale, and hence by Itô's formula,

$$\frac{1}{2}\psi''(x) + \left[\frac{1-2a}{x} + \frac{a}{x+1}\right]\psi'(x) = 0.$$

By considering this as a first-order equation in ψ' we obtain two linearly independent solutions: $\psi \equiv 1$ and

$$\psi(x) = \int_0^x \frac{dy}{y^{2-4a}(1+y)^{2a}} = \int_0^{\frac{x}{1+x}} \frac{du}{(1-u)^{2a} u^{2-4a}}.$$

We also know from Proposition 1.21 that $\psi(0) = 0, \psi(\infty) = 1$. Using (B.5), we get (6.22). □

If $\kappa = 6$ the two distributions in the above propositions agree. We therefore get the following corollary (recalling that the distribution of SLE_κ is symmetric about the imaginary axis).

COROLLARY 6.35. *Suppose γ is a chordal SLE_6 curve from 0 to ∞. Define the random variables*

$$Y = \sup\{x : T_x \leq T_{-1}\} = \inf\{x : x \in \partial K_{T_{-1}}\},$$

$$Z = \gamma(T_1) - 1 = \sup\{x : T_x = T_1\} - 1.$$

Then Y and Z have the same distribution,

$$\mathbf{P}\{Y \leq x\} = \mathbf{P}\{Z \leq x\} = \frac{\Gamma(2/3)}{\Gamma(1/3)^2} \int_0^{\frac{x}{1+x}} \frac{du}{u^{2/3}(1-u)^{2/3}}.$$

The following proposition calculates a "crossing exponent" for SLE_6 and will be used in the next chapter. It shows that the exponent for crossing a rectangle is the same for SLE_6 as for a Brownian excursion. This should not be surprising given the relationship between (whole-plane) SLE_6 and Brownian motion discussed in the previous section. Recall that $\mathcal{R}_L$ denotes the rectangular domain

$$\mathcal{R}_L = \{x + iy : 0 < x < L, 0 < y < \pi\},$$

with vertical boundaries $\partial_1 = [0, i\pi], \partial_2 = [L, L + i\pi]$ and horizontal boundaries $\partial_3 = [0, L], \partial_4 = [i\pi, L + i\pi]$.

PROPOSITION 6.36. *For every $0 \leq \theta < \pi/2$, there exist c_1, c_2 such that the following holds. Suppose $|y - (\pi/2)| \leq \theta$, and let γ be a chordal SLE_6 curve in $\mathcal{R}_L$ from iy to a point in $\partial \mathcal{R}_L \setminus \partial_1$. Then the probability that the curve reaches ∂_2 before hitting either of the horizontal boundaries is bounded between $c_1 e^{-L}$ and $c_2 e^{-L}$.*

PROOF. We fix θ, assume $|y - (\pi/2)| < \theta$, and allow constants to depend on θ. Without loss of generality, we may assume $L \geq 1$. Let $q_j, j = 3, 4$, be defined by

$$q_j = \lim_{\epsilon \to 0+} \epsilon^{-1} \mathbf{P}^{\epsilon + iy}\{B_\tau \in \partial_j\}.$$

Here B is a complex Brownian motion and $\tau = \tau_{\mathcal{R}_L}$ is the first time that the B leaves $\mathcal{R}_L$. A straightforward argument shows that the limit exists and that there exists $0 < q_- < q_+ < 1$ such that

(6.23) $$q_- \leq q_j \leq q_+.$$

The locality property for SLE_6 tells us that the probability we are interested in does not depend on which terminal point in $\partial \mathcal{R}_L \setminus \partial_1$ we choose. Let E_1, E_2 be the events

$$E_1 = \{\gamma \text{ hits } \partial_2 \cup \partial_3 \text{ before } \partial_4\},$$

$$E_2 = \{\gamma \text{ hits } \partial_3 \text{ before } \partial_2 \cup \partial_4\}.$$

Then $E_2 \subset E_1$ and the probability we are interested in is $\mathbf{P}(E_1) - \mathbf{P}(E_2)$. Let f_1 be a conformal transformation of $\mathcal{R}_L$ onto $\mathbb{H}$ with $f_1(iy) = 0, |f_1'(iy)| = 1, f_1(L) = \infty$. Let f_2 be the conformal transformation with $f_2(iy) = 0, |f_2'(iy)| = 1, f_2(L + i\pi) = \infty$. Let $x_{+,j} = f_j(0), x_{-,j} = -f_j(i\pi), r_j = x_{-,j}/x_{+,j}$. Then

$$\mathbf{P}(E_j) = \frac{\Gamma(2/3)}{\Gamma(1/3)^2} \int_{\frac{r_j}{r_j + 1}}^1 \frac{du}{u^{2/3}(1-u)^{2/3}}.$$

Using conformal invariance and (6.23), we can see that $r_j/(r_j+1)$ is bounded away from 0 and 1. Hence,
$$\mathbf{P}(E_1) - \mathbf{P}(E_2) \asymp \left[\frac{r_1}{1+r_1} - \frac{r_2}{1+r_2}\right].$$
But $r_j/(1+r_j)$ is the probability that a Brownian excursion in $\mathbb{H}$ starting at 0 exits the upper half plane at $(-\infty, -x_{j,-})$ given that it exits at $(-\infty, -x_{j,-}] \cup [x_{j,+}, \infty)$. By considering $\mathcal{R}_L$ we can see that $\mathbf{P}(E_1) - \mathbf{P}(E_2)$ is comparable to the probability that a Brownian excursion in $\mathcal{R}_L$ starting at iy exits $\mathcal{R}_L$ at ∂_2 given that it exits at $\partial_2 \cup \partial_3 \cup \partial_4$. We have seen that this is comparable to e^{-L}, see (2.10). $\square$

6.8. SLE_6 in an equilateral triangle

In this section, we will consider only SLE_6. It is useful to map $\mathbb{H}$ to the Jordan domain D whose boundary is the equilateral triangle with vertices $v_1 = 0, v_2 = 1$, and $v_3 = e^{i\pi/3}$. Such a map can be found explicitly; it is an example of what are called Schwarz-Christoffel transformations. One can check that
$$f(z) = \frac{\Gamma(2/3)}{\Gamma(1/3)^2} \int_0^z \frac{dw}{w^{2/3}(1-w)^{2/3}}$$
maps $\mathbb{H}$ conformally onto D with $f(0) = v_1, f(1) = v_2, f(\infty) = v_3$. A little care is needed in taking the roots in the expression above. The branches are taken so that
$$f(-x) = e^{i\pi/3} \frac{\Gamma(2/3)}{\Gamma(1/3)^2} \int_0^{\frac{x}{1+x}} \frac{du}{u^{2/3}(1-u)^{2/3}}, \quad 0 < x < \infty,$$
$$f(x) = \frac{\Gamma(2/3)}{\Gamma(1/3)^2} \int_0^x \frac{du}{u^{2/3}(1-u)^{2/3}}, \quad 0 \leq x \leq 1,$$
$$f(1+x) = 1 + e^{i2\pi/3} \frac{\Gamma(2/3)}{\Gamma(1/3)^2} \int_0^{\frac{x}{1+x}} \frac{du}{u^{2/3}(1-u)^{2/3}}, \quad 0 < x < \infty.$$

Corollary 6.35 implies the following proposition. Recall that the locality property implies that SLE_6 in D from v_1 to v_3 and SLE_6 in D from v_1 to v_2 have the same distribution up to the first time the curve hits the edge $[v_2, v_3]$.

PROPOSITION 6.37. *Let γ be an SLE_6 curve in D connecting v_1 and v_2. Let l^* denote the line segment $[v_2, v_3]$, and for $0 < r < 1$, let l_r denote the line segment $[v_1, re^{i\pi/3}]$. Let t_* be the first time t that $\gamma(t) \in l^*$. Then*

- *The probability that l_r is connected to l^* by a path avoiding $\gamma[0, \infty)$ is r.*
- *The distribution of $\gamma(t_*)$ is uniform on l^*.*

We now suppose γ is an SLE_6 curve in D from v_3 to v_2, and we let t_* be the first time t with $\gamma(t) \in [v_1, v_2] = [0, 1]$. Let V denote the "hull observed from v_1" obtained by filling everything to the "right" of $\gamma[0, t_*]$ in D. More precisely, V is the closure of $D \setminus U$ where U is the connected component of $D \setminus \gamma[0, t_*]$ whose boundary includes v_1. The next proposition describes the distribution of V.

PROPOSITION 6.38. *Let γ be an SLE_6 path in the equilateral triangle D from v_3 to v_2. Suppose $\eta : [0, 1] \to \overline{D}$ is a simple curve with $\eta(0) \in (v_1, v_3), \eta(1) \in (v_1, v_2)$ and $\eta(0, 1) \in D$. Let D' be the connected component of $D \setminus \eta(0, 1)$ whose boundary includes v_3. Then*
$$\mathbf{P}\{\gamma[0, t_*] \subset \overline{D'}\} = 1 - f(\eta(1)),$$

where $f : D' \to D$ is the conformal transformation with $f(v_2) = v_2, f(v_3) = v_3$, $f(\eta(0)) = v_1$.

PROOF. The locality property tells us that $f \circ \gamma[0, t_*]$ is an SLE_6 path in D from v_3 to v_2 stopped at the first time it reaches $[v_1, v_2]$. The event $\{\gamma[0, t_*] \subset \overline{D'}\}$ is the same as $\{\gamma(t_*) \in [\eta(1), v_2]\}$ which is the same as the event that $f \circ \gamma$ hits $[v_1, v_2]$ on $[f(\eta(1)), v_2]$. But we know the hitting distribution of $[v_1, v_2]$ by $f \circ \gamma$ is uniform. □

For the remainder of this section, let D be the infinite 60° wedge

$$D = \{re^{i\theta} : r > 0, \; \frac{\pi}{3} < \theta < \frac{2\pi}{3}\},$$

and let γ be an SLE_6 path in D from 0 to ∞. Let $\sigma_s = \inf\{t : \gamma(t) \in \mathcal{I}_s\}$ where $\mathcal{I}_s = \{z : \text{Im}(z) = s\}$, and let D_s denote the equilateral triangular domain $D \cap \{\text{Im}(z) < s\}$. The locality property and Proposition 6.37 imply that the distribution of $\gamma(\sigma_s)$ is uniform on the line segment $\mathcal{I}_s \cap D$. Let V_s denote the hull generated by $\gamma[0, \sigma_s]$ viewed by the vertex $(2/\sqrt{3})\, s\, e^{i\pi/3}$, i.e., $V_s = \overline{D_s \setminus U}$ where U is the connected component of $D_s \setminus \gamma[0, \sigma_s]$ whose boundary includes $(2/\sqrt{3})\, s\, e^{i\pi/3}$. Proposition 6.38 gives the distribution of V_s. We will now describe an analogue of Proposition 6.32, i.e., we will find a Brownian motion that generates a hull with the same distribution as V_s.

Let W_t be a Brownian motion in D that is reflected on ∂D in a particular way that we now describe. We do not reflect in the normal direction (angle $\pi/2$), but rather we do an "oblique" reflection. On the line $l_1 := \{\arg(z) = \pi/3\}$ we reflect at angle $\pi/3$, and on the line $l_2 := \{\arg(z) = 2\pi/3\}$ we reflect at angle $2\pi/3$. It is not difficult to show that this process is well defined away from the origin, and since the reflection angles push the process "away" from the origin, it is not difficult to see that the process does not return to the origin after time $t = 0$. Let $p(t, z, w)$ be the transition function for the process, i.e., the density (in w) of the distribution of W_t given $W_0 = z$. Then $p(t, z, w)$ satisfies (see Appendix C)

(6.24) $$\dot{p}(t, z, w) = \frac{1}{2}\Delta_z p(t, z, w), \quad z \in D,$$

(6.25) $$\dot{p}(t, z, w) = \frac{1}{2}\Delta_w p(t, z, w), \quad w \in D,$$

(6.26) $$\partial_{\mathbf{v}_1, z} p(t, z, w) = 0, \; z \in l_1, \quad \partial_{\mathbf{v}_2, z} p(t, z, w) = 0, \; z \in l_2,$$

(6.27) $$\partial_{x, w} p(t, z, w) = 0, \quad w \in l_1 \cup l_2,$$

where $\mathbf{v}_1 = e^{2i\pi/3}, \mathbf{v}_2 = e^{i\pi/3}$ and $\partial_{\mathbf{v}, z}$ denotes the directional derivative in the direction $\mathbf{v}$ in the variable z. Our particular choice of reflection angles was made so that (6.27) holds. We can give an explicit solution to these equations. In fact, if $q(t, y, y'), y, y' > 0$, is the density of a one-dimensional Brownian motion killed upon reaching the origin, then $p(t, x + iy, x' + iy') := q(t, y, y')\, 1\{x + iy' \in \overline{D}\}$ is a solution to (6.24) – (6.27). We can conclude the following.

- Suppose the distribution of W_0 is uniform on each line $\mathcal{I}_s$. Then for every $t > 0$, the distribution of W_t is uniform on each line $\mathcal{I}_s$. In particular, if $W_0 = 0$ and $\tau_s = \inf\{t : W_t \in \mathcal{I}_s\}$, then the distribution of $W(\tau_s)$ is uniform on $\mathcal{I}_s$.

Fix $s = \sqrt{3}/2$ so that the triangle D_s has side length 1. Let $\eta : [0,1] \to \overline{D}$ be a simple curve with $\eta(0) \in l_1, \eta(1) \in \mathcal{I}_s$ and $\eta(0,1) \subset D_s$. Let $\tau_\eta = \inf\{t : W_t \in \mathcal{I}_s \cup \eta[0,1]\}$. Let $\tilde{D} = \tilde{D}_\eta$ be the component of $D_s \setminus \eta[0,1]$ whose boundary includes the origin and let $f : \tilde{D} \to D_s$ be the conformal transformation fixing the vertices $0, e^{2i\pi/3}$ and with $f(\eta(0)) = e^{i\pi/3}$. Let l denote the line segment $[e^{2i\pi/3}, \eta(1)]$ and if $z \in \tilde{D}$, let $h(z) = \mathbf{P}^z\{W_\eta \in l\}$. Then from (6.24) and (6.26) we can see that h is the solution to the equation

$$\Delta h(z) = 0, z \in \tilde{D},$$

$$h(z) = 1, z \in l, \quad h(z) = 0, z \in \eta(0,1),$$

$$\partial_{\mathbf{v}_1} h(z) = 0, z \in [0, \eta(0)], \quad \partial_{\mathbf{v}_2} h(z) = 0, z \in l_2 \cap \partial D_s.$$

Let $\hat{h}(z) = h(f^{-1}(z))$. Then it is easy to check that $\hat{h}(z)$ satisfies

$$\Delta \hat{h}(z) = 0, \ z \in D_s,$$

$$\hat{h}(z) = 1, \ z \in [e^{2i\pi/3}, f(\eta(1))), \quad \hat{h}(z) = 0, \ z \in (f(\eta(1)), e^{i\pi/3}],$$

$$\partial_{\mathbf{v}_1}\hat{h}(z) = 0, z \in l_1 \cap \partial D_s, \quad \partial_{\mathbf{v}_2}\hat{h}(z) = 0, z \in l_2 \cap \partial D_s.$$

Since the distribution of $W(\tau_s)$ is uniform, this implies that

$$h(0) = \hat{h}(0) = \text{length}(\,[e^{2i\pi/3}, f(\eta(1))]\,).$$

Since these probabilities determine the distribution of the hull generated by $W(\tau_s)$ we have derived the following.

PROPOSITION 6.39. *Let D, D_s, W_t, τ_s be as in the previous two paragraphs. Let $V' = V'_s$ be the hull generated by $W[0, \tau_s]$, i.e., the closure of $D_s \setminus U$ where U is the connected component of $D_s \setminus W[0, \tau_s]$ whose boundary includes $(2/\sqrt{3})\, s\, e^{i\pi/3}$. Then the distribution of V' is the the same as the the distribution of the hull of SLE_6 in D from 0 to $(2/\sqrt{3})\, s\, e^{2i\pi/3}$ viewed by $(2/\sqrt{3})\, s\, e^{i\pi/3}$.*

6.9. Derivative estimates

Much can be learned about SLE_κ by considering the spatial derivative $g'_t(z)$. We start by considering the chordal case. By differentiating (6.1), we see that

$$\dot{g}'_t(z) = -\frac{2\, g'_t(z)}{(g_t(z) - \sqrt{\kappa}\, B_t)^2}, \quad t < T_z.$$

If $v_t = \log|g'_t(z)| = \text{Re}[\log g'_t(z)]$, then

$$\dot{v}_t(z) = -\text{Re}\left[\frac{2}{(g_t(z) - \sqrt{\kappa}B_t)^2}\right] = -\text{Re}\left[\frac{a}{\hat{g}_t(z)^2}\right],$$

where, as before, $\hat{g}_t(z) = [g_t(z) - \sqrt{\kappa}\, B_t]/\sqrt{\kappa}$ and $a = 2/\kappa$. Since $g'_0(z) = 1$, this implies

$$|g'_t(z)|^b = \exp\{b\, v_t(z)\} = \exp\left\{-ab \int_0^t \text{Re}\left[\frac{1}{\hat{g}_s(z)^2}\right] ds\right\}.$$

If $x \in \mathbb{R} \setminus \{0\}$ we can write this as

(6.28) $$g'_t(x)^b = \exp\left\{-ab \int_0^t \frac{ds}{\hat{g}_s(x)^2}\right\}.$$

Note that for fixed t, $g'_t(x)$ is increasing in x on $(0, \infty)$ where we define $g'_t(x) = 0$ if $t \geq T_x$.

Consider the case when $z = x \in (0, \sqrt{\kappa})$, i.e., $\hat{g}_0(x) \in (0,1)$. Note that $g'_t(x)$ is positive and decreasing in t for $0 \leq t < T_x$. Let σ be the first time that $\hat{g}_t(x) \in \{0, 1\}$. Then using (1.11) we can see that for $\alpha \geq 0$,

$$\mathbf{E}[|g'_\sigma(x)|^b; \hat{g}_t(x) = 1] = \mathbf{E}[|g'_\sigma(x)|^b; T_x > \sigma] = (x/\sqrt{\kappa})^q,$$

where

(6.29) $$q = q(a, b) = \frac{1 - 2a + \sqrt{(1-2a)^2 + 8ab}}{2}.$$

Of course, σ is a rather unnatural stopping time in the context of the chordal SLE_κ. We can also get results for fixed times or a more natural stopping time, but then the formulas start to involve hypergeometric functions. In the case of fixed time, we have already done the calculation.

PROPOSITION 6.40. *For every $\kappa > 0$ and $b > 0$, $\mathbf{E}[g'_t(x)^b] = \psi(|x|/\sqrt{\kappa t})$ where $\psi(x) = \psi(x; a, ab)$ is as in (1.12). In particular as $t \to \infty$, $\mathbf{E}[g'_t(1)^b] \sim c t^{-q/2}$ where q is as in (6.29). This is also true for $b = 0$ if $\kappa > 4$.*

PROOF. Since $\hat{g}_t(x)$ is a Bessel-$(2a+1)$ process starting at $x/\sqrt{\kappa}$, this follows from (6.28) and (1.12). □

Now assume that $\kappa > 4$, $x, y > 0$, and let

(6.30) $$\psi(x, y) = \mathbf{E}[g'_{T_{-y}}(x)^b] = \mathbf{E}\left[\exp\left\{-ab \int_0^{T_{-y}} \frac{dt}{\hat{g}_t(x)^2}\right\}; T_x > T_{-y}\right].$$

By scaling we can write $\psi(x, y) = \psi(x/(x+y))$ where $\psi(r) = \psi(1, r)$.

REMARK 6.41. We have derived a scaling rule for g'_t. It is worth noting that we could have derived this from the scaling rule for SLE, Proposition 6.5. If $x \in \mathbb{R}$ and $r > 0$, then the distribution of $r^{-1} g_{r^2 t}(rx)$ is the same as the distribution of $g_t(x)$. Hence, $g'_{r^2 t}(rx)$ and $g'_t(x)$ have the same distribution. If $y > 0$, we can similarly see that $g'_{T_{-y}}(1)$ has the same distribution as $g'_{T_{-1}}(1/y)$.

PROPOSITION 6.42. *If $\kappa > 4$ and $b \geq 0$, there is a $c = c(\kappa, b)$ such that as $y \to \infty$,*

$$\mathbf{E}[g'_{T_{-1}}(1/y)^b] = \mathbf{E}[g'_{T_{-y}}(1)^b] \sim c y^{-q}$$

where $q = q(2/\kappa, b)$ as in (6.29).

PROOF. The first equality follows by scaling, so we need only show that

$$\mathbf{E}[g'_{T_{-y}}(1)^b] \sim c y^{-q}.$$

We will use (6.30). Let $X_t, Y_t, Z_t, \sigma, r(t), \tilde{Z}_t, \tilde{B}_t, \tilde{\mathcal{F}}_t$ be as in the proof of Proposition 6.33. Note that $\sigma = T_{-y}$ and $Z_\sigma = 0$ if and only if $T_x < T_{-y}$. Let

$$K_t = \exp\left\{-ab \int_0^{r(t)} \frac{ds}{X_s^2}\right\},$$

and note that $K_t \psi(\tilde Z_t) = \mathbf{E}[K_\sigma \mid \tilde{\mathcal{F}}_t]$ is a martingale. Also $\dot K_t = -ab\, X_{r(t)}^{-2}\, \dot r(t)\, K_t = -ab\, \tilde Z_t^{-2}\, K_t$. Therefore,

$$\frac{d[K_t \psi(\tilde Z_t)]}{K_t} = \left[-\frac{ab}{\tilde Z_t^2} \psi(\tilde Z_t) + \psi'(\tilde Z_t) \left[\frac{a}{\tilde Z_t} - \frac{a}{1-\tilde Z_t} \right] + \frac{1}{2} \psi''(\tilde Z_t) \right] dt + d\tilde B_t.$$

Hence ψ satisfies the equation

$$u^2 (1-u)\, \psi''(u) + 2a\, u\, (1-2u)\, \psi'(u) - 2ab\, (1-u)\, \psi(u) = 0.$$

There are two linearly independent solutions[3] to this equation which can be written as $x^{\alpha_\pm} F_\pm(x)$, where F is smooth near zero with $F_\pm(0) = 1$ (the solutions can be written in terms of hypergeometric functions) and

$$\alpha_\pm = \frac{1}{2} - a \pm \frac{\sqrt{(1-2a)^2 + 8ab}}{2}.$$

Since the solution must equal zero at zero, we choose the solution that looks like x^{α_+}. □

6.10. Crossing exponent for SLE_6

The crossing exponents for chordal SLE_6, which are very closely related to the derivative estimates, are used to find the intersection exponents for planar Brownian motion. Let $\mathcal{R}_L = (0,L) \times (0,\pi)$ be the rectangle as in §3.7 with vertical boundaries $\partial_1 = [0, i\pi], \partial_2 = \partial_{2,L} = [L, L+i\pi]$. Let γ be an SLE_6 path from $i\pi$ to $L+i\pi$ in $\mathcal{R}_L$ and let t_* be the first time that $\gamma(t) \in \partial_2$. Locality (Proposition 6.14) tells us that the distribution of $\gamma[0, t_*]$ is the same if we take an SLE_6 path from $i\pi$ to L. Let D be the connected component of $\mathcal{R}_L \setminus \gamma[0, t_*]$ whose boundary includes 0 and let $\mathcal{L}^*$ denote the π-extremal distance between $\partial_1 \cap \partial D$ and $\partial_2 \cap \partial D$ in D. If $\partial_2 \cap \partial D = \emptyset$, then $\mathcal{L}^* = \infty$. We use the convention $e^{-0 \cdot \mathcal{L}^*} = I\{\mathcal{L}^* < \infty\}$.

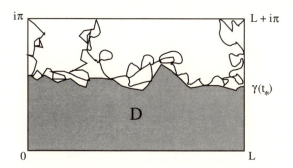

PROPOSITION 6.43. *If $b \geq 0$, then*

$$\mathbf{E}[\exp\{-b\, \mathcal{L}^*\}] \asymp e^{-\eta(b)\, L}, \quad L \to \infty$$

where

$$\eta(b) = b + q(1/3, b) = b + \frac{1}{6}[1 + \sqrt{1 + 24b}].$$

---

[3]Good tools for finding these solutions are Maple and Mathematica.

PROOF. We fix $b \geq 0$, let $q = q(1/3, b)$, and allow all constants in this proof to depend on b. We first transform this problem to $\mathbb{H}$. Let $f_L : \mathcal{R}_L \to \mathbb{H}$ be the conformal transformation with $f_L(i\pi) = 0, f_L(0) = 1, f_L(L) = \infty$; note that $|f_L(L + i\pi)| \asymp e^L$. One way to see this is to note that the excursion measure in $\mathcal{R}_L$ of the set of paths from ∂_1 to ∂_2 is comparable to e^{-L} which is comparable to the excursion measure in $\mathbb{H}$ of paths from $[0, 1]$ to $(-\infty, -e^L]$.

Since π-extremal distance is a conformal invariant, scaling of SLE_6 shows that the proposition is equivalent to the following statement. Let γ be an SLE_6 path in $\mathbb{H}$ from 0 to ∞ and let $\mathcal{L} = \mathcal{L}(x)$ be the π-extremal distance in D between $l_1 = l_{1,x} := [0, x] \cap \partial D$ and $l_2 := (-\infty, -1] \cap \partial D$, where D denotes the unbounded component of $\mathbb{H} \setminus \gamma[0, T_{-1}]$. Then,

$$\mathbf{E}[\exp\{-b\mathcal{L}\}] = \mathbf{E}[\exp\{-b\mathcal{L}\}; T_{-1} < T_x] \asymp x^{b+q}, \quad x \to 0+.$$

We will assume $x \leq 1$. By conformal invariance, $\mathcal{L}$ is the π-extremal distance between $g_{T_{-1}}(l_1)$ and $g_{T_{-1}}(l_2)$ in $\mathbb{H}$, and hence we can see that

$$e^{-\mathcal{L}} \asymp \frac{\text{length}[g_{T_{-1}}(l_1)]}{\text{dist}[g_{T_{-1}}(l_1), g_{T_{-1}}(l_2)]}.$$

By considering $\text{cap}_{\mathbb{H}}(\gamma[0, T_{-1}])$ (see (3.13) for definition), we can see that the denominator is comparable to $\text{rad}(\gamma[0, T_{-1}])$, which is at least 1. Since $g_t'(x)$ is increasing in x for $x > 0$,

$$\frac{x}{2} g_{T_{-1}}'(\frac{x}{2}) \leq \text{length}[g_{T_{-1}}(l_1)] \leq x\, g_{T_{-1}}'(x).$$

The upper bound for the proposition now follows from Proposition 6.42,

$$\mathbf{E}[\exp\{-b\mathcal{L}\}] \leq c\, x^b\, \mathbf{E}[g_{T_{-1}}'(x)^b; T_{-1} < T_x] \leq c\, x^{b+q}.$$

The lower bound will follow similarly if we show that for some constants c_1, c_2,

$$\mathbf{E}[g_{T_{-1}}'(x)^b; T_{-1} < T_x; \text{rad}(\gamma[0, T_{-1}]) \leq c_1] \geq c_2\, x^q.$$

Using Lemma 4.13, we can see it suffices to find c_1, c_2 such that

(6.31) $\quad \mathbf{E}[g_{T_{-1}}'(x)^b; T_{-1} < T_x; T_{-1} \leq c_1; \text{rad}(B[0, T_{-1}]) \leq c_1] \geq c_2\, x,$

where $U_t = -\sqrt{6}\, B_t$ is the driving function. Let $X_t = \hat{g}_t(x), Y_t = \hat{g}_t(-1)$ and recall that

$$dX_t = \frac{1}{3X_t} + dB_t, \quad dY_t = \frac{1}{3Y_t} dt + dB_t,$$

and

$$g_{T_{-1}}'(x)^b = 1\{T_{-1} < T_x\} \exp\left\{-\frac{b}{3} \int_0^{T_{-1}} \frac{ds}{X_s^2}\right\}.$$

It is a straightforward exercise to show that there is a constant c such that for every $x > 0$,

(6.32) $\quad \mathbf{E}[g_{T_{-1}}'(x)^b; T_{-1} \leq c; \text{rad}(B[0, 1]) \leq c] = \rho(c, x) > 0,$

and that $\rho(c, x)$ is increasing with x. Proposition 6.42 tells us that $\mathbf{E}[g_{T_{-1}}'(x)^b] \asymp x^q$. If $x < 2^{-n}$, let $\sigma_n = \inf\{t \geq 0 : X_t \in \{0, 2^{-n}\}\}$, and let V_n be the event

$$\{X_{\sigma_n} = 2^{-n}; \sigma_n - \sigma_{n+1} \geq n\, 2^{-2n} \text{ or } \text{diam}(B[\sigma_{n+1}, \sigma_n]) \geq n\, 2^{-n}\}.$$

Note that if $\sigma_n - \sigma_{n+1} \geq n\, 2^{-2n}$, then
$$\exp\left\{-b \int_{\sigma_{n+1}}^{\sigma_n} \frac{ds}{X_s^2}\right\} \geq e^{-bn}.$$
Also, there is a constant c' such that if $x \in (0, 2^{-n})$, then
$$\mathbf{P}^x\{\operatorname{diam}(B[0, \sigma_n]) \leq 2 \cdot 2^{-n}\}$$
$$\geq \mathbf{P}^x\{B_{2^{-2n}} \geq 2^{-n}; \operatorname{diam}(B[0, 2^{-2n}]) \leq 2 \cdot 2^{-n}\} \geq c'.$$
Hence by iterating (using the stopping times t_j defined to be the minimum of σ_n and the first time t that $\operatorname{diam}(B_{[t_{j-1}, t]}) \geq 2 \cdot 2^{-n}$),
$$\mathbf{P}^x\{\operatorname{diam}(B[0, \sigma_n]) \geq n\, 2^{-n}\} \leq \mathbf{P}\{t_{n/2} < \sigma_n\} \leq (1 - c')^{n/2}.$$
Therefore, by splitting the path $X[0, \sigma]$ into $X[0, \sigma_{n+1}], X[\sigma_{n+1}, \sigma_n], X[\sigma_n, \sigma]$, we can see that for some c and some $\rho < 1$,
$$\mathbf{E}[g'_{T_{-1}}(x)^b; V_n] \leq c\, \rho^n\, x^q.$$
Hence, for some $n > 0$, and all $x \leq 2^{-n}$,
$$\mathbf{E}[g'_{T_{-1}}(x)^b; \bigcap_{k=n}^{\infty} V_k^c] \geq c\, x^q.$$
Note that on the event $\bigcap_{k=n}^{\infty} V_k^c$,
$$\sigma_n \leq \sum_{k=n}^{\infty} k\, 2^{-2k}, \quad \operatorname{diam}(B[0, \sigma_n]) \leq \sum_{k=n}^{\infty} k\, 2^{-k}.$$
Combining this with (6.32) gives the estimate. $\square$

COROLLARY 6.44. *There exist constants c_1, c_2 such that if γ is an SLE_6 curve in $\mathcal{R}_L$ from $i\pi$ to $L + i\pi$, then*
$$c_1\, e^{-L/3} \leq \mathbf{P}\{\gamma[0, \infty) \cap [0, L] = \emptyset\} \leq c_2\, e^{-L/3}.$$

PROOF. This is Proposition 6.43 with $b = 0$. Note that $\gamma[0, \infty) \cap [0, L] = \emptyset$ if and only if $\gamma[0, t_*] \cap [0, L] = \emptyset$. $\square$

In Chapter 8 we will use the crossing exponent to calculate the half-plane Brownian intersection exponent. We will need a stronger version of Proposition 6.43 that uses the locality property of SLE_6.

PROPOSITION 6.45. *For every $b \geq 0$, there exist c_1, c_2 such that the following is true. Suppose γ is an SLE_6 curve from $i\pi$ to $L + i\pi$ in $\mathcal{R}_L$. Let $\hat{\gamma} : [0, 1] \to \overline{\mathcal{R}}_L$ be a simple curve with $\hat{\gamma}(0) \in [0, i\pi], \hat{\gamma}(1) \in [L, L + i\pi), \gamma(0, 1) \subset \mathcal{R}_L$, and let D be the connected component of $\mathcal{R}_L \setminus \hat{\gamma}[0, 1]$ whose boundary includes $[i\pi, L + i\pi]$. Let D_γ be the connected component of $D \setminus \gamma[0, \infty)$ whose boundary includes $\hat{\gamma}[0, 1]$. Then,*
$$c_1 \exp\{-\eta\, L(D; \partial_1, \partial_2)\} \leq \mathbf{E}[\exp\{-b\, L(D_\gamma; \partial_1, \partial_2)\}]$$
$$\leq c_2 \exp\{-\eta\, L(D; \partial_1, \partial_2)\},$$
where $\eta = \eta(b) = b + [1 + \sqrt{1 + 24b}]/6$ and $L(D_\gamma; \partial_1, \partial_2)$ is defined to be ∞ if either $\gamma(0, \infty) \cap \hat{\gamma}(0, 1) \neq \emptyset$ or $\gamma[0, \infty) \not\subset \overline{D}$.

PROOF. The locality property implies that the conditional distribution on γ given that it stays in $\overline{D}$ is the same as SLE_6 from $i\pi$ to $L + i\pi$ in D. The result then follows from conformal invariance of π-extremal distance. $\square$

6.11. Derivative estimates, radial case

Suppose g_t is radial SLE_κ in $\mathbb{D}$ going from 1 to 0, and let $h_t(z) = -i \log g_t(e^{iz})$ (defined locally) as in §6.5 with driving function $U_t = -B_{\kappa t}$ where B_t is a standard Brownian motion. Let $\hat{h}_t(z) = h_{t/\kappa}(z)$, so that

$$\dot{\hat{h}}_t(z) = \frac{a}{2} \cot \left[\frac{\hat{h}_t(z) + B_t}{2} \right], \quad \hat{h}_0(z) = z, \tag{6.33}$$

where $a = 2/\kappa$. Fix $x \in (0, 2\pi)$ and let $Y_t = h_{t/\kappa}(x) + B_t$. Then Y_t satisfies

$$dY_t = \frac{a}{2} \cot \left(\frac{Y_t}{2} \right) dt + dB_t, \tag{6.34}$$

where B_t is a standard Brownian motion starting at the origin. This equation is valid until time $T = T_{e^{ix}}$ which is the first time t such that $Y_t \in \{0, 2\pi\}$.

Note that if $x \in (0, 2\pi)$, then $|g'_t(e^{ix})| = h'_t(x) = \hat{h}'_{2t/a}(x)$. Differentiating (6.33) with respect to z gives

$$\dot{\hat{h}}'_t(x) = -\frac{a \hat{h}'_t(z)}{4 \sin^2(Y_t/2)}.$$

Since $\hat{h}'_0(z) = 1$, this implies that

$$\hat{h}'_t(x) = \exp \left\{ -\frac{a}{2} \int_0^t \frac{ds}{2 \sin^2(Y_s/2)} \right\}, \quad t < T_x.$$

We now fix $b \geq 0$; if $a \geq 1/2$, we require $b > 0$. All constants in this section may depend on a, b. Let

$$\psi(t, x) = \psi(t, x; a, b) = \mathbf{E}[\hat{h}'_t(x)^b; T_x > t],$$

$$q = q(a, b) = \frac{1 - 2a + \sqrt{(1 - 2a)^2 + 8ab}}{2},$$

$$\lambda = \lambda(a, b) = \frac{q + 2ab}{8} = \frac{1 - 2a + 4ab + \sqrt{(1 - 2a)^2 + 8ab}}{16}. \tag{6.35}$$

PROPOSITION 6.46. *There exist c_1, c_2 such that for every $x \in (0, \pi)$ and every $t \geq 1$,*

$$c_1 e^{-\lambda t} x^q \leq \psi(t, x) \leq c_2 e^{-\lambda t} x^q.$$

Hence, if $t \geq a/2, x \in (0, \pi)$,

$$\mathbf{E}[|g'_t(e^{ix})|^b] = \mathbf{E}[\hat{h}'_{2t/a}(x)^b] \asymp x^q e^{-2\lambda t/a}.$$

PROOF. See (1.22). $\square$

COROLLARY 6.47. *There exist c_1, c_2 such that if γ is a radial SLE_κ curve from 1 to 0 in $\mathbb{D}$, $z \in \partial \mathbb{D}$, and*

$$\sigma_r = \inf\{t : |\gamma(t)| = r\},$$

then if $r \leq e^{-a/2}$,

$$c_1 |z - 1|^q r^{2\lambda/a} \leq \mathbf{E}[|g'_{\sigma_r}(z)|^b; \sigma_r < T_z] \leq c_2 |z - 1|^q r^{2\lambda/a},$$

where $\lambda = \lambda(a,b)$ is as in (6.35). In particular, if $\kappa > 4$,
$$\mathbf{P}\{\sigma_r < T_z\} \asymp |z-1|^q \, r^{(\kappa-4)/8}.$$

PROOF. This is an immediate consequence of Proposition 6.46 and (4.22). □

The next proposition discusses a "crossing exponent" for radial SLE_κ.

PROPOSITION 6.48. *For every $\kappa > 0, b > 0$ (or $\kappa \geq 4, b = 0$), there exist c_1, c_2 such that if γ is a radial SLE_κ path from 1 to 0 in $\mathbb{D}$; $\sigma_r = \inf\{t : |\gamma(t)| = r\}$; and D_r is the connected component of $\{r < |z| < 1\} \setminus \gamma[0, \sigma_r]$ whose boundary contains $r\,\partial \mathbb{D}$, then*
$$c_1 \, r^{\kappa\lambda} \leq \mathbf{E}[\exp\{-b\,L(D_r; r\partial\mathbb{D}, \partial\mathbb{D} \cap \partial D)\}] \leq c_2 \, r^{\kappa\lambda},$$
where $\lambda = \lambda(2/\kappa, b)$ is as in (6.35) and $\mathbf{E}[\exp\{-b\,L(D_r; r\partial\mathbb{D}, \emptyset)\}]$ is defined to be 0, even for $b = 0$.

The hard work in proving the proposition was done in §1.11. In order to use the result in that section we will need to prove two simple lemmas about π-extremal distance.

LEMMA 6.49. *There exist c_1, c_2 such that if $\theta \in (0, \pi)$ and*
$$\mathcal{L}(\theta) = \mathcal{L}(\mathcal{A}; l, (1/8)\,\partial\mathbb{D})$$
is the π-extremal distance of the collection of curves connecting $\{|z| = 1/8\}$ and $l := \{e^{i\theta'} : -\theta \leq \theta' \leq \theta\}$ in the annulus $\mathcal{A} := \{1/8 < |z| < 1\}$, then $c_1\theta \leq e^{-\mathcal{L}(\theta)} \leq c_2\theta$.

PROOF. Without loss of generality we assume that $\theta < \pi/2$. Let $D = \mathcal{A} \setminus \{\arg(z) = \pi\}$. Clearly, $L(\theta) \leq L(D; l, (1/8)\,\mathbb{D})$. We claim that $L(\theta) \geq L(D; l, l_1)$, where $l_1 = (1/8)\partial\mathbb{D} \cup \{re^{i\pi} : 1/8 \leq r \leq 1\}$. To see this, note that any ρ satisfying (3.17) for D, l, l_1 also satisfies (3.17) for $\mathcal{A}, l, (1/8)\,\partial\mathbb{D}$.

Let W_t denote a complex Brownian motion and let $\tau = \inf\{t : W_t \notin D\}$. Let $x = 1 - \sqrt{\theta}$. Using the Poisson kernel in $\mathbb{D}$, we can see that $\mathbf{P}^x\{W_\tau \in l\} \geq c\sqrt{\theta}$; and a straightforward estimate shows that $\mathbf{P}^x\{|W_\tau| = 1/8\} \geq c\sqrt{\theta}$. Hence $\Theta(D; l, (1/8)\partial\mathbb{D}) \geq c\sqrt{\theta}$, where Θ is defined as in Proposition 3.69. Similarly, any z with $\mathbf{P}^z\{W_\tau \in l\} \geq c\sqrt{\theta}$ must be within distance $c_2\sqrt{\theta}$ of 1. For such z, it is easy to show that $\mathbf{P}^z\{W_\tau \in l_1\} \leq c_3\sqrt{\theta}$. Hence $\Theta(D; l, l_1) \leq c\sqrt{\theta}$. The result now follows from Proposition 3.69. □

LEMMA 6.50. *There exist c_1, c_2 such that the following holds. Suppose $0 < r < 1$ and $\gamma : [0,1] \to \mathbb{C}$ is a curve with $|\gamma(0)| = 1$, $|\gamma(1)| = r$, $r < |\gamma(t)| \leq 1, 0 \leq t < 1$. For $s \leq r$, let D_s be the connected component of $\{s < |z| < 1\} \setminus \gamma[0,1]$ whose boundary includes $s\,\partial\mathbb{D}$. Let $l = \partial D_s \cap \partial\mathbb{D}$ (which is independent of s for $s \leq r$), and let $L_s = L(D_s; l, s\,\partial\mathbb{D})$. Let $\tilde{D}$ be the component of $\mathbb{D} \setminus \gamma[0,1]$ containing the origin and let $g : \tilde{D} \to \mathbb{D}$ be a conformal transformation with $g(0) = 0$. Then*
$$c_1\,e^{-L_r/50} \leq \mathrm{length}[g(l)] \leq c_2\,e^{-L_r}.$$

PROOF. Let $f = g^{-1}$, and note that Corollary 3.19 implies $r/4 \leq |f'(0)| \leq r$. By applying the Growth Theorem (Theorem 3.23) to $f(z)/f'(0)$ we can see that
$$\left(\frac{8}{9}\right)^2 \frac{1}{4}|z|\,r \leq |f(z)| \leq \left(\frac{8}{7}\right)^2 |z|\,r, \qquad |z| \leq 1/8.$$

In particular, f maps the circle of radius $1/8$ into the closed annulus $\{2r/81 \leq |z| \leq 8r/49\}$. Hence by conformal invariance,
$$L_r \leq L(\{1/8 < |z| < 1\}; g(l), (1/8)\,\partial\mathbb{D}) \leq L_{r/50}.$$
Now use Lemma 6.49. □

PROOF OF PROPOSITION 6.48. Let
$$R_t = \text{length}[g_t(\partial\mathbb{D})] = \int_0^{2\pi} |g_t'(e^{ix})|\,dx = \int_0^{2\pi} \hat{h}'_{\kappa t}(x)\,dx.$$
Then,
$$c_1 \exp\{-b\,L(D_{50r}; 50\,r\partial\mathbb{D}, \partial\mathbb{D} \cap \partial D)\} \leq R_{\sigma_r} \leq R_{-\log r},$$
$$R_{-\log(r/4)} \leq R_{\sigma_r} \leq c_2 \exp\{-b\,L(D_r; r\partial\mathbb{D}, \partial\mathbb{D} \cap \partial D)\}.$$
Hence it suffices to show that $\mathbf{E}[R_t^b] \asymp e^{-\kappa\lambda t}$. This is done in Proposition 1.36. □

CHAPTER 7

More results about SLE

7.1. Introduction

In this chapter we will discuss some fine properties of the conformal maps g_t given by chordal SLE_κ. In particular, we will discuss the following:
- How do we prove that the SLE_κ curve γ exists, i.e., that there is a curve γ such that the domain of g_t is the unbounded component of $\mathbb{H} \setminus \gamma[0,t]$?
- What is the fractal dimension of the path $\gamma[0,t]$?
- Is the map $f_t = g_t^{-1}$ Hölder continuous?

In the previous chapter, much was derived by considering the real-valued processes $t \mapsto g_t(x)$ for $x \in \mathbb{R} \setminus \{0\}$. For the study in this chaper we consider the $\mathbb{H}$-valued processes $t \mapsto g_t(z)$ for $z \in \mathbb{H}$ which we will write as two real-valued processes, $\mathrm{Re}[g_t(z)]$ and $\mathrm{Im}[g_t(z)]$.

Suppose g_t is chordal SLE_κ with driving function $U_t = -\sqrt{\kappa} B_t$ where B_t is a standard Brownian motion. Then g_t satisfies
$$\dot{g}_t(z) = \frac{2}{g_t(z) + \sqrt{\kappa} B_t}, \qquad g_0(z) = z.$$

It will be convenient to let
$$a = \frac{2}{\kappa}, \qquad h_t(z) = \frac{g_t(\sqrt{\kappa}\, z)}{\sqrt{\kappa}}.$$

Note that h_t satisfies the Loewner equation
$$(7.1) \qquad \dot{h}_t(z) = \frac{a}{h_t(z) + B_t}, \qquad h_0(z) = z.$$

If K_t denotes the hulls generated by h_t (i.e., the domain of h_t is $\mathbb{H} \setminus K_t$), then K_t has the distribution of the hulls of $SLE_{2/a}$ with time parametrized so that $\mathrm{hcap}(K_t) = at$.

If $z_0 = x_0 + i y_0 \in \mathbb{H}$, let $Z_t = h_t(z_0) + B_t$, so that
$$(7.2) \qquad dZ_t = \frac{a}{Z_t}\, dt + dB_t, \qquad Z_0 = z_0.$$

7.1.1. Some computations. Most of the proofs in this section will involve functions of Z_t, so we will start by stating some SDEs satisfied by these functions. These equations are derived easily from (7.1) and (7.2) using Itô's formula.

If we write $Z_t = X_t + i Y_t$, then (7.2) can be written as
$$dX_t = \frac{a X_t}{X_t^2 + Y_t^2}\, dt + dB_t, \qquad dY_t = -\frac{a Y_t}{X_t^2 + Y_t^2}\, dt,$$

Hence, we also get
$$d\log Y_t = -\frac{a}{X_t^2 + Y_t^2}\, dt,$$

$$d|Z_t|^2 = d(X_t^2 + Y_t^2) = \frac{(2a+1)X_t^2 + (1-2a)Y_t^2}{X_t^2 + Y_t^2}\,dt + 2X_t\,dB_t.$$

(7.3)
$$d\begin{bmatrix} X_t \\ Y_t \end{bmatrix} = 2a\,\frac{X_t}{Y_t}\,\frac{1}{X_t^2 + Y_t^2}\,dt + \frac{1}{Y_t}\,dB_t.$$

By using Itô's formula on the real and imaginary parts, one can see that if $\psi : \mathbb{H} \to \mathbb{C}$ is an analytic function,

$$d[\psi(Z_t)] = \left[\frac{a\,\psi'(Z_t)}{Z_t} + \frac{\psi''(Z_t)}{2} \right] dt + \psi'(Z_t)\,dB_t.$$

For example,

$$d[\log Z_t] = \left(a - \frac{1}{2}\right) \frac{1}{Z_t^2}\,dt + \frac{1}{Z_t}\,dB_t.$$

Taking real and imaginary parts of the last equation gives

$$d[\log |Z_t|] = \left(a - \frac{1}{2}\right) \frac{X_t^2 - Y_t^2}{(X_t^2 + Y_t^2)^2}\,dt + \frac{X_t}{X_t^2 + Y_t^2}\,dB_t,$$

$$d[\arg(Z_t)] = (1 - 2a) \frac{X_t Y_t}{(X_t^2 + Y_t^2)^2}\,dt - \frac{Y_t}{X_t^2 + Y_t^2}\,dB_t.$$

REMARK 7.1. The last equation shows that $\arg(Z_t)$ is a martingale if and only if $a = 1/2$, i.e., $\kappa = 4$. Recall (see (2.8)) that $\arg(Z_t)/\pi$ is the probability that a complex Brownian motion starting at Z_t hits $(-\infty, 0)$ before hitting $(0, \infty)$. Schramm and Sheffield [78] used this observation to define a discrete process called the *harmonic explorer* that converges to SLE_4.

If $L_t = \log h_t'(z_0)$, then differentiating (7.1) gives

$$L_t = -\int_0^t \frac{a}{Z_s^2}\,ds.$$

In particular,

(7.4) $\qquad |h_t'(z_0)| = \exp\{\mathrm{Re}[\log h_t'(z)]\} = \exp\left\{ a \int_0^t \frac{Y_s^2 - X_s^2}{(X_s^2 + Y_s^2)^2}\,ds \right\}.$

7.1.2. A useful time change. It is sometimes useful to consider the processes above under the time change $\sigma(t)$, where

$$\dot{\sigma}(t) = X_t^2 + Y_t^2, \quad t = \int_0^{\sigma(t)} \frac{ds}{X_s^2 + Y_s^2}.$$

This random time change depends on both B_t and the initial position z_0. We let $\tilde{Z}_t = Z_{\sigma(t)}, \tilde{X}_t = X_{\sigma(t)}, \tilde{Y}_t = Y_{\sigma(t)}$, etc. Note that $\tilde{Y}_t$ evolves deterministically,

(7.5) $\qquad d\tilde{Y}_t = -a\tilde{Y}_t\,dt, \quad \tilde{Y}_t = y_0\,e^{-at},$

and $\tilde{X}_t$ satisfies

$$d\tilde{X}_t = a\,\tilde{X}_t\,dt + \sqrt{\tilde{X}_t^2 + \tilde{Y}_t^2}\,d\tilde{B}_t,$$

where $\tilde{B}_t$ denotes the standard Brownian motion

$$\tilde{B}_t = \int_0^{\sigma(t)} \frac{1}{\sqrt{X_t^2 + Y_t^2}}\,dB_t.$$

Let
$$\tilde{h}_t(z) = h_{\sigma(t)}(z), \quad \tilde{K}_t = \frac{\tilde{X}_t}{\tilde{Y}_t}, \quad \tilde{N}_t = \frac{\tilde{X}_t^2}{\tilde{X}_t^2 + \tilde{Y}_t^2} = \frac{\tilde{K}_t^2}{\tilde{K}_t^2 + 1}.$$

Then using (7.3), (7.4), and Itô's formula, we get

(7.6) $$d\tilde{K}_t = 2a\,\tilde{K}_t\,dt + \sqrt{\tilde{K}_t^2 + 1}\,d\tilde{B}_t,$$

$$d\tilde{N}_t = (1 - \tilde{N}_t)\,[4(a-1)\tilde{N}_t + 1]\,dt + 2\sqrt{\tilde{N}_t\,(1 - \tilde{N}_t)}\,dB_t,$$

(7.7) $$|\tilde{h}'_t(z_0)| = \exp\left\{a\int_0^t \frac{Y_s^2 - X_s^2}{X_s^2 + Y_s^2}\,ds\right\} = e^{at}\exp\left\{-2a\int_0^t \tilde{N}_s\,ds\right\}.$$

7.1.3. Negative times. If $B_{t,-}$ is another standard Brownian motion, independent of B_t, we can extend B_t to be a two-sided Brownian motion by setting $B_{-t} = B_{t,-}$. We can then define h_t for negative t by using (7.1). Let $h_{t,-} = h_{-t}$ so that $h_{t,-}$ satisfies the "backwards" Loewner equation

$$\dot{h}_{t,-}(z) = -\dot{h}_{-t}(z) = -\frac{a}{h_{t,-}(z) + B_{t,-}}.$$

We also define Z_t for negative t, and write $Z_{t,-} = Z_{-t} = h_{t,-}(z_0) + B_{t,-} = X_{t,-} + iY_{t,-}$. Here $h_{0,-}(z) = z$ and $Z_{0,-} = z_0 = x_0 + iy_0$. Then

$$dZ_{t,-} = -\frac{a}{Z_{t,-}}\,dt + dB_{t,-},$$

(7.8) $$dX_{t,-} = -\frac{a\,X_{t,-}}{X_{t,-}^2 + Y_{t,-}^2}\,dt + dB_{t,-}, \quad dY_{t,-} = \frac{a\,Y_{t,-}}{X_{t,-}^2 + Y_{t,-}^2}\,dt,$$

$$d\left[\frac{X_{t,-}}{Y_{t,-}}\right] = -2a\,\frac{X_{t,-}}{Y_{t,-}}\,\frac{1}{X_{t,-}^2 + Y_{t,-}^2}\,dt + \frac{1}{Y_{t,-}}\,dB_{t,-}.$$

(7.9) $$|h'_{t,-}(z_0)| = \exp\left\{a\int_0^t \frac{X_{s,-}^2 - Y_{s,-}^2}{(X_{s,-}^2 + Y_{s,-}^2)^2}\,ds\right\}.$$

We can do the same time change as in the previous subsection, i.e., $\dot{\sigma}_-(t) = X_{t,-}^2 + Y_{t,-}^2$, and we write $\tilde{X}_{t,-}, \tilde{Y}_{t,-}, \tilde{K}_{t,-}, \tilde{N}_{t,-}$ for the corresponding processes. Note that $\tilde{Y}_{t,-} = y_0\,e^{at}$. We get

$$d\tilde{K}_{t,-} = -2a\tilde{K}_{t,-}\,dt + \sqrt{\tilde{K}_{t,-}^2 + 1}\,d\tilde{B}_{t,-},$$

$$d\tilde{N}_{t,-} = (1 - \tilde{N}_{t,-})\,[-4(a+1)\tilde{N}_{t,-} + 1]\,dt + 2\sqrt{\tilde{N}_{t,-}\,(1 - \tilde{N}_{t,-})}\,dB_{t,-}.$$

(7.10) $$|\tilde{h}'_{t,-}(z_0)| = \exp\left\{a\int_0^t \frac{\tilde{X}_{s,-}^2 - \tilde{Y}_{s,-}^2}{\tilde{X}_{s,-}^2 + \tilde{Y}_{s,-}^2}\,ds\right\} = e^{-at}\,\exp\left\{2a\int_0^t \tilde{N}_{s,-}\,ds\right\}.$$

In the next two sections, we will need estimates on $|h'_{t,-}(z_0)|$. We will now derive the critical estimate (Corollary 7.3) by considering a particular martingale.

PROPOSITION 7.2. *If r, b satsify*
$$r^2 - (2a+1)r + ab = 0,$$
then
$$M_t := \tilde{Y}_{t,-}^{b-(r/a)} (|\tilde{Z}_{t,-}|/\tilde{Y}_{t,-})^{2r} |\tilde{h}'_{t,-}(z_0)|^b$$
is a martingale. In particular,
$$\mathbf{E}\left[(|\tilde{Z}_{t,-}|/\tilde{Y}_{t,-})^{2r} |\tilde{h}'_{t,-}(z_0)|^b\right] = (|z_0|/y_0)^{2r} e^{t(r-ab)}.$$
If $r \geq 0$, then for all $\lambda > 0$,

(7.11) $$\mathbf{P}\{|\tilde{h}'_{t,-}(z_0)| \geq \lambda\} \leq \lambda^{-b} (|z_0|/y_0)^{2r} e^{t(r-ab)}.$$

PROOF. By (7.10) and the relation $\tilde{Y}_{t,-} = y_0 e^{at}$,
$$M_t = y_0^{b-(r/a)} e^{-rt} (1 - \tilde{N}_{t,-})^{-r} \exp\{2ab \int_0^t \tilde{N}_{s,-} \, ds\}.$$
Therefore, Itô's formula and the product formula give
$$dM_t = 2r \sqrt{\tilde{N}_t} M_t \, d\tilde{B}_t,$$
which shows that M_t is a martingale. In particular,
$$\mathbf{E}[M_t] = \mathbf{E}[M_0] = y_0^{b-(r/a)} (|z_0|/y_0)^{2r}.$$
But,
$$M_t = y_0^{b-(r/a)} e^{t(ab-r)} (|\tilde{Z}_{t,-}|/\tilde{Y}_{t,-})^{2r} |\tilde{h}'_{t,-}(z_0)|^b$$
If $r \geq 0$, then $(|\tilde{Z}_{t,-}|/Y_{t,-})^{2r} \geq 1$, and hence the last inequality follows from the Markov inequality. □

COROLLARY 7.3. *For every $0 \leq r \leq 2a+1$, there is a $c = c(r,a) < \infty$ such that for all $0 \leq t \leq 1, 0 < y_0 \leq 1, e \leq \lambda \leq y_0^{-1}$,*
$$\mathbf{P}\{|h'_t(z_0)| \geq \lambda\} \leq c \lambda^{-b} (|z_0|/y_0)^{2r} \delta(y_0, \lambda),$$
where $b = [(2a+1)r - r^2]/a \geq 0$ and
$$\delta(y_0, \lambda) = \begin{cases} \lambda^{(r/a)-b}, & r < ab, \\ -\log(\lambda y_0), & r = ab, \\ y_0^{b-(r/a)}, & r > ab. \end{cases}$$

PROOF. Using (7.8), we see that $\dot{Y}_{t,-} \leq a/Y_{t,-}$ and hence $Y_{t,-} \leq \sqrt{2at + y_0^2} \leq \sqrt{2a+1}$. Therefore,
$$\mathbf{P}\{|h'_{t,-}(z_0)| \geq \lambda\} \leq \mathbf{P}\{\sup_{0 \leq t \leq T} |\tilde{h}'_{t,-}(z_0)| \leq \lambda\},$$
where $T = [\log \sqrt{2a+1} - \log y_0]/a$ is the time with $\tilde{Y}_{T,-} = \sqrt{2a+1}$. Using (7.10) we see that $|\tilde{h}'_{t+s,-}(z_0)| \leq e^{as} |\tilde{h}'_{t,-}(z_0)|$. Therefore,
$$\mathbf{P}\{\sup_{0 \leq t \leq T} |\tilde{h}'_{t,-}(z_0)| \geq e^a \lambda\} \leq \sum_{j=0}^{\lfloor T \rfloor} \mathbf{P}\{|\tilde{h}'_{j,-}(z_0)| \geq \lambda\}.$$

The Schwarz lemma tells us that $|\tilde{h}'_{t,-}(z_0)| \leq \operatorname{Im}[\tilde{h}'_{t,-}(z_0)]/y_0 = e^{at}$; hence (7.11) gives

$$\begin{aligned}
\mathbf{P}\{\sup_{0\leq t\leq T} |\tilde{h}'_{t,-}(z_0)| \geq e^a \lambda\} &\leq \sum_{(1/a)\log\lambda \leq j \leq T} \mathbf{P}\{|\tilde{h}'_{j,-}(z_0)| \geq \lambda\} \\
&\leq \lambda^{-b}(|z_0|/y_0)^{2r} \sum_{(1/a)\log\lambda \leq j \leq T} e^{j(r-ab)} \\
&\leq c\lambda^{-b}(|z_0|/y_0)^{2r} \theta(y_0, \lambda).
\end{aligned}$$

□

7.2. The existence of the path

In this section, we prove the following. Most of the work has already been done.

THEOREM 7.4. *If $\kappa \neq 8$, chordal SLE_κ is generated by a path w.p.1.*

REMARK 7.5. This result is also true for $\kappa = 8$ but requires a different proof. It was established in [57] by giving a measure on discrete curves that approach SLE_8.

LEMMA 7.6. *Let $\hat{f}_t(z) = h_t^{-1}(z - B_t)$. If $s \geq 0$, then the distribution of the function $z \mapsto \hat{f}_s(z) + B_s$ is the same as the distribution of $z \mapsto h_{-s}(z)$. In particular, $z \mapsto \hat{f}'_s(z)$ has the same distribution as $z \mapsto h'_{-s}(z)$.*

PROOF. Let $h_t^*(z) = h_{s+t} \circ h_s^{-1}(z - B_s) + B_s$. Then

$$h_0^*(z) = z, \quad h_{-s}^* = \hat{f}_s(z - B_s) + B_s, \quad \dot{h}_t^*(z) = \frac{a}{h_t^*(z) + (B_{t+s} - B_s)}.$$

But $B_{t+s} - B_s$ has the same distribution as B_t. □

PROOF OF THEOREM 7.4. By scaling it suffices to show the result for $t \in [0, 1]$. By Proposition 4.29 amd Lemma 4.33, it suffices to show that w.p.1 there exists an $\epsilon > 0$ and a (random) constant c such that

$$|\hat{f}'_{k2^{-2j}}(i2^{-j})| \leq c\, 2^{j-\epsilon}, \quad j = 1, 2, 3, \ldots, \quad k = 0, 1, \ldots, 2^{2j},$$

$$|B_t - B_s| \leq c\,|t-s|^{1/2}\,|\log\sqrt{|t-s|}|, \quad 0 \leq t \leq 1.$$

The second inequality follows from the modulus of continuity of Brownian motion (Corollary 1.38). For the first inequality, it suffices by Lemma 7.6 and the Borel-Cantelli Lemma to find c, ϵ such that for all $0 \leq t \leq 1$,

(7.12) $$\mathbf{P}\{|h'_t(i2^{-j})| \geq 2^{j-\epsilon}\} \leq c\,2^{-(2+\epsilon)j}.$$

Let $r = a + (1/4) < 2a + 1$ and

$$b = \frac{(1+2a)r - r^2}{a} = a + 1 + \frac{3}{16a}$$

as in Corollary 7.3. Note that $r < ab$. From this corollary, we see that

$$\mathbf{P}\{|h'_t(i2^{-j})| \geq 2^{j-\epsilon}\} \leq c\,2^{-j[2b-(r/a)](1-\epsilon)}.$$

But, $2b - (r/a) = 2a + 1 + (8a)^{-1} > 2$ provided that $a \neq 1/4$. Hence if $a \neq 1/4$, we can find a positive ϵ such that (7.12) holds. □

7.3. Hölder continuity

We define the Hölder exponent $\alpha_0 = \alpha(\kappa) = \alpha(2/a)$ of SLE_κ to be the supremum of all α such that if $t > 0$, w.p.1 the function $z \mapsto g_t^{-1}(z)$ is Hölder α-continuous. By scaling, this is independent of the choice of t, and g_t^{-1} is Hölder α-continuous if and only if $\hat{f}_t$ is Hölder α-continuous.

PROPOSITION 7.7. *For every $\kappa \neq 4$, $\alpha_0(\kappa) > 0$. Moreover,*
$$\lim_{\kappa \to \infty} \alpha_0(\kappa) = 1, \quad \lim_{\kappa \to 0+} \alpha_0(\kappa) = \frac{1}{2}.$$

REMARK 7.8. The exact solution of the Loewner equation with a constant driving function (see (4.13)) shows that $\alpha_0(0) = 1/2$. It is expected that $\alpha_0(4) = 0$, i.e., SLE_4 paths do not give a Hölder continuous g_t^{-1}.

PROOF. By scaling we may assume that $t \leq 1, -1 \leq \mathrm{Re}(z), \mathrm{Re}(w) \leq 1, 0 < \mathrm{Im}(z), \mathrm{Im}(w) \leq 1$. We first claim that it suffices to find a c such that

(7.13) $\quad |\hat{f}_t'(\frac{j}{2^n} + \frac{i}{2^n})| \leq c\, 2^n\, 2^{-\alpha n}, \quad n = 0, 1, 1, \ldots, \quad j = -2^n, \ldots, 2^n.$

To see this, note that (7.13) and the Distortion Theorem (see the comment after (4.29)) imply that
$$|\hat{f}_t'(x + iy)| \leq c_1 \mathrm{Im}(y)^{\alpha - 1}, \quad -1 \leq x \leq 1, \quad 0 < y \leq 1,$$
and by integrating this we get the result.

For any $0 < r \leq 2a + 1$ let $b = [(1 + 2a)r - r^2]/a \geq 0$. Then, using Corollary 7.3,

$$\mathbf{P}\{|\hat{f}_t'(\frac{j}{2^n} + \frac{i}{2^n})| \geq 2^n\, 2^{-\alpha n}\} \leq c\,[1 + 2^{2n}]^r\, 2^{-n[2b - (r/a)](1-\alpha)}, \quad \text{if } r < ab,$$

$$\mathbf{P}\{|\hat{f}_t'(\frac{j}{2^n} + \frac{i}{2^n})| \geq 2^n\, 2^{-\alpha n}\} \leq c\,[1 + 2^{2n}]^r\, 2^{-nb(1-\alpha)}\, n, \quad \text{if } r = ab,$$

$$\mathbf{P}\{|\hat{f}_t'(\frac{j}{2^n} + \frac{i}{2^n})| \geq 2^n\, 2^{-\alpha n}\} \leq c\,[1 + 2^{2n}]^r\, 2^{-nb(1-\alpha)}\, 2^{n[(r/a) - b]}, \quad \text{if } r > ab.$$

Hence, if
$$\alpha < \frac{2b - (r/a) - 2r - 1}{b + \max\{0, b - (r/a)\}},$$
then
$$\sum_{n=1}^\infty \sum_{j=-2^n}^{2^n} \mathbf{P}\{|\hat{f}_t'(\frac{j}{2^n} + \frac{i}{2^n})| \geq 2^n\, 2^{-\alpha n}\} < c \sum_{n=1}^\infty 2^{-n\epsilon} < \infty,$$
for some $c, \epsilon > 0$. By the Borel-Cantelli Lemma, we get (7.13)

If $a < 1/4$, we can choose $r = 2a$ so that $b = 2$ and
$$2b - \frac{r}{a} - 2r - 1 = 1 - 4a > 0.$$

If $a \geq 1/6$, let $r = (a/2) + (1/4)$. Then $b = (3a/4) + (3/4) + (3/16a)$,
$$2b - \frac{r}{a} - 2r - 1 = \frac{a}{2} - \frac{1}{2} + \frac{1}{8a}.$$

Note that the right hand side is strictly positive unless $a = 1/2$.

If we let $r = \sqrt{a}, b = 2\sqrt{a} - 1 + a^{-1/2}$. then we see that the paths are Hölder α-continuous for all

$$\alpha < \frac{2\sqrt{a} - 3 + a^{-1/2}}{2\sqrt{a} - 1 + a^{-1/2} + \max\{0, 2\sqrt{a} - 1\}}.$$

By letting $a \to \infty$, we see that $\alpha_0(0+) = 1/2$, and by letting $a \to 0$, we see that $\alpha_0(\infty) = 1$. □

7.4. Dimension of the path

Let γ denote the chordal SLE_κ path parametrized so that $\text{hcap}[\gamma[0,t]] = at = (2/\kappa)t$. Then $g_{\gamma[0,t]} = h_t$ as defined in §7.1. In this section we will study the dimension of $\gamma[0,\infty)$. One simple definition of dimension is the box dimension, $\dim_b$, defined roughly by saying that for each R, the number of disks of radius ϵ needed to cover $\gamma[0,\infty) \cap \{|z| \leq R\}$ grows like $\epsilon^{-\dim_b}$ as $\epsilon \to 0+$. It is known that the box dimension of a set is no smaller than the Hausdorff dimension. In this section we will show that the *expected* number of needed disks decays like $\epsilon^{-\dim(\kappa)}$ where $\dim(\kappa) = \min\{1 + (\kappa/8), 2\}$. It is easy to see that this will follow from the estimate

$$\mathbf{P}\{\mathcal{B}(z,\epsilon) \cap \gamma[0,\infty) \neq \emptyset\} \asymp \epsilon^{2-\dim(\kappa)}, \quad \epsilon \to 0+, z \in \mathbb{H}.$$

For $x \in \mathbb{R}$, define the following random variable

$$\Delta_x = \text{dist}[x + i, \gamma[0,\infty) \cup \mathbb{R}].$$

Note that scaling implies that $\text{dist}[x + iy, \gamma[0,\infty)]$ has the same distribution as $y\Delta_{x/y}$. Also, the event $\{\mathcal{B}(x+i,\epsilon) \cap \gamma[0,\infty) \neq \emptyset\}$ is the same as the event $\{\Delta_x < \epsilon\}$.

THEOREM 7.9. *If $\kappa \geq 8$, then w.p.1 $\Delta_x = 0$ for all $x \in \mathbb{R}$. If $\kappa < 8$, then for every $M \geq 0$ there exist positive constants $c = c_\kappa$ and $c'_M = c'_{M,\kappa}$ such that,*

$$c'_{|x|} \epsilon^{1-(\kappa/8)} \leq \mathbf{P}\{\Delta_x \leq \epsilon\} \leq c\epsilon^{1-(\kappa/8)}, \quad 0 \leq |x| \leq M.$$

We will use the notation from §7.1. Using Proposition 4.38, we see that it suffices to find constants so that

$$(7.14) \qquad c'_{|x|} e^{-r(4a-1)/(4a)} \leq \mathbf{P}\{D(x) \geq r\} \leq c e^{-r(4a-1)/(4a)},$$

where

$$D(x) = \lim_{t \to T_{x+i}-} \log \frac{|h'_t(x+i)|}{\text{Im}[h_t(x+i)]}.$$

Using (4.33), we see that

$$D(x) = 2a \int_0^{T_{x+i}} \frac{Y_s^2}{(X_s^2 + Y_s^2)^2} \, ds,$$

where $Z_t = X_t + iY_t$ has initial condition $Z_0 = x + i$.

PROPOSITION 7.10. *If $\kappa \geq 8$, then w.p.1 $D(x) = \infty$. If $\kappa < 8$, w.p.1 $D(x) < \infty$ and has characteristic function*

$$\mathbf{E}[e^{ibD_\infty(x)}] =$$

$$\frac{\Gamma(a + \sqrt{(a-\tfrac{1}{2})^2 - iab})\,\Gamma(a - \sqrt{(a-\tfrac{1}{2})^2 - iab})}{\Gamma(\tfrac{1}{2})\,\Gamma(2a - \tfrac{1}{2})} F(\alpha_+, \alpha_-, \tfrac{1}{2}, \tfrac{x^2}{x^2+1}),$$

where F denotes the hypergeometric function and

$$\alpha_\pm = \frac{1}{2} - a \pm \sqrt{(a - \frac{1}{2})^2 - iab}.$$

PROOF. Let

$$D_t = D_t(x) = 2a \int_0^t \frac{Y_s^2}{(X_s^2 + Y_s^2)^2} \, ds,$$

and recall the time change

$$\dot\sigma(t) = X_t^2 + Y_t^2.$$

Let $\tilde Y_t = Y_{\sigma(t)}, \tilde X_t = X_{\sigma(t)}, \tilde D_t = D_{\sigma(t)}, \tilde K_t = \tilde X_t/\tilde Y_t$, as before, and let $\tilde C_t = \log \tilde K_t$. From (7.6) and Itô's formula, we get

(7.15) $$d\tilde C_t = [2a - \frac{1}{2} - \frac{1}{2} e^{-2\tilde C_t}] \, dt + \sqrt{1 + e^{-2\tilde C_t}} \, d\tilde B_t.$$

Also,

$$\dot{\tilde D}_t = \frac{2a Y_t^2}{X_t^2 + Y_t^2} = \frac{2a}{1 + e^{2\tilde C_t}},$$

and hence, since $D_0(x) = 0$,

$$D(x) = \int_0^\infty \frac{2a \tilde Y_s^2}{\tilde X_s^2 + \tilde Y_s^2} \, ds = \int_0^\infty \frac{2a \, ds}{\tilde K_s^2 + 1} = \int_0^\infty \frac{2a \, ds}{e^{2\tilde C_s} + 1}.$$

Note that the drift term in (7.15) is negative if $a \le 1/4$. Hence, w.p.1 for every $T > 0$, there is a $t > T$ such that $C_t \le 0$ for $t \le s \le t + 1$. This shows that w.p.1 $D(x) = \infty$ for $a \le 1/4$.

For the remainder of the proof, assume that $a > 1/4$. Define another change of time,

$$\dot{\hat\sigma}(t) = \frac{1}{\tilde K_t^2 + 1},$$

and let $\hat K_t = \tilde K_{\hat\sigma(t)}$. Then

$$d\hat K_t = \frac{2a \hat K_t}{1 + \hat K_t^2} \, dt + d\hat B_t,$$

for a standard Brownian motion $\hat B_t$, and

$$D(x) = 2a \int_0^\infty \frac{dt}{(1 + \hat K_t^2)^2}.$$

Let $\phi(b|x) = \mathbf{E}[e^{ibD(x)}]$. By (1.34), we can see that $\psi(x) := \phi(b|x)$ is the solution to

$$\frac{1}{2}\psi''(x) + \frac{2ax}{1 + x^2}\psi'(x) + \frac{abi}{(1 + x^2)^2}\psi(x) = 0$$

that is bounded for $x \ge 0$ and such that $\psi(\infty) = 1$. Letting $\psi(x) = H(x^2/(x^2 + 1))$ we get the hypergeometric equation

$$u(1 - u)H''(u) + [\frac{1}{2} + 2(a - 1)u] H'(u) + \frac{1}{2}abi H(u) = 0.$$

A bounded solution of this equation is (see (B.4))

$$H(u) = c\, F(\alpha_+, \alpha_-, \tfrac{1}{2}, u),$$

where $\alpha_\pm = (1/2) - a \pm \sqrt{(a-\tfrac{1}{2})^2 - iab}$. Since $H(1) = 1$, we can use (B.6) to conclude that

$$H(u) = \frac{\Gamma(a + \sqrt{(a-\tfrac{1}{2})^2 - iab})\, \Gamma(a - \sqrt{(a-\tfrac{1}{2})^2 - iab})}{\Gamma(1/2)\, \Gamma(2a - (1/2))}\, F(\alpha_+, \alpha_-, \tfrac{1}{2}, u).$$

$\square$

LEMMA 7.11. *Suppose X is a random variable with characteristic function ϕ. Suppose for some $u, \lambda, \epsilon > 0$,*

$$\phi(t) = \frac{u\,\lambda}{\lambda - it} + v(t),$$

where v is an analytic function on $\{|z| < \lambda + 2\epsilon\}$. Then,

$$\mathbf{P}\{X \geq x\} = u\, e^{-\lambda x} + o(e^{-(\lambda+\epsilon)x}), \quad x \to \infty.$$

PROOF. Write the distribution of X as $\mu_1 + \mu_2$ where $\mu_1 = u\lambda e^{-\lambda x} 1_{x>0}\, dx$ and μ_2 is a signed measure with

$$\int_{-\infty}^{\infty} e^{itx}\, \mu_2(dx) = v(t).$$

Analyticity of v implies that $|\mu_2|[x, \infty) = o(e^{-(\lambda+\epsilon)x})$ and hence

$$\mathbf{P}\{X \geq x\} = \mu_1[x, \infty) + \mu_2[x, \infty) = u\, e^{-\lambda x} + o(e^{-(\lambda+\epsilon)x}).$$

$\square$

Proof of Theorem 7.9. We will do the case $x = 0$; given this, the other cases can be handled easily. Let $D = D(0)$, $a = 2/\kappa$, $\lambda = 1 - (\kappa/8) = (4a-1)/4$. Recall (see, e.g. [**64**, (1.1.4)]) that

$$\Gamma(z) = \sum_{k=0}^{\infty} \frac{(-1)^k}{k!}\, \frac{1}{z+k} + \int_1^\infty e^{-t} t^{z-1}\, dt.$$

In particular,

$$\Gamma(a - z)\, \Gamma(a + z) = \frac{2a}{a^2 - z^2} + H(z),$$

where H is analytic on $\{|z| < a+1\}$. Therefore,

$$\mathbf{E}[e^{itD}] = \frac{c}{\lambda - it} + v(t),$$

where v is analytic in $\{|z| < \lambda + \epsilon\}$ for some $\epsilon > 0$. Lemma 7.11 then implies (7.14). $\square$

CHAPTER 8

Brownian intersection exponent

8.1. Dimension of exceptional sets

The first major application of the Schramm-Loewner evolution was to compute the planar Brownian intersection exponents from which the Hausdorff dimension of exceptional sets of the Brownian path can be determined. In this chapter we will discuss this application. Let B_t denote a complex Brownian motion starting at the origin. Let K_t be the hull generated by B_t, i.e., $\mathbb{C} \setminus K_t$ is the unbounded component of $\mathbb{C} \setminus B[0, t]$.

DEFINITION 8.1.

- V_s^{cut} is the set of *cut times* of $B[0, s]$, i.e., the set of $t \in [0, s]$ such that $B[0, t) \cap B(t, s] = \emptyset$.
- V_s^{fron} is the set of *frontier times* of $B[0, s]$, i.e., the set of $t \in [0, s]$ such that $B_t \in \partial K_s$.
- V_s^{pion} is the set of *pioneer times* in $[0, s]$, i.e., the set of $t \in [0, s]$ such that $B_t \in \partial K_t$.

We write just $V^{\text{cut}}, V^{\text{fron}}, V^{\text{pion}}$ for $V_1^{\text{cut}}, V_1^{\text{fron}}, V_1^{\text{pion}}$. We call elements of $B(V_s^{\text{cut}})$ *cut points*, elements of $B(V_s^{\text{pion}})$ *pioneer points*, and $B(V_s^{\text{fron}}) = \partial K_s$ is called the *frontier* or *outer boundary*. Note that a point is a "pioneer" if it was on the frontier at some time. In this section, we will show that the proof of the following theorem reduces to certain estimates about Brownian probabilities. The remainder of this chapter will concern these and related estimates. The proofs of the estimates will use SLE_6.

THEOREM 8.2. *For every $s > 0$, w.p.1,*

$$\dim_h[B(V_s^{\text{cut}})] = \frac{3}{4}, \quad \dim_h[B(V_s^{\text{fron}})] = \frac{4}{3}, \quad \dim_h[B(V_s^{\text{pion}})] = \frac{7}{4},$$

where $\dim_h$ denotes Hausdorff dimension.

We start by stating Theorem 8.3 which gives some particular cases of a more general result proved in this chapter. It will be useful to set up some notation that will be used throughout the chapter. Let $B_t^1, B_t^2, \ldots$ denote independent complex Brownian motions. We assume that they start at the origin, unless stated otherwise. If we need to specify the starting points we will use notation $\mathbf{E}^{z_1,\ldots,z_k}$ and $\mathbf{P}^{z_1,\ldots,z_k}$ to denote expectations and probabilities assuming $B_0^1 = z_1, \ldots, B_0^k = z_k$. Let

$$T_n^j = \inf\{t \geq 0 : |B_t^j| \geq e^n\},$$

and let $\mathcal{G}_n$ denote the σ-algebra generated by $\{B_t^j : j = 1, 2, \ldots; 0 \leq t \leq T_n^j\}$. Here n can take on any nonnegative real value. Let $\Gamma_{m,n}^j = B^j[T_m^j, T_n^j]$ and $\Gamma_n^j =$

$\Gamma^j_{0,n} = B^j[T^j_0, T^j_n]$. Let $\mathcal{D}^j_n$ denote the connected component of $e^n \mathbb{D} \setminus (\Gamma^1_n \cup \cdots \cup \Gamma^j_n)$ containing the origin. Let $\Upsilon^j_n, \Upsilon^j_{n,-}$ be the events

$$\Upsilon^j_n = \left\{ B^j[T^j_{n-(1/2)}, T^j_n] \subset \{z : \mathrm{Re}(z) \geq e^{n-1}\} \right\},$$

$$\Upsilon^j_{n,-} = \left\{ B^j[T^j_{n-(1/2)}, T^j_n] \subset \{z : \mathrm{Re}(z) \leq -e^{n-1}\} \right\}.$$

If B is another complex Brownian motion, we write $T_n, \Upsilon_n, \Upsilon_{n,-}$ for the corresponding quantities and we write $\tilde{\mathcal{G}}_n$ for the σ-algebra generated by $\{B_t : 0 \leq t \leq T_n\}$.

THEOREM 8.3. *There exist c_1, c_2 such that for every $n \geq 0$,*

$$c_1 e^{-5n/4} \leq \mathbf{P}\{\Gamma^1_n \cap \Gamma^2_n = \emptyset; \Upsilon^1_n \cap \Upsilon^2_{n,-}\}$$
$$\leq \mathbf{P}\{\Gamma^1_n \cap \Gamma^2_n = \emptyset\} \leq c_2 e^{-5n/4},$$

$$c_1 e^{-2n/3} \leq \mathbf{P}\{-e^n \in \partial \mathcal{D}^2_n; \Upsilon^1_n \cap \Upsilon^2_n\}$$
$$\leq \mathbf{P}\{\partial \mathcal{D}^2_n \cap \partial [e^n \mathbb{D}] \neq \emptyset\} \leq c_2 e^{-2n/3},$$

$$c_1 e^{-n/4} \leq \mathbf{P}\{-e^n \in \partial \mathcal{D}^1_n; \Upsilon^1_n\}$$
$$\leq \mathbf{P}\{\partial \mathcal{D}^1_n \cap \partial [e^n \mathbb{D}] \neq \emptyset\} \leq c_2 e^{-n/4}.$$

We will now show how to derive Theorem 8.2 from Theorem 8.3. We will make use of facts about Hausdorff dimension found in §A. Since the three cases are similar, we will consider only one case, V^{cut}_s. By scaling we may choose $s = 1$, and using Theorem A.7, we can see that it suffices to show that w.p.1 $\dim_h[V^{\mathrm{cut}}] = 3/8$. Note that if $j \leq n-1$,

$$\mathbf{P}\{\Gamma^1_n \cap \Gamma^2_n = \emptyset \mid \mathcal{G}_j\} \leq 1\{\Gamma^1_j \cap \Gamma^2_j = \emptyset\} \mathbf{P}\{\Gamma^1_{j+1,n} \cap \Gamma^2_{j+1,n} = \emptyset \mid \mathcal{G}_j\}$$
$$\leq c 1\{\Gamma^1_j \cap \Gamma^2_j = \emptyset\} \mathbf{P}\{\Gamma^1_{n-j-1} \cap \Gamma^2_{n-j-1} = \emptyset\}$$
(8.1) $$\leq c e^{-5(n-j)/4} 1\{\Gamma^1_j \cap \Gamma^2_j = \emptyset\}.$$

The second inequality above uses the Harnack inequality and Brownian scaling. The next proposition shows that Theorem 8.3 implies a similar result for fixed times:

$$\mathbf{P}\{B^1[1,t] \cap B^2[1,t] = \emptyset\} \asymp t^{-5/8}, \quad t \to \infty.$$

Since the rotational symmetry of Brownian motion implies that T^j_0 and $B^j(T^j_0)$ are independent random variables, it suffices to prove that

$$\mathbf{P}\{B^1[T^1_0, t] \cap B^2[T^2_0, t] = \emptyset\} \asymp t^{-5/8}, \quad t \to \infty.$$

The proposition gives a slightly stronger version of this.

PROPOSITION 8.4. *There exist c_1, c_2 such that for every $n \geq 1$,*

(8.2) $\quad \mathbf{P}\{\Gamma^1_n \cap \Gamma^2_n = \emptyset; \Upsilon^1_n \cap \Upsilon^2_{n,-}; T^1_n - T^1_0 \geq e^{2n}, T^2_n - T^2_0 \geq e^{2n}\} \geq c_1 e^{-5n/4},$

(8.3) $\quad \mathbf{P}\{B^1[T^1_0, e^{2n} \wedge T^1_n] \cap B^2[T^2_0, e^{2n} \wedge T^2_n] = \emptyset\} \leq c_2 e^{-5n/4}.$

8.1. DIMENSION OF EXCEPTIONAL SETS

PROOF. Let V_n^1 be the event that
$$\mathrm{diam}(B^1[T_{n-1}^1, T_{n-1}^1 + e^{2n}]) \leq e^n \, e^{-3},$$
$$\mathrm{Re}[B^1(T_{n-1}^1 + 2\,e^{2n}) - B^1(T_{n-1}^1)] \geq e^n,$$
and
$$B^1[T_{n-1}^1, T_{n-1}^1 + 2\,e^{2n}] - B^1(T_{n-1}^1) \subset \{\mathrm{Re}(z) \geq -e^{-3}e^n; |\mathrm{Im}(z)| \leq e^{-3}e^n\}.$$
Then $\mathbf{P}(V_n^1) \geq p > 0$ for some p independent of n. Let V_n^2 be the same event with T_{n-1}^1 replaced with T_{n-1}^2 and B^1 replaced with $\tilde{B}_t^2 := -\mathrm{Re}[B_t^2] + i\mathrm{Im}[B_t^2]$. Then $\mathbf{P}(V_n^2) \geq p$. By the strong Markov property,
$$\mathbf{P}\{\Gamma_n^1 \cap \Gamma_n^2 = \emptyset; \Upsilon_n^1 \cap \Upsilon_{n,-}^2; T_n^1 - T_0^1 \geq e^{2n}, T_n^2 - T_0^2 \geq e^{2n}\}$$
$$\geq \mathbf{P}\{\Gamma_{n-1}^1 \cap \Gamma_{n-1}^2; \Upsilon_{n-1}^1 \cap \Upsilon_{n-1,-}^2 \cap V_n^1 \cap V_n^2\}$$
$$\geq p^2 \, \mathbf{P}\{\Gamma_{n-1}^1 \cap \Gamma_{n-1}^2; \Upsilon_{n-1}^1 \cap \Upsilon_{n-1,-}^2\}.$$
Hence, the first inequality follows from the lower bound in Theorem 8.3.

It suffices to prove the second inequality for integer n. We first note that it is easy and standard to check that there exist c, r such that for every b'
$$(8.4) \qquad \mathbf{P}\{T_n^1 - T_{n-1}^1 \geq b' \, e^{2n} \mid \mathcal{G}_{n-1}\} \leq c \, e^{-rb'}.$$
We claim that there exist c, α such that for every $b \geq 1$ and every $n \geq 1$,
$$(8.5) \qquad \mathbf{P}\{\Gamma_{n-1}^1 \cap \Gamma_{n-2}^2 = \emptyset; T_n^1 - T_0^1 \geq b e^{2n}\} \leq c \, e^{-\alpha b} \, e^{-5n/4}.$$
To see this, first note that if $T_n^1 - T_0^1 \geq b e^{2n}$, then there exists an $m \in \{1, 2, \ldots, n\}$ such that
$$T_m^1 - T_{m-1}^1 \geq b \, 2^{m-n-1} \, e^{2n} \geq b \, 2^{n-m-1} \, e^{2m}.$$
But using (8.1), (8.4) and the strong Markov property, we see that
$$\mathbf{P}\{\Gamma_{n-1}^1 \cap \Gamma_{n-1}^2 = \emptyset; T_m^1 - T_{m-1}^1 \geq 2^{n-m-1} b e^{2m}\} \leq c \, e^{-5n/4} \, \exp\{-rb2^{n-m-1}\}.$$
Summing over m gives (8.5). By symmetry (8.5) also holds if we replace $T_n^1 - T_0^1$ with $T_n^2 - T_0^2$. But
$$\mathbf{P}\{B^1[T_0^1, e^{2n} \wedge T_n^1] \cap B^2[T_0^2, e^{2n} \wedge T_n^2] = \emptyset\}$$
$$\leq \mathbf{P}\{B^1[T_0^1, T_n^1] \cap B^2[T_0^2, T_n^2] = \emptyset\}$$
$$+ \sum_{j=1}^{2} \sum_{m=1}^{\infty} \mathbf{P}\{\Gamma_{n-m}^1 \cap \Gamma_{n-m}^2 = \emptyset; T_{n-m}^j \leq e^{2n} \leq T_{n-m+1}^j\}.$$
So (8.5) implies the second inequality. $\square$

We will now show that if B is a Brownian motion, then
$$\mathbf{P}\{\mathrm{dim}_h(V^{\mathrm{cut}}) \leq \frac{3}{8}\} = 1.$$
$$\mathbf{P}\{\mathrm{dim}_h(V^{\mathrm{cut}} \cap [\frac{1}{4}, \frac{3}{4}]) = \frac{3}{8}\} > 0.$$
Note that there is a positive probability that $V^{\mathrm{cut}} \cap [1/4, 3/4] = \emptyset$. Let $I(j, n)$ be the interval $[(j-1)2^{-n}, j2^{-n}]$ and let $J(j, n)$ be the indicator function of the event

$B[0, (j-1)2^{-n}] \cap B[j2^{-n}, 1] = \emptyset$. By Propositions A.13 and A.16, it suffices to show that

(8.6) $\qquad \mathbf{E}[J(j,n)] \asymp 2^{-5n/8}, \quad \frac{1}{4}2^n \leq j \leq \frac{3}{4}2^n,$

(8.7) $\qquad \mathbf{E}[J(j,n)\,J(k,n)] \leq c\,2^{-5n/8}\,(k-j+1)^{-5/8}, \quad \frac{1}{4}2^n \leq j \leq k \leq \frac{3}{4}2^n.$

But, (8.6) follows immediately from (8.2) and (8.3), and it is not difficult to derive (8.7) from (8.3) by bounding $\mathbf{E}[J(j,n)\,J(k,n)]$ by the probability that

$$B[\frac{j}{2^n} - \frac{k-j}{2\cdot 2^n}, \frac{j-1}{2^n}] \cap B[\frac{j}{2^n}, \frac{j}{2^n} + \frac{k-j}{2\cdot 2^n}] = \emptyset,$$

$$B[\frac{k}{2^n} - \frac{k-j}{2\cdot 2^n}, \frac{k-1}{2^n}] \cap B[\frac{k}{2^n}, \frac{k}{2^n} + \frac{k-j}{2\cdot 2^n}] = \emptyset,$$

$$B[0, \frac{j}{2^n} - \frac{k-j}{2\cdot 2^n}] \cap B[\frac{k}{2^n} + \frac{k-j}{2\cdot 2^n}, 1] = \emptyset.$$

We finish this section by sketching the argument that shows that $\dim_h(V^{\text{cut}}) = 3/4$ w.p.1. Let R_n be the event that

$$\dim_h(V^{\text{cut}} \cap B[T_{n-(1/2)}, T_{n-(1/4)}]) = 3/4,$$

$$B[T_{n-(1/2)}, T_n] \cap \mathcal{B}(0, e^{n-(3/4)}) = \emptyset,$$

$$B[T_{n-(1/8)}, T_n] \cap \mathcal{B}(0, e^{n-(1/4)}) = \emptyset.$$

Note that R_n depends only on $B[T_{n-1}, T_n]$. A similar argument as the previous paragraph shows that $\mathbf{P}(R_n) = c > 0$ for some c independent of n. For $n \geq 1$, let A_n be the event

$$A_n = \{T_{-n} \leq 1; B[T_{-n}, 1] \cap \mathcal{B}(0, e^{-n-(1/8)}) = \emptyset\}.$$

Then $\mathbf{P}(A_n) \asymp n^{-1}$, and it is not difficult to show that w.p.1, A_n occurs infinitely often. From this, we can show that $R_{-n} \cap A_n$ occurs infinitely often. But on the event $R_{-n} \cap A_n$,

$$\dim_h(V^{\text{cut}} \cap [\mathcal{B}(0, e^{-n}) \setminus \mathcal{B}(0, e^{-n-1})]) = 3/4.$$

8.2. Subadditivity

We will be using subadditivity in our arguments, so it will be useful to review a well known fact.

LEMMA 8.5. *Suppose b_n is a sequence of real numbers such that for each n, m, $b_{n+m} \leq b_n + b_m$. Then*

$$\lim_{n \to \infty} n^{-1} b_n = \inf_n n^{-1} b_n.$$

PROOF. If N is a positive integer, we can write any positive integer n as $mN + r$ where m, r are nonnegative integers with $r < N$. Let $\bar{b}_N = \max\{b_0, \ldots, b_{N-1}\}$. Since $b_n \leq m\,b_N + b_r$, we get

$$\limsup_{n \to \infty} n^{-1} b_n \leq \limsup_{n \to \infty} n^{-1} [m\,b_N + \bar{b}_N] = N^{-1} b_N.$$

Since this holds for every N, the lemma follows. $\square$

LEMMA 8.6. *Suppose r_n is a sequence of positive numbers and α_1, α_2 are positive numbers such that for all n, m,*

(8.8)
$$\alpha_1 r_n r_m \leq r_{n+m} \leq \alpha_2 r_n r_m.$$

Then the limit

(8.9)
$$\beta := \lim_{n \to \infty} n^{-1} \log r_n,$$

exists. Moreover, for all n,

$$\alpha_2^{-1} e^{\beta n} \leq r_n \leq \alpha_1^{-1} e^{\beta n}.$$

PROOF. Let $b_n = \log r_n + \log \alpha_2$, $\tilde{b}_n = -\log r_n - \log \alpha_1$. Then $b_{n+m} \leq b_n + b_m$, $\tilde{b}_{n+m} \leq \tilde{b}_n + \tilde{b}_m$. The previous lemma tells us that $n^{-1} b_n$ has a limit, which must be the same as the β in (8.9), and $b_n \geq n\beta$. Similarly, $-\tilde{b}_n \geq -n\beta$. □

REMARK 8.7. Note that if r_n satisfies (8.8), so does $r_n e^{\beta_1 n}$. Hence establishing an inequality such as (8.8) does not help determine the exponent β.

8.3. Half-plane or rectangle exponent

If $k_1, \ldots, k_j$ are positive integers, D a domain whose boundary is a finite union of Jordan curves, and ∂, ∂' disjoint analytic arcs on ∂D we let

$$\mu_{D,\partial,\partial'}^{k_1,\ldots,k_j} = \mu_D(\partial, \partial')^{\otimes k_1} \times \cdots \times \mu_D(\partial, \partial')^{\otimes k_j}.$$

Here $\mu_D(\partial, \partial')$ is the Brownian excursion measure restricted to paths going from ∂ to ∂' as in (5.10). Although $\mu_{D,\partial,\partial'}^{k_1,\ldots,k_j}$ is the same as $\mu_{D,\partial,\partial'}^{k_1+\cdots+k_j}$, we write it this way to distinguish between the paths. We imagine having j colors and coloring the first k_1 paths color 1, the next k_2 paths color 2, and so on. We write just $\mu_L^{k_1,\ldots,k_j}$ if $D = \mathcal{R}_L, \partial = \partial_1, \partial' = \partial_2$ as in §3.7. Let $F(L) = |\mu_{\mathcal{R}_L}(\partial_1, \partial_2)|$ where $|\cdot|$ denotes total mass. In (5.11) we noted that there is a $c_0 > 0$ such that

$$F(L) = c_0 e^{-L} + O(e^{-3L}), \quad L \to \infty.$$

Suppose that a sample is chosen from $\mu_L^{k_1,\ldots,k_j}$ and let us write $\gamma^l, l = 1, \ldots, j$, for the union of all the paths of color l. Let $D_{l,+}$ (resp., $D_{l,-}$) be the connected component of $D \setminus \gamma^l$ whose boundary contains πi (resp., contains 0). Let $D^1 = D_{1,+}, D^{j+1} = D_{j,-}$ and for $2 \leq l \leq j$, let $D^l = D_{l-1,-} \cap D_{l,+}$. Let $\mathcal{L}^l = L(D^l; \partial_1, \partial_2)$ where L denotes π-extremal distance. Conformal invariance tells us that $|\mu_{D^l}(\partial_1, \partial_2)| = F(\mathcal{L}^l)$. Let

$$\Phi_L(\lambda_1, k_1, \lambda_2, k_2, \ldots, k_j, \lambda_{j+1}) = \mu_L^{k_1,\ldots,k_j}\left[\prod_{l=1}^{j+1} F(\mathcal{L}^l)^{\lambda_l}\right].$$

Here we use the shorthand notation $\mu(f) = \int f\, d\mu$, and, as before, $F(\infty)^\lambda = 0$ even if $\lambda = 0$. We can also define

$$\Phi_{D,\partial,\partial'}(\lambda_1, k_1, \lambda_2, k_2, \ldots, k_j, \lambda_{j+1})$$

similarly, and conformal invariance of excursion measure shows that this is a conformal invariant.

We will now give a lower bound on Φ_L which will show that it decays no faster than exponentially in L. Suppose we divide $\mathcal{R}_L$ into $(2j+1)$ rectangles, $R_1^*, R_1, R_2^*, R_2, \ldots, R_j, R_{j+1}^*$ each of length L and height $\pi/(2j+1)$ ordered from highest to lowest. Note that scaling shows that the π-extremal distance in these

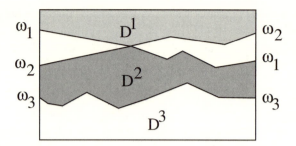

FIGURE 8.1. An example with $j=2, k_1=2, k_2=1$

rectangles between the vertical boundaries is $(2j+1)L$. The $\mu_l^{k_1,\ldots,k_j}$ measure of the set of excursions such that the paths of color l stay in R_l is bounded below by $c^{k_1+\cdots+k_j} \exp\{-(k_1+\cdots+k_j)(2j+1)L\}$, and on this set of paths, $F(\mathcal{L}^l) \geq ce^{-(2j+1)L}$. Hence,

(8.10)
$$\Phi_L(\lambda_1, k_1, \ldots, k_j, \lambda_{j+1}) \geq c^{\lambda_1+k_1+\cdots+k_j+\lambda_{j+1}} e^{-(\lambda_1+k_1+\lambda_2+\cdots+k_j+\lambda_{j+1})(2j+1)L}.$$

By symmetry we know that

$$\Phi_{D,\partial,\partial'}(\lambda_1, k_1, \ldots, k_j, \lambda_{j+1}) = \Phi_{D,\partial,\partial'}(\lambda_{j+1}, k_j, \ldots, k_1, \lambda_1).$$

In fact, as the next proposition goes, Φ_L is invariant under all permutations.

PROPOSITION 8.8. *If $k_1, \ldots, k_j$ are positive integers, $\lambda_1, \ldots, \lambda_{j+1}$ are nonnegative real numbers and $\sigma: \{1,\ldots,j\} \to \{1,\ldots,j\}, \theta: \{1,\ldots,j+1\} \to \{1,\ldots,j+1\}$ are permutations then*

$$\Phi_L(\lambda_1, k_1, \ldots, k_j, \lambda_{j+1}) = \Phi_L(\lambda_{\theta(1)}, k_{\sigma(1)}, \ldots, k_{\sigma(j)}, \lambda_{\theta(j+1)}).$$

PROOF. If $m < j$,

$$\mu_L^{k_1,\ldots,k_j}\left[\prod_{l=1}^{j+1} F(\mathcal{L}^l)^{\lambda_l}\right]$$

$$= (\mu_L^{k_1,\ldots,k_m} \times \mu_L^{k_{m+1},\ldots,k_j})\left[\prod_{l=1}^{j+1} F(\mathcal{L}^l)^{\lambda_l}\right]$$

$$= \mu_L^{k_1,\ldots,k_m}\left[(\prod_{l=1}^{m} F(\mathcal{L}^l)^{\lambda_l}) \Phi_{D_m,-,\partial_1,\partial_2}(\lambda_{m+1}, k_{m+1}, \ldots, k_j, \lambda_{j+1})\right]$$

$$= \mu_L^{k_1,\ldots,k_m}\left[(\prod_{l=1}^{m} F(\mathcal{L}^l)^{\lambda_l}) \Phi_{D_m,-,\partial_1,\partial_2}(\lambda_{j+1}, k_j, \ldots, k_m, \lambda_{m+1})\right]$$

$$= \mu_L^{k_1,k_m,k_j,k_{j-1},\ldots,k_{m+1}}\left[\prod_{l=1}^{j+1} F(\mathcal{L}^l)^{\tilde{\lambda}_l}\right],$$

where $\tilde{\lambda}_l = \lambda_l$ if $l \leq m$ and $\tilde{\lambda}_l = \lambda_{j+2+m-l}$ if $l \geq m+1$. This gives

$$\Phi_L(\lambda_1, k_1, \ldots, k_j, \lambda_{j+1}) = \Phi_L(\lambda_1, k_1, \ldots, k_m, \lambda_{j+1}, k_j, \ldots, k_{m+1}, \lambda_{m+1}).$$

Repeated use of this gives the proposition. □

8.3. HALF-PLANE OR RECTANGLE EXPONENT

DEFINITION 8.9. The *rectangle or half-plane Brownian intersection exponent*
$$\tilde\xi = \tilde\xi(\lambda_1, k_1, \ldots, k_j, \lambda_{j+1})$$
is defined by
$$\Phi_L(\lambda_1, k_1, \ldots, k_j, \lambda_{j+1}) \asymp e^{-\tilde\xi L} \asymp F(L)^{\tilde\xi}, \quad L \to \infty.$$

Of course, one needs to prove that such an exponent $\tilde\xi$ exists. We will show existence in the next section by proving the following proposition. The proof of this proposition will not give the value of $\tilde\xi$; we will show in this section how to use SLE_6 to determine its value. Note that $\tilde\xi(0, k, 0) = k$ since $|\mu_L| \asymp e^{-L}$.

PROPOSITION 8.10. *For every $M < \infty$ there exist c_1, c_2 such that if $\lambda_1, \lambda_2, k \leq M$, there exists $\tilde\xi = \tilde\xi(\lambda_1, M, \lambda_2)$ such that for all $L \geq 1$,*
$$c_1 e^{-\tilde\xi L} \leq \Phi_L(\lambda_1, k, \lambda_2) \leq c_2 e^{-\tilde\xi L}.$$

PROOF. See §8.3.1. Note that by Lemma 8.6 it suffices to show that for each $M < \infty$, there exist c_1, c_2 such that for all $\lambda_1, \lambda_2, k \leq M$ and all positive integers L, L',

$$\begin{aligned}
c_1\, \Phi_L(\lambda_1, k, \lambda_2)\, \Phi_{L'}(\lambda_1, k, \lambda_2) &\leq \Phi_{L+L'}(\lambda_1, k, \lambda_2) \\
&\leq c_2\, \Phi_L(\lambda_1, k, \lambda_2)\, \Phi_{L'}(\lambda_1, k, \lambda_2).
\end{aligned} \quad (8.11)$$

We emphasize that c_1, c_2 depend only on M and not on λ_1, k, λ_2. □

PROPOSITION 8.11. *For every $M < \infty$, there exist c_1, c_2 such that if $k_1, \ldots, k_j$ are positive integers with $k_1 + \cdots + k_j \leq M$ and $0 \leq \lambda_0, \ldots, \lambda_{j+1} \leq M$ are nonnegative real numbers, then for all $L \geq 1$,*
$$c_1 e^{-\tilde\xi L} \leq \Phi(\lambda_1, k_1, \ldots, k_j, \lambda_{j+1}) \leq c_2 e^{-\tilde\xi L},$$
where $\tilde\xi = \tilde\xi(\lambda_1, k_1, \ldots, k_j, \lambda_{j+1})$ is defined recursively by
$$\tilde\xi(\lambda_1, k_1, \ldots, k_j, \lambda_{j+1}) = \tilde\xi(\lambda_1, k_1, \ldots, k_{j-1}, \tilde\xi(\lambda_j, k_j, \lambda_{j+1})).$$

PROOF. Note that
$$\begin{aligned}
& \Phi_L(\lambda_1, k_1, \ldots, k_j, \lambda_{j+1}) \\
&= (\mu_L^{k_1,\ldots,k_{j-1}} \times \mu_L^{k_j}) \left[\prod_{l=1}^{j+1} F(\mathcal{L}^l)^{\lambda_l} \right] \\
&= \mu_L^{k_1,\ldots,k_{j-1}} \left[(\prod_{l=1}^{j-1} F(\mathcal{L}^l)^{\lambda_l})\, \Phi_{D_{j-1,-}}(\lambda_j, k_j, \lambda_{j+1}) \right] \\
&\asymp \mu_L^{k_1,\ldots,k_{j-1}} \left[(\prod_{l=1}^{j-1} F(\mathcal{L}^l)^{\lambda_l})\, F(L(D_{j-1,-}; \partial_1, \partial_2))^{\tilde\xi(\lambda_j, k_j, \lambda_{j+1})} \right] \\
&= \Phi_L(\lambda_1, k_1, \ldots, k_{j-1}, \tilde\xi(\lambda_j, k_j, \lambda_{j+1})).
\end{aligned}$$
□

REMARK 8.12. If k is a positive integer, we write just $\tilde\xi(k)$ for $\tilde\xi(0, k, 0) = k$. More generally, if $k_1, \ldots, k_j$ are positive integers we write
$$\tilde\xi(k_1, \ldots, k_j) = \tilde\xi(0, k_1, 0, k_2, \ldots, 0, k_j, 0).$$

The $\mu_L^{k_1,\ldots,k_j}$ measure of the set of excursions so that for each $1 \leq l_1 < l_2 \leq j$, the paths of color l_1 do not intersect the paths of color l_2 and lie "above" them is comparable to $\exp\{-L\tilde{\xi}(k_1,\ldots,k_j)\}$ as $L \to \infty$. There may appear to be some ambiguity in the notation, e.g., $\tilde{\xi}(k_1, k_2, k_3)$ could mean either $\tilde{\xi}(0, k_1, 0, k_2, 0, k_3, 0)$ or $\tilde{\xi}(\lambda_1, k_2, \lambda_2)$ with $\lambda_1 = k_1$ and $\lambda_2 = k_3$. However, it is easy to check that these two exponents are the same. In fact,

$$\Phi_L(k_1, k_2, k_3) = \Phi_l(0, k_1, 0, k_2, 0, k_3, 0).$$

Note that Proposition 8.8 implies that

$$\tilde{\xi}(k_1,\ldots,k_j) = \tilde{\xi}(k_{\sigma(1)},\ldots,k_{\sigma(j)}),$$

which is not an obvious relation.

REMARK 8.13. We can define $\tilde{\xi}(a_1,\ldots,a_j)$ for any collection of nonnegative reals $a_1,\ldots,a_j$ by: $\tilde{\xi}(a_1) = a_1$; $\tilde{\xi}(a_1, a_2)$ is defined by the relation

(8.12) $$\tilde{\xi}(0, 1, \tilde{\xi}(a_1, a_2)) = \tilde{\xi}(a_1, 1, a_2);$$

and then inductively by

$$\tilde{\xi}(a_1,\ldots,a_j) = \tilde{\xi}(a_1,\ldots,a_{j-2}, \tilde{\xi}(a_{j-1}, a_j)).$$

The definition in (8.12) is legitimate once we show that $\tilde{\xi}(0, 1, \lambda)$ is continuous and strictly increasing in λ; this is not difficult. It is straightforward to check that this definition is consistent with our previous definitions (say, when j is odd and a_l is an integer for even l), $\tilde{\xi}$ is a symmetric function of its variables, and

$$\tilde{\xi}(a_1,\ldots,a_j) = \tilde{\xi}(a_1,\ldots,a_m, \tilde{\xi}(a_{m+1},\ldots,a_j)).$$

REMARK 8.14. Suppose $k_1,\ldots,k_j$ are integers. Let

$$\tilde{\xi}^\#(k_1,\ldots,k_j) = \tilde{\xi}(k_1,\ldots,k_j) - (k_1 + \cdots + k_j).$$

This is the intersection exponent for the normalized excursion measure $[\mu_L^{k_1,\ldots,k_j}]^\#$. In other words, given that we have k_1 excursions of color 1, k_2 excursions of color 2, etc., connecting ∂_1 and ∂_2 in $\mathcal{R}_L$, the probability that no paths of different colors intersect is comparable to $\exp\{-L\tilde{\xi}^\#(k_1,\ldots,k_j)\}$.

PROPOSITION 8.15. For all $\lambda_1, \lambda_2 \geq 0$ and $0 < r < 1$,

(8.13) $$1 + (\lambda_1^{1/2} + \lambda_2^{1/2})^2 \leq \tilde{\xi}(\lambda_1, 1, \lambda_2) \leq (\lambda_1^{1/2} + \lambda_2^{1/2})^2 \frac{1}{1-r} + \frac{1}{r}.$$

The limit

(8.14) $$U(\lambda) := \lim_{n \to \infty} n^{-1}\, \xi(\underbrace{\lambda, \lambda, \cdots, \lambda}_{n})^{1/2},$$

exists and for all $\lambda_1,\ldots,\lambda_j$,

(8.15) $$U(\tilde{\xi}(\lambda_1,\ldots,\lambda_j)) = U(\lambda_1) + \cdots + U(\lambda_j).$$

PROOF. If we split $\mathcal{R}_L$ into two disjoint simply connected domains D_1, D_2, then

$$\frac{1}{L} \geq \frac{1}{L(D_1; \partial_1, \partial_2)} + \frac{1}{L(D_2; \partial_1, \partial_2)}.$$

This can be seen from (3.17). In particular

(8.16) $\qquad \lambda_1 L(D_1; \partial_1, \partial_2) + \lambda_2 L(D_2; \partial_1, \partial_2) \geq (\lambda_1^{1/2} + \lambda_2^{1/2})^2 L,$

with equality if D_1, D_2 are rectangles of horizontal length L and vertical heights $a\pi$ and $(1-a)\pi$ where $a = \lambda_1^{1/2}/(\lambda_1^{1/2} + \lambda_2^{1/2})$. The first inequality in (8.13) is obtained by choosing the excursion of "color 2" first (i.e., the "1" in $\tilde{\xi}(\lambda_1, 1, \lambda_2)$) and using (8.16). For the second inequality, we divide $\mathcal{R}_L$ into three rectangles of vertical heights $(1-r)a\pi, r\pi, (1-r)(1-a)\pi$, respectively, and consider those excursions of color 2 that stay in the middle rectangle. Plugging $r = 1/\sqrt{\lambda}$ into (8.13), we see that

$$1 + \lambda \leq \tilde{\xi}(1, \lambda) \leq \lambda + O(\sqrt{\lambda}).$$

Also,

(8.17) $\qquad \lim_{n \to \infty} n^{-1} \tilde{\xi}(n\lambda_1, n\lambda_2) = \lim_{n \to \infty} n^{-1} \tilde{\xi}(n\lambda_1, 1, n\lambda_2) = (\lambda_1^{1/2} + \lambda_2^{1/2})^2.$

Let $b_n = \tilde{\xi}(\underbrace{1, 1, \ldots, 1}_{n-1})^{1/2}$. We can write

$$b_{n+m} = \tilde{\xi}(\underbrace{1, 1, \ldots, 1}_{n-1}, 1, \underbrace{1, 1, \ldots, 1}_{m-1})^{1/2} = \tilde{\xi}(\tilde{\xi}(\underbrace{1, \ldots, 1}_{n-1}), 1, \tilde{\xi}(\underbrace{1, \ldots, 1}_{m-1}))^{1/2},$$

and hence the first inequality in (8.13) shows that $b_{n+m} \geq b_n + b_m$. Therefore, by Lemma 8.5 the limit in (8.14) exists for $\lambda = 1$. For other $\lambda > 0$, note that if $\tilde{\xi}(\underbrace{\lambda, \cdots, \lambda}_{j}) \leq \tilde{\xi}(\underbrace{1, \ldots, 1}_{m})$,, then

$$\limsup_{n \to \infty} n^{-1} \tilde{\xi}(\underbrace{\lambda, \cdots, \lambda}_{n}) \leq \frac{m}{j} U(1),$$

and if $\tilde{\xi}(\underbrace{\lambda, \cdots, \lambda}_{j}) \geq \tilde{\xi}(\underbrace{1, \ldots, 1}_{m})$,

$$\liminf_{n \to \infty} n^{-1} \tilde{\xi}(\underbrace{\lambda, \cdots, \lambda}_{n}) \geq \frac{m}{j} U(1).$$

Also, using the facts that $\tilde{\xi}(\underbrace{\lambda, \cdots, \lambda}_{j}) \to \infty$ and $b_{n+1}/b_n \to 1$, we can see that the set of numbers $\tilde{\xi}(\underbrace{\lambda, \cdots, \lambda}_{j})/\tilde{\xi}(\underbrace{1, \cdots, 1}_{m})$ is dense in the positive reals and hence the limit in (8.14) exists with

$$U(\lambda) = U(1) \inf\{m/j : \tilde{\xi}(\underbrace{\lambda, \cdots, \lambda}_{j}) \leq \tilde{\xi}(\underbrace{1, \cdots, 1}_{m})\}.$$

Since

$$\tilde{\xi}(\tilde{\xi}(\underbrace{\lambda_1, \ldots, \lambda_1}_{n}), \tilde{\xi}(\underbrace{\lambda_2, \ldots, \lambda_2}_{n})) = \tilde{\xi}(\underbrace{\tilde{\xi}(\lambda_1, \lambda_2), \ldots, \tilde{\xi}(\lambda_1, \lambda_2)}_{n}),$$

we can conclude (8.15) using (8.17). $\qquad\square$

Unfortunately, the proof of the last proposition does not tell us how to compute the function U. Let

$$(8.18) \qquad \eta(t,x) = x + \frac{t^2 + t\sqrt{1+24x}}{6},$$

and let $\eta(x) = \eta(1,x)$ so that η agrees with the η in Proposition 6.43. Note that $\eta(0,x) = x$ and that η satisfies $\eta(t+s,x) = \eta(t,\eta_s(x))$. The next proposition relates the SLE_6 exponent η to the Brownian exponent $\tilde{\xi}$.

PROPOSITION 8.16. *For every positive integer k and nonnegative real λ_1, λ_2,*

$$\eta(\tilde{\xi}(\lambda_1, k, \lambda_2)) = \tilde{\xi}(\eta(\lambda_1), k, \lambda_2).$$

PROOF. Let γ be an SLE_6 curve from πi to $L + \pi i$ in $\mathcal{R}_L$ and let D be the domain "under" γ in $\mathcal{R}_L$ as in Proposition 6.43. Let $\gamma_1, \ldots, \gamma_k$ be excursions from ∂_1 to ∂_2 in $\mathcal{R}_L$ under the measure μ_L^k and let $\gamma^1 = \gamma_1 \cup \cdots \cup \gamma_k$. Let D_+, D_- be the domains above and below γ^1, respectively, as described in this section, let $\mathcal{D} = D \cap D_+$, and let $\mathcal{L}_1 = L(\mathcal{D}; \partial_1, \partial_2), \mathcal{L}_2 = L(D_-; \partial_1, \partial_2)$. We will consider

$$(\mathbf{P} \times \mu_L^k)\left[e^{-\lambda_1 \mathcal{L}_1 - \lambda_2 \mathcal{L}_2}\right],$$

where $\mathbf{P}$ denotes the probability measure under which the SLE_6 path is defined. Note that given γ,

$$\mu_L^k[e^{-\lambda_1 \mathcal{L}_1 - \lambda_2 \mathcal{L}_2}] \asymp \exp\{-\tilde{\xi}(\lambda_1, k, \lambda_2)\, L(D; \partial_1, \partial_2)\}.$$

Hence, by Proposition 6.43,

$$(\mathbf{P} \times \mu_L^k)\left[e^{-\lambda_1 \mathcal{L}_1 - \lambda_2 \mathcal{L}_2}\right] \asymp e^{-\eta(\tilde{\xi}(\lambda_1, k, \lambda_2))\, L}.$$

However, if we fix γ^1, then Proposition 6.45 states that

$$(8.19) \qquad \mathbf{P}[e^{-\lambda_1 \mathcal{L}}] \asymp \exp\{-\eta(\lambda_1)\, L(D_+; \partial_1, \partial_2)\}.$$

Hence,

$$(\mathbf{P} \times \mu_L^k)\left[e^{-\lambda_1 \mathcal{L}_1 - \lambda_2 \mathcal{L}_2}\right] \asymp e^{-\tilde{\xi}(\eta(\lambda_1), k, \lambda_2)\, L}.$$

This gives the proposition. □

REMARK 8.17. The step in the last proof that uses $\kappa = 6$ is (8.19) because the proof of Proposition 6.45 uses the locality property.

Proposition 8.16 allows us to determine a number of values of $\tilde{\xi}$. For example,

$$\tilde{\xi}(1/3, 1, 0) = \tilde{\xi}(\eta(0), 1, 0) = \eta(\tilde{\xi}(0, 1, 0)) = \eta(1) = 2,$$

$$\tilde{\xi}(1, 1) = \tilde{\xi}(\eta(1/3), 1, 0) = \eta(\tilde{\xi}(1/3, 1, 0)) = \eta(2) = 10/3,$$

$$\tilde{\xi}(1, \tilde{\xi}(1/3, 1/3)) = \tilde{\xi}(1/3, 1, 1/3) = \eta(\tilde{\xi}(1, 1/3)) = \eta(2) = 10/3.$$

Since $\lambda \mapsto \tilde{\xi}(1, \lambda)$ is strictly increasing, the last two lines give $\tilde{\xi}(1/3, 1/3) = 1$. Let $\tilde{\xi}_n = \tilde{\xi}(\underbrace{1/3, 1/3, \ldots, 1/3}_{n})$ and note that $\tilde{\xi}_0 = 0, \tilde{\xi}_1 = 1/3 = \eta(0), \tilde{\xi}_2 = 1 = \eta(1/3), \tilde{\xi}_3 = \tilde{\xi}(1/3, \tilde{\xi}(1/3, 1/3)) = \tilde{\xi}(1/3, 1) = 2 = \eta(1)$. Also if $n \geq 3$, $\tilde{\xi}_n = \tilde{\xi}(1/3, 1, \tilde{\xi}_{n-3}) = \eta(\tilde{\xi}(1, \tilde{\xi}_{n-3})) = \eta(\tilde{\xi}_{n-1})$. By induction we get $\tilde{\xi}_n = n(n+1)/6$ and

$$\tilde{\xi}(\underbrace{1, 1, \ldots, 1}_{n}) = \tilde{\xi}_{2n} = \frac{n(2n+1)}{3}.$$

COROLLARY 8.18. *For all $\lambda \geq 0$,*

$$\tilde{\xi}(1/3, \lambda) = \eta(\lambda) = \lambda + \frac{1}{6} + \frac{1}{6}\sqrt{1 + 24\lambda},$$

$$\tilde{\xi}(1, \lambda) = \eta \circ \eta(\lambda) = \lambda + \frac{2}{3} + \frac{1}{3}\sqrt{1 + 24\lambda},$$

and, more generally,

$$\tilde{\xi}(\tilde{\xi}_n, \lambda) = \eta(n, \lambda),$$

where $\tilde{\xi}_n = \underbrace{(1/3, \ldots, 1/3)}_{n}$ and $\eta(t, \lambda)$ is as in (8.18).

PROOF. By Proposition 8.16,

$$\tilde{\xi}(1, \tilde{\xi}(1/3, \lambda)) = \tilde{\xi}(1/3, 1, \lambda) = \eta(\tilde{\xi}(0, 1, \lambda)) = \tilde{\xi}(0, 1, \eta(\lambda)) = \tilde{\xi}(1, \eta(\lambda)).$$

Since $\lambda \mapsto \tilde{\xi}(1, \lambda)$ is strictly increasing this shows that $\tilde{\xi}(1/3, \lambda) = \eta(\lambda)$. Also,

$$\tilde{\xi}(1, \lambda) = \tilde{\xi}(1/3, 1/3, \lambda) = \tilde{\xi}(1/3, \tilde{\xi}(1/3, \lambda)) = \eta \circ \eta(\lambda),$$

and similarly $\tilde{\xi}(\tilde{\xi}_n, \lambda) = \tilde{\xi}(1/3, \tilde{\xi}_{n-1}, \lambda) = \eta(\tilde{\xi}(\tilde{\xi}_{n-1}, \lambda))$. □

The crossing exponent we have computed for SLE_6 is not sufficient for finding all of the values of U; however, a more general "two-sided" SLE_6 crossing exponent can be computed which can then be used to find U.

PROPOSITION 8.19.

$$U(s) = \frac{\sqrt{24s + 1} - 1}{\sqrt{24}}.$$

PROOF. See [53]. □

In particular,

$$\tilde{\xi}(k, \lambda) = \eta(t_k, \lambda),$$

where $t_k = (\sqrt{1 + 24k} - 1)/2 = \sqrt{6}\, U(k)$. The relation $t_{\tilde{\xi}(j,k)} = t_j + t_k$ can be seen as the "flow" relation $\eta(t_j + t_k, \lambda) = \eta(t_j, \eta(t_k, \lambda))$.

8.3.1. Proof of Proposition 8.10. We will give a proof of Proposition 8.10 in the case $\lambda_1 = 0, k = 1, \lambda_2 = \lambda \in [0, M]$. The other cases are essentially the same, but the notation is a little nicer if we restrict to this case. We fix $M < \infty$ and allow all constants $c, c_1, c_2, \ldots$ in this section to depend on M; however, they are not allowed to depend on λ. We will write just $\tilde{\xi}$ for $\tilde{\xi}(0, 1, \lambda) = \tilde{\xi}(1, \lambda)$ and Φ_L for $\Phi_L(0, 1, \lambda)$. As pointed out in (8.11), it suffices to find constants c_1, c_2 such that for all positive integers L, L',

$$c_1 \Phi_L \Phi_{L'} \leq \Phi_{L+L'} \leq c_2 \Phi_L \Phi_{L'}.$$

We will set up some notation. Let $\mu_L = \mu_L^1$ be the excursion measure on $\mathcal{R}_L$ restricted to curves connecting ∂_1 and ∂_2. Let γ denote an excursion, and let $D = D_-^1$ be the connected component of $\mathcal{R}_L \setminus \gamma$ that contains the origin on its boundary. We write just F_L for $F(L(D; \partial_1, \partial_2))$. Recall that $F_L \asymp \exp\{-L(D; \partial_1, \partial_2)\}$. This implies that $F_L^\lambda \asymp \exp\{-\lambda L(D; \partial_1, \partial_2)\}$, but note that in order to choose the implicit constants in the $\asymp$ notation uniformly in λ we must restrict to $0 \leq \lambda \leq M$.

When we use the $\asymp$ notation in this section we will require that the constants are uniform in $\lambda \leq M$.

Let $B_t = X_t + iY_t$ be a complex Brownian motion. Let
$$S_L = \inf\{t : X_t = L\}, \quad S_0^* = \sup\{t \leq S_1 : X_t = 0\},$$

$$\tau_L = \inf\{t > S_0^* : B_t \notin \mathcal{R}_L\}, \quad \hat{B}_t = B_{t+S_0^*},$$

$$T^* = \inf\{t > S_1 : X_t = 0\}.$$

and let V_L be the event that $\tau_L = S_L$. Let $\mathcal{F}_L$ be the σ-algebra generated by $\{B_t : t \leq S_L\}$. Let $\mathcal{D}_L = \{z : 0 < \text{Re}(z) < L\}$ and let ν_L denote the excursion measure on $\mathcal{D}_L$ restricted to excursions going from $\{\text{Re}(z) = 0\}$ to $\{\text{Re}(z) = L\}$. Note that μ_L can be obtained from ν_L by restricting ν_L to excursions γ with $\gamma(0, t_\gamma) \subset \mathcal{R}_L$. If $L \geq 1$ and we start a Brownian motion with Lebesgue measure on $\{\text{Re}(z) = 0\}$ (this is an infinite initial measure, but this is well defined), then the distribution of $\hat{B}_t, 0 \leq t \leq S_L - S_0^*$, restricted to the event $\{S_L < T^*\}$ is ν_L. In our proof, we will use the following simple facts about this construction that we leave to the reader.

- The initial part $X_t, 0 \leq t \leq S_0^*$, and the latter part $X_t, S_0^* \leq t < \infty$, are independent.
- The conditional distribution of $B_t, S_0^* \leq t < \infty$, given $B_{S_0^*}$ is independent of $B_t, 0 \leq t \leq S_0^*$, and is equal to a translation of the conditional distribution given $B_{S_0^*} = 0$.
- There is a c such that if $B_0 = iy, y \in \mathbb{R}$, then the distribution of $B_t, S_0^* \leq t \leq S_L$, restricted to the event $S_L < T^*$ is bounded above by $c\nu_L$.
- There is a c such that if $\epsilon > 0$ and $B_0 = 0$, then the distribution of $B_t, S_0^* \leq t \leq S_L$ restricted to the event

(8.20) $$\{B[0, S_1] \subset \mathcal{B}(0, \epsilon) \cup \{\text{Re}(z) > 0\}\}$$

is bounded below by $c\epsilon\nu_{L,\epsilon}$ where $\nu_{L,\epsilon}$ denotes ν_L restricted to excursions whose starting point is in $\mathcal{B}(0, \epsilon/2)$. Note that the gambler's ruin estimate shows that the probability of the event (8.20) is greater than $c'\epsilon$.

On the event V_L, let D_L be the connected component of $\mathcal{R}_L \setminus B[S_0^*, S_L]$ whose boundary contains 0 and let $\mathcal{L}_L = L(D_L; \partial_1, \partial_2)$. Then,

$$\Phi_L = \int_{-\infty}^{\infty} \mathbf{E}^{iy}[F_L^\lambda; V_L] \, dy.$$

While this is an integral over all of $\mathbb{R}$, it is easy to see that the expectations are $O(|y|^{-2})$ for all $L \geq 1$, and hence the integral is finite. If $L < L'$, define $F_{L,L'}$ to be the same as F_L for the translated Brownian motion, $W_t := W_{t+S_L} - Li$. Then the second fact above can be used to see that $\mathbf{E}[F_{L,L'}^\lambda \mid \mathcal{F}_L] \leq c\Phi_{L'-L}$. Since $\log F_{L,L'}$ is equal to an appropriate π-extremal distance, up to an additive error of $O(1)$, we can see that

$$\Phi_{L+L'} \leq c_2 \Phi_L \Phi_{L'}.$$

This is enough to conclude the existence of the exponent $\tilde{\xi} := -\lim_{L \to \infty} \log \Phi_L / L$ and to give the inequality $\Phi_L \geq c_2^{-1} e^{-\tilde{\xi} L}$.

8.3. HALF-PLANE OR RECTANGLE EXPONENT

The bound $\Phi_{L+L'} \geq c_1 \Phi_L \Phi_{L'}$ takes more work to establish. Consider the set of excursions γ between ∂_1 and ∂_2 in $\mathcal{R}_L$ such that

$$(8.21) \qquad \gamma[0, t_\gamma] \cap \left\{ z : \operatorname{Re}(z) \geq L - \frac{1}{2} \right\} \subset \left\{ z : |\operatorname{Im}(z) - \frac{\pi}{2}| \leq \frac{1}{5} \right\}.$$

We will show that there is a constant such that for all L, the μ_L measure of such excursions is at least $cF(L)$. By symmetry this is also true for the set of excursions such that

$$(8.22) \qquad \gamma[0, t_\gamma] \cap \left\{ z : \operatorname{Re}(z) \leq \frac{1}{2} \right\} \subset \left\{ z : |\operatorname{Im}(z) - \frac{\pi}{2}| \leq \frac{1}{5} \right\}.$$

For fixed L, L' we consider excursions derived from the Brownian motion such that the following holds: the part from $\{\operatorname{Re}(z) = 0\}$ to $\{\operatorname{Re}(z) = L\}$ satisfies (8.21); the Brownian motion between the first and last visits to $\{\operatorname{Re}(z) = L\}$ before $T_{L+L'}$ stays in $\mathcal{B}(B_{S_L}, 1/5)$; and the excursion derived (by translation by $-L$) from $\{\operatorname{Re}(z) = L\}$ to $\{\operatorname{Re}(z) = L + L'\}$ satisfies (8.22). The path of the corresponding excursion from $\{\operatorname{Re}(z) = 0\}$ to $\{\operatorname{Re}(z) = L + L'\}$ consists of these two excursions and a subset of $\mathcal{B}(B_{S_L}, 1/5)$. Using Lemma 5.14 we can see that $F_{L+L'} \geq cF_L F_{L'}$ on this event. Hence we get $\Phi_{L+L'} \geq c \Phi_L \Phi_{L'}$ which finishes the proof.

In proving the estimate, we will prove a stronger fact. If $L \geq 1$, define the event

$$\tilde{\Upsilon}_L = \left\{ B[S_{L-(1/2)}, S_L] \subset \{x + iy : L - \frac{3}{4} \leq x \leq L, |\frac{\pi}{2} - y| \leq \frac{1}{5}\} \right\}.$$

LEMMA 8.20 (Separation Lemma). *There is a $c > 0$ such that for all $L \geq 1$,*

$$\mathbf{E}[F_{L+1}^\lambda 1_{\tilde{\Upsilon}_{L+1}} \mid \mathcal{F}_L] \geq c\, \mathbf{E}[F_{L+1}^\lambda \mid \mathcal{F}_L].$$

PROOF. We will start by sketching the outline of the argument, and then we will do the necessary estimates. Let J_L be the supremum of all $\epsilon > 0$ such that:

- $|Y_{S_L} - \frac{\pi}{2}| \geq \frac{\pi}{2} - 4\epsilon$,
- $Y_t \geq \frac{1}{2} Y_{S_L}$, $S_{L-\epsilon} \leq t \leq S_L$.

Note that $J_L > 0$ if $F_L > 0$. Let

$$r_n = 1 - \sum_{m=n}^\infty u\, m^2 2^{-m} \quad \text{where} \quad u = [4 \sum_{m=1}^\infty m^2 2^{-m}]^{-1}.$$

Let

$$h_m = \inf \frac{\mathbf{E}[F_{L+s}^\lambda 1_{\tilde{\Upsilon}_{L+s}} 1\{J_L \geq 2^{-m}\} \mid \mathcal{F}_L]}{\mathbf{E}[F_{L+s}^\lambda 1\{J_L \geq 2^{-m}\} \mid \mathcal{F}_L]},$$

where the infimum is over all $L \geq 1$ and all $r_n \leq s \leq 1$. The first estimate is to show there exist c, α such that

$$(8.23) \qquad \mathbf{E}\left[F_{L+s}^\lambda 1_{\tilde{\Upsilon}_{L+s}} 1\{J_L \geq 2^{-m}\} \mid \mathcal{F}_L\right] \geq c 2^{-m\alpha} F_L^\lambda 1\{J_L \geq 2^{-m}\}.$$

In particular, $h_m > 0$ for each m. The second estimate is to show there exists a summable sequence δ_m such that

$$(8.24) \qquad h_{m+1} \geq h_m [1 - \delta_m].$$

This implies that $\inf h_m > 0$ and gives the lemma.

To make the notation a little easier, we will establish (8.23) only in the (harder) case where $Y_{S_L} \leq \pi/2$. Let us write ϵ for $c_1 \, 2^{-m}$ for some constant c_1 we will choose later (independent of m, L). Let U be the domain that is bounded by the following curves: the line segment from $B_{S_L} - \epsilon i$ to $(L + (1/8)) + [(\pi/2) - (1/10)]i$; the line segment from $B_{S_L} + \epsilon i$ to $(L + (1/8)) + [(\pi/2) + (1/10)]i$; the line segment from $(L + (1/8)) + [(\pi/2) - (1/10)]i$ to $(L + (1/8)) + [(\pi/2) + (1/10)]i$; and the part of the circle of radius ϵ about B_{S_L} contained in $\{\operatorname{Re}(z) \leq L\}$. This is a "wedge-like" region, and by comparison with Brownian motion in a wedge[1], we can see that $\mathbf{P}\{B[S_L, S_{L+(1/8)}) \subset U \mid \mathcal{F}_L\} \geq c\epsilon^\beta \geq \tilde{c}(2^{-m})^\beta$. Let K_L be the event that $B[S_L, S_{L+1}) \subset U$ and for every $3/4 \leq s \leq 1$, the event $\tilde{\Upsilon}_{L+s}$ holds. Then it is easy to see by the strong Markov property that there is a c (depending on c_1 still to be chosen) such that

$$\mathbf{P}[K_L \cap \{J_L \geq 2^{-m}\} \mid \mathcal{F}_L] \geq c \, 2^{-m\beta} \mathbf{1}\{J_L \geq 2^{-m}\}.$$

But a similar wedge estimate combined with Lemma 5.14 shows that if we choose c_1 sufficiently small, then on the event $K_L \cap \{J_L \geq 2^{-m}\}$, $F_{L+1} \geq c_2 \, 2^{-\beta' m} F_L$. This gives (8.23).

To get (8.24), fix m and assume that $J_L \geq 2^{-m-1}$. Let $\sigma_m = \sigma_{m,L}$ denote the first time of the form $l 2^{-m}$, $l = 0, 1, 2, \ldots$ such that either either $F_{L+\sigma} = 0$ or $J_{L+\sigma} \geq 2^{-m}$. It is easy to see that there is a $\rho > 0$ such that $\mathbf{P}\{\sigma_m \leq 4 \cdot 2^{-m} \mid \mathcal{F}_L\} \geq \rho$. By iterating this we get

$$\mathbf{P}\{\sigma_m \geq u m^2 2^{-m} \mid \mathcal{F}_L\} \leq c e^{-\beta m^2}.$$

Therefore, using (8.23),

$$\mathbf{E}\left[F_{L+s}^\lambda \mathbf{1}\{\sigma_m \geq u m^2 2^{-m}\} I\{J_m \geq 2^{-m}\} \mid \mathcal{F}_L\right]$$

$$\leq c e^{-\beta m^2} \mathbf{E}\left[F_{L+s}^\lambda I\{J_m \geq 2^{-m}\} \mid \mathcal{F}_L\right]$$

(with a different c, β). But,

$$\mathbf{E}\left[F_{L+s}^\lambda \tilde{\Upsilon}_{L+s} \mathbf{1}\{\sigma_m < u m^2 2^{-m}\} I\{J_m \geq 2^{-m}\}\right]$$

$$\geq h_m \mathbf{E}\left[F_L^\lambda \mathbf{1}\{u \sigma_m < m^2 2^{-m}\} \mid \mathcal{F}_L\right].$$

This gives

$$h_{m+1} \geq h_m \left[1 - O(e^{-m^2 \beta})\right],$$

which gives (8.24). □

8.4. Whole-plane or annulus exponent

In this section we will discuss the intersection exponent $\xi(k, \lambda)$ where k is a positive integer and λ is a nonnegative real number. The definitions could be given in terms of the excursion measure on annuli and perhaps are cleaner in this framework; however, we will choose to describe them in terms of Brownian motions. Our approach has two advantages: it is the Brownian versions that are needed to prove results such as Theorem 8.3 and many of these definitions have direct analogues for three-dimensional Brownian motions. (In contrast, the exponent $\tilde{\xi}$

---

[1] Brownian motion in a wedge $\{re^{i\theta} : 0 < r < \infty, 0 < \theta < \theta_0\}$ can be studied by conformally mapping to $\mathbb{H}$ using $z \mapsto z^{\pi/\theta_0}$.

8.4. WHOLE-PLANE OR ANNULUS EXPONENT

of the previous section does not have a three-dimensional analogue.) Throughout this section we will fix an $M < \infty$ as well as $k, \lambda \leq M$. Constants $c, c_1, c_2, \ldots$ are allowed to depend on M, but do not depend on k, λ. We will use the notation set up before the statement of Theorem 8.3. For ease we will assume that $B^1, \ldots, B^k$ are defined on a probability space $(\Omega, \mathbf{P}, \mathcal{G})$ and that another Brownian motion B is defined on the space $(\tilde{\Omega}, \tilde{\mathbf{P}}, \tilde{\mathcal{G}})$. We let $\bar{\Omega} = \Omega \times \tilde{\Omega}$ be the product space; note that $B, B^1, \ldots, B^k$ are independent on $\bar{\Omega}$.

We define the following random variables on Ω: if $n \geq 0$,

$$Z_n = \tilde{\mathbf{P}}\{B[0, T_n] \cap (\Gamma_n^1 \cup \cdots \cup \Gamma_n^k) = \emptyset\},$$

$$Z_n^* = \tilde{\mathbf{P}}\{B[0, T_n] \cap (\Gamma_n^1 \cup \cdots \cup \Gamma_n^k) = \emptyset; \Upsilon_{n,-}\}.$$

Note that $Z_n^* \leq Z_n$.

DEFINITION 8.21. The *(whole-plane or annulus) Brownian intersection exponent* $\xi = \xi(k, \lambda)$ is defined by

$$\mathbf{E}[Z_n^\lambda] \asymp e^{-\xi n}, \quad n \to \infty.$$

The random variable Z_n is a function of the random path $\Gamma_n^1 \cup \cdots \cup \Gamma_n^k$. If λ is a positive integer, then Z_n^λ denotes the probability that λ Brownian motions (independent of $B^1, \ldots, B^k$) reach the circle of radius e^n without intersecting $\Gamma_n^1 \cup \cdots \cup \Gamma_n^k$. In particular,

$$\bar{\mathbf{P}}\{B[0, T_n] \cap B^1[T_0, T_n] = \emptyset\} = \mathbf{E}[Z_n].$$

The existence of the exponent $\xi(k, \lambda)$ comes from the following proposition.

PROPOSITION 8.22. *For every $M < \infty$ there exist c_1, c_2 such that for every $k, \lambda \leq M$, there is a $\xi = \xi(k, \lambda)$ such that for all $n \geq 0$,*

$$c_1 e^{-\xi n} \leq \mathbf{E}[(Z_n^*)^\lambda; \Upsilon_n^1 \cap \cdots \cap \Upsilon_n^k] \leq \mathbf{E}[Z_n^\lambda] \leq c_2 e^{-\xi n}.$$

Note that Theorem 8.3 follows from this proposition and the (difficult) fact that $\xi(1, 1) = 5/4, \xi(2, 0) = 2/3, \xi(1, 0) = 1/4$. The proof of Proposition 8.22 is similar to that of Proposition 8.10 so we will not give it. In the proof, one uses a separation lemma proved in a same way as Lemma 8.20. It is sufficiently useful that we will state it here.

LEMMA 8.23 (Separation Lemma). *There is a c such that for every $n \geq 0$,*

$$\mathbf{E}[(Z_{n+1}^*)^\lambda; \Upsilon_{n+1}^1 \cap \cdots \cap \Upsilon_{n+1}^k \mid \mathcal{G}_n] \geq c \, \mathbf{E}[Z_{n+1}^\lambda \mid \mathcal{G}_n].$$

In particular,

$$\mathbf{E}[(Z_n^*)^\lambda; \Upsilon_n^1 \cap \cdots \cap \Upsilon_n^k] \geq c \, \mathbf{E}[Z_n^\lambda].$$

We will now interpret the intersection exponent as an eigenvalue of an operator. For ease, we will consider the case $k = 1$, but this can easily be generalized to other values of k. Let $\mathcal{Y}$ be the set of curves $\gamma : [0, t_\gamma] \to \overline{\mathbb{D}}$ with $\gamma(0) = 0; 0 < |\gamma(t)| < 1, 0 < t < t_\gamma; |\gamma(t_\gamma)| = 1$. (For $k > 1$, we would choose the space of k-tuples of such curves.) There are a number of natural metrics to put on $\mathcal{Y}$; we will not specify at the moment which one we choose. Let $\mathcal{Y}^*$ be the subset of $\mathcal{Y}$ such that the connected component of $\mathbb{D} \setminus \gamma[0, t_\gamma]$ whose boundary contains $\partial \mathbb{D}$ also has the origin on its boundary. Suppose B^1 is a Brownian motion starting at $\gamma(t_\gamma)$. For $n \geq 0$, define $\hat{\gamma}_n$ by $t_{\hat{\gamma}_n} = t_\gamma + T_n^1$ and

$$\hat{\gamma}_n(t) = \begin{cases} \gamma(t), & 0 \leq t \leq t_\gamma \\ B^1_{t-t_\gamma}, & t_\gamma \leq t \leq t_\gamma + T_n^1. \end{cases}$$

Let $\gamma_n \in \mathcal{Y}$ be obtained from $t_{\hat{\gamma}_n}$ by Brownian scaling, i.e., $t_{\gamma_n} = e^{-2n} t_{\hat{\gamma}_n}$, $\gamma_n(t) = e^{-n} \hat{\gamma}_n(e^{2n} t)$. If $\gamma_0 \in \mathcal{Y}^*$, let

$$Z_n = Z_n(\gamma_n) = \mathbf{P}\{B[0, T_n] \cap \hat{\gamma}_n(0, t_{\hat{\gamma}_n}] = \emptyset \mid B[0, T_0] \cap \gamma(0, t_\gamma] = \emptyset\}$$

Of course, what we have written here does not make sense since we are conditioning on a set of probability zero. To be precise, we fix γ and then choose a sample $B[0, T_0]$ from the normalized excursion measure of excursions from 0 to $\partial \mathbb{D}$ in $\mathbb{D} \setminus \gamma[0, t_\gamma]$. Then we extend one of these excursions to $B[0, T_n]$ by starting a Brownian motion at the endpoint and stopping it when it reaches the disk of radius e^n. Note that if $\gamma \in \mathcal{Y}^*, \gamma_n \notin \mathcal{Y}^*$, then $Z_n(\gamma_n) = 0$. If $\gamma \notin \mathcal{Y}^*$, then $\gamma_n \notin \mathcal{Y}^*$ and we define $Z_n(\gamma_n) = 0$. The map $\gamma \mapsto \gamma_n$ gives a Markovian semigroup from $\mathcal{Y}$ to $\mathcal{Y}$ whose invariant distribution is the distribution of a Brownian motion stopped when it reaches $\partial \mathbb{D}$. Note that this gives measure zero to $\mathcal{Y}^*$.

For each $\lambda \geq 0$, this Markovian semigroup gives a subMarkovian transformation $\mathcal{P}_{n,\lambda} : \mathcal{Y}^* \to \mathcal{Y}^*$, where γ_n is "not killed" with probability $Z_n(\gamma_n)^\lambda$ (again we use the convention $0^0 = 0$, i.e., if $\gamma \in \mathcal{Y}^*$, then $Z_n(\gamma_n)^0$ is the indicator function that $\gamma_n \in \mathcal{Y}$). If we start with an initial $\gamma \in \mathcal{Y}^*$ then the total mass of the measure, $\nu_m(\gamma)$, induced by $\mathcal{P}_{n,\lambda}$ decays like $e^{-\xi(1,\lambda)n}$. In fact, using the proofs of Proposition 8.22 and Lemma 8.23, it can be shown that there exist c_1, c_2 such that for $n \geq 1$,

$$c_1 \, e^{-\xi(1,\lambda)n} \, |\nu_1(\gamma)| \leq |\nu_n(\gamma)| \leq c_2 \, e^{-\xi(1,\lambda)n} \, |\nu_1(\gamma)|.$$

We will now phrase the intersection exponent in terms of excursion measure. Let $\mathcal{A}_n = \{1 < |z| < e^n\}$ and let $\hat{\mu}_n$ denote the Brownian excursion measure on $\mathcal{A}_n$ restricted to the curves that start on the unit circle and end at the circle of radius e^n. This can be obtained in a number of different ways. One way is to consider the excursion measure on $\{|z| > 1\}$, restrict to those paths that reach $e^n \, \partial \mathbb{D}$, and then truncate the paths. Another way is to start with the excursion measure on $\{z : 0 < \text{Re}(z) < n\}$ restricted to curves that start on $[0, 2\pi i)$ and end on $\{\text{Re}(z) = n\}$ and taking the image under the exponential map (the exponential map is not one-to-one but the image of the measure is excursion measure on the annulus). If $\gamma^1, \ldots, \gamma^k$ denote k excursions from $\partial \mathbb{D}$ to $e^n \partial \mathbb{D}$, let

$$L_n = L(\mathcal{A}_n \setminus (\gamma^1 \cup \cdots \cup \gamma^k); \partial \mathbb{D}, e^n \, \partial \mathbb{D}).$$

Typically, if n is large this will be ∞; if it is not ∞, then with very large probability there is a single component, D_n, of $\mathcal{A} \setminus (\gamma^1 \cup \cdots \cup \gamma^k)$ whose boundary intersects both circles, in which case $L_n = L(D_n; \partial \mathbb{D}, e^n \, \partial \mathbb{D})$. However, it is possible to have as many as k different components with boundaries intersecting both circles. The intersection exponent $\xi = \xi(k, \lambda)$ satisfies

$$\mathbf{E}[\exp\{-\lambda L_n\}] \asymp e^{-\xi n} \quad n \geq 1,$$

where the implicit constants can be chosen uniformly for $k, \lambda \leq M$.

Let γ denote a radial SLE_6 path in $e^n \, \mathbb{D}$ from $e^n \, \partial \mathbb{D}$ to the origin; we will choose the starting point uniformly on $r^n \, \partial \mathbb{D}$. Let $\tau = \inf\{t : |\gamma(t)| = 1\}$ and write just γ for $\gamma[0, \tau]$ (the particular parametrization of γ will not be relevant, just the set $\gamma = \gamma[0,\tau]$.) Let $\mathcal{D}$ denote the connected component of $\mathcal{A}_n \setminus \gamma$ whose boundary includes $\partial \mathbb{D}$, and let $\mathcal{L} = L(\mathcal{D}; \partial \mathbb{D}, e^n \, \partial \mathbb{D})$. Proposition 6.48 tells us that

$$\mathbf{E}[\exp\{-b\mathcal{L}\}] \asymp e^{-\rho(b)n}, \quad n \geq 1,$$

where

$$\rho(b) = \frac{b}{2} + \frac{1}{8} + \frac{1}{8}\sqrt{24b+1}.$$

PROPOSITION 8.24. $\xi(1,1) = 5/4$.

PROOF. Let γ be a radial SLE_6 on the probability space $(\Omega, \mathbf{P})$ and γ^1 an excursion in $\mathcal{A}_n$ chosen according to $\hat{\mu}_n$. We claim that

$$(\mathbf{P} \times \hat{\mu}_n)\{\gamma \cap \gamma^1 = \emptyset\} \asymp e^{-5n/4}, \quad n \geq 1.$$

To see this, note that

$$(\mathbf{P} \times \hat{\mu}_n)\{\gamma \cap \gamma^1 = \emptyset\} = \mathbf{P}[\mu_{\mathcal{D}}(\partial \mathbb{D}, e^n \partial \mathbb{D})] \asymp \mathbf{P}[\exp\{-\mathcal{L}\}] \asymp e^{-\rho(1)n} = e^{-5n/4}.$$

But if γ^1 is fixed and D_n, L_n are as defined as above, the probability that γ stays in D_n is comparable to e^{-L_n}. (Here we have used Proposition 6.36, which is an estimate for *chordal* SLE_6 and Proposition 6.23 which relates chordal and radial SLE_6.) Hence,

$$(\mathbf{P} \times \hat{\mu}_n)\{\gamma \cap \gamma^1 = \emptyset\} \asymp \hat{\mu}_n[e^{-L_n}] \asymp e^{-\xi(1,1)n}.$$

$\square$

If $\gamma^1, \gamma^2, \gamma^3$ are three mutually nonintersecting excursions in $\mathcal{A}_n$ connecting the inner and outer boundary circles, let D_n denote the connected component of $\mathcal{A}_n \setminus (\gamma^1 \cap \gamma^3)$ whose boundary intersects both circles and such that γ^2 does not intersect D_n. Define the exponent $\xi(1,1,1,\lambda)$ by the relation

$$(\hat{\mu}_n \times \hat{\mu}_n \times \hat{\mu}_n)\left[\exp\{-\lambda L(D_n; \partial \mathbb{D}, e^n \, \partial \mathbb{D})\}; \gamma^j \cap \gamma^k = \emptyset, j \neq k\right] \asymp e^{-\xi(1,1,1,\lambda)n}.$$

PROPOSITION 8.25. *Using the notation of Corollary 8.18,*

$$\xi(1,1,1,\lambda) = \rho(\tilde{\xi}(1,1,\lambda)) = \rho(\tilde{\xi}(\xi_4,\lambda)) = \rho(\eta_4,\lambda) = \frac{\lambda}{2} + \frac{7}{3} + \frac{11}{24}\sqrt{1+24\lambda}.$$

PROOF. By using the crossing estimate for chordal SLE_κ as in the previous proposition, we can see that

$$(\hat{\mu}_n \times \mathbf{P} \times \hat{\mu}_n)[\exp\{-\lambda L(D_n; \partial \mathbb{D}, e^n \, \partial \mathbb{D})\}; \gamma^j \cap \gamma^k = \emptyset, j \neq k]$$

$$\asymp (\hat{\mu}_n \times \hat{\mu}_n \times \hat{\mu}_n)[\exp\{-\lambda L(D_n; \partial \mathbb{D}, e^n \, \partial \mathbb{D})\}; \gamma^j \cap \gamma^k = \emptyset, j \neq k].$$

Here we mean that γ^2 on the left hand side has been chosen to be a radial SLE_6 as above rather than an excursion. But by fixing the SLE_6 first, we can see that

$$\xi(1,1,1,\lambda) = \rho(\tilde{\xi}(1,1,\lambda)) = \rho(\tilde{\xi}(\tilde{\xi}_4,\lambda)) = \rho(\tilde{\xi}(4,\lambda)) = \eta(4,\lambda).$$

$\square$

COROLLARY 8.26. *If $\lambda \geq 10/3$, then*

$$\xi(2,\lambda) = \frac{\lambda}{2} + \frac{11}{24} + \frac{5}{24}\sqrt{24\lambda+1}.$$

PROOF. Choose $\tilde{\lambda}$ so that $\lambda = \tilde{\xi}(10/3, \lambda) = \tilde{\xi}(1/3, 1, \lambda)$. Then
$$\xi(2, \lambda) = \xi(2, \tilde{\xi}(1/3, 1, \tilde{\lambda})) = \xi(2, 1/3, 1, \tilde{\lambda}) =$$
$$\xi(\tilde{\xi}(2, 1/3), 1, \tilde{\lambda}) = \xi(\tilde{\xi}(1, 1), 1, \tilde{\lambda}) = \xi(1, 1, 1, \tilde{\lambda}).$$

□

Our methods up to this point only determine $\xi(2, \lambda)$ for $\lambda \geq 10/3$. In order to extend this to $\lambda \geq 0$, we need the following proposition. The proof, which we omit, does not use the Schramm-Loewner evolution. Instead, one analyzes the family of operators described above, parametrized by λ which can be extended to be complex, and shows that this is analytic in a neighborhood of any $\lambda > 0$.

PROPOSITION 8.27. *For each positive integer k, the function $\lambda \mapsto \xi(k, \lambda)$ is real analytic for $\lambda > 0$ and is right continuous at $\lambda = 0$.*

PROOF. See [54]. □

COROLLARY 8.28. *For all $\lambda \geq 0$,*
$$\xi(1, \lambda) = \frac{\lambda}{2} + \frac{1}{8} + \frac{1}{8}\sqrt{24\lambda + 1},$$
$$\xi(2, \lambda) = \frac{\lambda}{2} + \frac{11}{24} + \frac{5}{24}\sqrt{24\lambda + 1}.$$
In particular, $\xi(1, 0) = 1/4, \xi(2, 0) = 2/3$.

PROOF. This follows from Corollary 8.26 and Proposition 8.27 since an analytic function in a domain D that vanishes on a line segment must be zero. This establishes the corollary for $\lambda > 0$ and continuity at $\lambda = 0+$ establishes it for $\lambda = 0$. □

CHAPTER 9

Restriction measures

9.1. Unbounded hulls in $\mathbb{H}$

We call a domain $D \subset \mathbb{H}$ a *right-domain* if it is simply connected and $\partial D \cap \mathbb{R} = [0, \infty)$; similarly, a simply connected $D \subset \mathbb{H}$ is a a *left-domain* if $\partial D \cap \mathbb{R} = (-\infty, 0]$, or, equivalently, if its reflection about the imaginary axis is a right-domain. Let $\mathcal{J}_+$ denote the set of closed sets K such that $K = \overline{K \cap \mathbb{H}}$ and $\mathbb{H} \setminus K$ is a right-domain. Let $\mathcal{J}$ denote the set of closed sets K such that $K = \overline{K \cap \mathbb{H}}$ and such that $\mathbb{H} \setminus K$ is the disjoint union of a right-domain and a left-domain. If $K \in \mathcal{J}$, let K^+ denote the corresponding element of $\mathcal{J}_+$ obtained from the filling on the left, i.e., K^+ is the closure of the union of K and the left-domain in $\mathbb{H} \setminus K$.

EXAMPLE 9.1. Suppose $\gamma^1, \ldots, \gamma^n : (0, \infty) \to \mathbb{H}$ are curves with
$$\lim_{t \to 0+} \gamma^k(t) = 0, \quad \lim_{t \to \infty} \gamma^k(t) = \infty.$$
Let $D = \mathbb{H} \setminus [\gamma^1(0, \infty) \cup \cdots \cup \gamma^n(0, \infty)]$. Let D^+ be the connected component of D whose boundary includes the positive real axis, and let D^- be the connected component of D whose boundary contains the negative real axis. Then D^+ is a right-domain, D^- is a left-domain, $\mathbb{H} \setminus (D^+ \cup D^-) \in \mathcal{J}$ and $\mathbb{H} \setminus D^+ \in \mathcal{J}_+$. We call $\mathbb{H} \setminus (D^+ \cup D^-)$ the *hull generated by* $\gamma^1, \ldots, \gamma^n$.

Recall the definition of $\mathcal{Q}$ from §3.4 and $\mathcal{Q}_+, \mathcal{Q}_-, \mathcal{Q}_\pm$ from §5.3. Let $\mathcal{V}_+$ (resp., $\mathcal{V}$) denote the set of subsets of $\mathcal{J}$, $\mathcal{V}_+ = \{V_A^+ : A \in \mathcal{Q}_+\}$ (resp., $\mathcal{V} = \{V_A : A \in \mathcal{Q}_\pm\}$), where
$$V_A^+ := \{K \in \mathcal{J}_+ : K \cap A = \emptyset\}, \quad V_A = \{K \in \mathcal{J} : K \cap A = \emptyset\}.$$
Note that $V_A^+ \cap V_{A'}^+ = V_{\tilde{A}}^+$ where $\tilde{A}$ is the compact hull generated by $A \cup A'$, i.e., the complement in $\mathbb{H}$ of the unbounded component of $\mathbb{H} \setminus (A \cup A')$. Hence $\mathcal{V}^+$ is a π-system (see, e.g., [**13**, Section 3] for definitions). Similarly, $\mathcal{V}$ is a π-system. By a measure on $\mathcal{J}_+$ (resp,. $\mathcal{J}$) we will mean a measure on $\sigma(\mathcal{V}_+)$ (resp., $\sigma(\mathcal{V})$). To specify a probability measure on $\sigma(\mathcal{V}_+)$ or $\sigma(\mathcal{V})$ it suffices to give the values on $\mathcal{V}_+$ or $\mathcal{V}$, respectively [**13**, Theorem 3.3]. Note that any measure on $\mathcal{J}$ induces a measure on $\mathcal{J}_+$ by the map $K \mapsto K^+$.

Recall that if $A \in \mathcal{Q}_\pm$, then Φ_A denotes the unique conformal transformation of $\mathbb{H} \setminus A$ onto $\mathbb{H}$ with $\Phi_A(0) = 0$ and such that $\Phi_A(z) \sim z$ as $z \to \infty$. If ν^+ is a measure on $\mathcal{J}_+$ supported on V_A^+, let $\Phi_A \circ \nu^+$ denote the measure
$$\Phi_A \circ \nu^+(V_{A'}^+) = \nu^+\{K : \Phi_A(K \cap \mathbb{H}) \cap A' = \emptyset\}.$$
We define $\Phi_A \circ \nu$ similarly if $A \in \mathcal{Q}_\pm$ and ν is a measure supported on V_A. A measure ν on $\mathcal{J}_+$ (resp., $\mathcal{J}$) is called *scale-invariant* if it is invariant under the map $K \mapsto rK$ for every $r > 0$. This implies that $\nu(V_{rA}^+) = \nu(V_A^+)$ for all $A \in \mathcal{Q}_+$ (resp., $\nu(V_{rA}) = \nu(V_A)$ for all $A \in \mathcal{Q}_\pm$).

9. RESTRICTION MEASURES

DEFINITION 9.2.

- A probability measure $\mathbf{P}^+$ on $\mathcal{J}_+$ is a *right-restriction measure* if it is scale-invariant and for every $A \in \mathcal{Q}_+$,
$$\Phi_A \circ \nu_A^+ = \mathbf{P}^+,$$
where ν_A^+ is the conditional probability distribution given $K \subset V_A^+$.
- A probability measure $\mathbf{P}$ on $\mathcal{J}$ is a *restriction measure* if it is scale-invariant and for every $A \in \mathcal{Q}_\pm$,
$$\Phi_A \circ \nu_A = \mathbf{P},$$
where ν_A is the conditional probability distribution given $K \subset V_A$.
- A probability measure $\mathbf{P}$ on $\mathcal{J}$ is a *right-restriction measure* if it is scale-invariant and for every $A \in \mathcal{Q}_+$,
$$\Phi_A \circ \nu_A = \mathbf{P},$$
where ν_A is the conditional probability distribution given $K \subset V_A$. Equivalently, $\mathbf{P}$ is a right-restriction measure if the induced measure on $\mathcal{J}_+$ is a right-restriction measure.

We can define an associative multiplication on $\mathcal{J}$ by $A' \cdot A = \Phi_A^{-1}(A') \cup A$, or equivalently
$$\Phi_{A' \cdot A} = \Phi_{A'} \circ \Phi_A.$$
Under this multiplication $\mathcal{J}$ is a nonabelian semigroup. A probability measure $\mathbf{P}$ is a restriction measure if and only if the map $A \mapsto \mathbf{P}(V_A)$ is a homomorphism from $\mathcal{J}$ to the abelian semigroup $[0, 1]$ with the usual product. Similarly, $\mathcal{J}_+$ is a nonabelian semigroup, and $\mathbf{P}^+$ is a right-restriction measure if and only if $A \mapsto \mathbf{P}^+(V_A^+)$ is a homomorphism. If $\alpha \geq 0$, the chain rule tells us that $A \mapsto \Phi'_A(0)^\alpha$ is a homomorphism. The next propositions tell us that these are the only possibilities for restriction or right-restriction measures.

PROPOSITION 9.3. *If $\mathbf{P}^+$ is a right-restriction measure on $\mathcal{J}_+$, then there exists $0 \leq \alpha < \infty$ such that for each $A \in \mathcal{Q}_+$,*
$$\mathbf{P}^+(V_A^+) = \Phi'_A(0)^\alpha. \tag{9.1}$$

PROOF. We fix $\mathbf{P}^+$ and allow constants in this proof to depend on $\mathbf{P}^+$. We will first show that there is a c such that for all ϵ,
$$\mathbf{P}^+(V_{\mathcal{B}_\epsilon}^+) \geq 1 - c\,\epsilon^2, \tag{9.2}$$
where $\mathcal{B}_\epsilon = \mathcal{B}(1,\epsilon) \cap \mathbb{H}$. To see this, first note that there is an $r > 0$ such that $\mathbf{P}(V_{\mathcal{B}_r}^+) \geq 1/2$, since $1 = \mathbf{P}(V_\emptyset^+) = \lim_{r \to 0+} \mathbf{P}^+(V_{\mathcal{B}_r}^+)$. (Here we use the fact that for each $K \in \mathcal{J}_+$, there is an $\epsilon > 0$ such that $K \cap \mathcal{B}_\epsilon = \emptyset$.) We fix such an r. We will prove the result for $\epsilon < r/4$. The map $\Phi_{\mathcal{B}_\epsilon}$ can be determined exactly,
$$\Phi_{\mathcal{B}_\epsilon}(z) = z + \frac{\epsilon^2}{z-1} + \epsilon^2.$$
In particular, if $|z-1| \geq r/4$, $|\Phi_{\mathcal{B}_\epsilon}(z) - z| \leq C\,\epsilon^2$ where $C = 1 + (4/r)$. Let k_ϵ be the largest integer such that $k_\epsilon C\,\epsilon^2 \leq r/2$; note that $k_\epsilon \sim (r/2C)\epsilon^{-2}$ as $\epsilon \to 0+$. Let
$$\Phi^{(\epsilon)} = \underbrace{\Phi_{\mathcal{B}_\epsilon} \circ \cdots \circ \Phi_{\mathcal{B}_\epsilon}}_{k_\epsilon}.$$

Note that $|\Phi^{(\epsilon)}(z) - 1| \geq r/2$ for $|z - 1| \geq r$; hence $\Phi^{(\epsilon)} = \Phi_{A^+}$ for some $A = A_\epsilon$ contained in $\mathcal{B}_r$. Hence, by monotonicity and right-restriction,
$$\frac{1}{2} \leq \mathbf{P}(V_{\mathcal{B}_r}^+) \leq \mathbf{P}(V_A^+) \leq \mathbf{P}(V_{\mathcal{B}_\epsilon}^+)^{k_\epsilon}.$$
This implies (9.2). By scale-invariance we see that for every $0 < x_1 < x_2 < \infty$, there is a $c_* = c_*(x_1, x_2, \mathbf{P}^+)$ such that

(9.3) $\qquad\qquad \mathbf{P}^+(V_A) \leq c_* \epsilon$ if $A = [x_1, x_2] \times (0, \epsilon]$

(we can cover this set with $O(1/\epsilon)$ balls of radius ϵ).

We claim that it suffices to establish (9.1) for all A of the form $A = \gamma(0, t_\gamma]$ where $\gamma : [0, t_\gamma] \to \mathbb{C}$ is a simple curve with $\gamma(0) = 1$ and $\gamma(0, t_\gamma] \subset \mathbb{H}$. To see this, note that if this is true, scale-invariance and $\Phi_{rA}(z) = r\Phi_A(z/r)$ imply (9.1) for $r\gamma$. Also, by taking limits, we see that (9.1) holds for $A = \gamma(0, t_\eta)$ if $\gamma(0), \gamma(t_\gamma) \in \mathbb{R}$ and $\gamma[0, t_\gamma)$ is simple. Finally, Lemma 5.19 can be used to show that (9.1) holds for all $A \in \mathcal{Q}_+$. Without loss of generality, we may assume that γ is parametrized by half-plane capacity, i.e., $\mathrm{hcap}(\gamma(0, t]) = 2t$.

Let us first consider the case $\gamma(t) = 1 + 2\sqrt{t}i$, $A_t = \gamma(0, t] = (1, 1 + 2\sqrt{t}i]$. Let us write Φ_t for Φ_{A_t}. Then, (see (4.13)),

(9.4) $\qquad \Phi_t(z) = \sqrt{(z-1)^2 + 4t} + \sqrt{1 + 4t}, \quad \Phi_t'(0) = \dfrac{1}{\sqrt{1+4t}}.$

(Note that the branch of the square root is such that $\Phi_t(0) = 0, \Phi_t'(0) > 0$.) Let $F(t) = \mathbf{P}(V_{A_t}^+)$. Then the restriction property and scale-invariance imply that

$$\begin{aligned} F(s + t) &= F(s)\, \mathbf{P}(V^+_{\Phi_s([1+2\sqrt{s}i, 1+2\sqrt{s+t}i])}) \\ &= F(s)\, \mathbf{P}(V^+_{[\sqrt{1+4s}, \sqrt{1+4s}+2\sqrt{t}i]}) \end{aligned}$$

(9.5) $\qquad\qquad\qquad = F(s)\, F\left(\dfrac{t}{1+4s}\right).$

Let $\alpha = -(1/2)\lim_{\epsilon \to 0+} \log F(\epsilon)/\epsilon$; the limit exists since $\log F$ is nonincreasing and concave by (9.5). Using (9.5), we see that $\lim_{\epsilon \to 0+} \log F(s+\epsilon)/\epsilon = -2\alpha/(1+4s)$. Integrating, and using $F(0) = 1$, gives
$$F(t) = (1 + 4t)^{-\alpha/2} = \Phi_t'(0)^\alpha.$$

For general γ parametrized by half-plane capacity, let $F_\gamma(t) = \mathbf{P}(V^+_{\gamma[0,t]})$ and let $\Phi_{t,\gamma} = \Phi_{\gamma[0,t]}$. We can write $\Phi_{t,\gamma}(z) = g_t(z) - g_t(0)$, where g_t satisfies the Loewner equation

(9.6) $\qquad\qquad \dot{g}_t(z) = \dfrac{2}{g_t(z) - U_t}, \quad g_0(z) = z,$

for some continuous real-valued function U_t with $U_0 > 0$. We will show that

(9.7) $\qquad\qquad F_\gamma'(0) = -\dfrac{2\alpha}{U_0^2},$

where the derivative is interpreted as a right-derivative. The same argument shows that $F_\gamma'(t) = -2\alpha F_\gamma(t)/(U_t - g_t(0))^2$ (as a right-derivative), and using Lemma 4.3 we see this is true as a two-sided derivative. Hence

$$F_\gamma(t) = \exp\left\{-\alpha \int_0^t \dfrac{2\,ds}{(U_t - g_t(0))^2}\right\} = g_t'(0)^\alpha = \Phi_{t,\gamma}'(0)^\alpha.$$

(The middle equality can be obtained by differentiating (9.6) with respect to z.) By scale-invariance, it suffices to prove (9.7) when $U_0 = \gamma(0) = 1$. Let $d_t = \text{diam}(\gamma[0,t])$ and let Φ_t be as in (9.4). Let $\hat{\gamma}(t) = 1 + 2\sqrt{t}i$ denote the straight line so that $\Phi_t = \Phi_{t,\hat{\gamma}}$. It follows from Proposition 3.46 that

$$(9.8) \qquad |\Phi_{t,\gamma}(z) - \Phi_t(z)| \leq c\,t\,d_t, \quad |z-1| \geq 1/2.$$

Using the semigroup notation, define the hulls U_t, U_t' (for t sufficiently small) by

$$\mathcal{B}_{1/2} = U_t \cdot \gamma(0,t] = U_t' \cdot \hat{\gamma}(0,t].$$

From (9.3) and (9.8) and the right-restriction property, we see that $\mathbf{P}^+(V_{U_t}^+) = \mathbf{P}^+(V_{U_t'}^+) + O(t\,d_t)$ and hence

$$F_\gamma(t) = F_{\hat{\gamma}}(t) + O(t\,d_t) = F_{\hat{\gamma}}(t) + o(t), \quad t \to 0+.$$

Since (9.7) holds for $\hat{\gamma}$, it also holds for γ. □

PROPOSITION 9.4. *If $\mathbf{P}$ is a restriction measure on $\mathcal{J}$, then there exists $0 \leq \alpha < \infty$ such that for each $A \in \mathcal{Q}$,*

$$(9.9) \qquad \mathbf{P}(V_A) = \Phi_A'(0)^\alpha.$$

PROOF. Since $\mathbf{P}$ generates a right-restriction measure, the previous proposition implies that there is an α_+ such that for all $A \in \mathcal{Q}_+$, $\mathbf{P}(V_A) = \Phi_A'(0)^{\alpha_+}$. By symmetry, we can also conclude there is an α_- such that for all $A \in \mathcal{Q}_-$, $\mathbf{P}(V_A) = \Phi_A'(0)^{\alpha_-}$. If $A \in \mathcal{Q}$, we can write $A = A_1 \cdot A_2 = A_3 \cdot A_4$ where $A_1, A_4 \in \mathcal{Q}_+, A_2, A_3 \in \mathcal{Q}_-$. It suffices, therefore, to show that $\alpha_+ = \alpha_-$.

Consider $A = A^\epsilon = A^+ \cup A^-$ where

$$A^+ = \{e^{i\theta} : 0 < \theta \leq \frac{\pi}{2} - \epsilon\} \in \mathcal{Q}_+, \quad A^- = \{e^{i\theta} : \frac{\pi}{2} + \epsilon \leq \theta < \pi\} \in \mathcal{Q}_-.$$

Define $A_1 \in \mathcal{Q}_+, A_3 \in \mathcal{Q}_-$ by $A = A_1 \cdot A^- = A_3 \cdot A^+$. By symmetry, $\Phi_{A_1}'(0) = \Phi_{A_3}'(0) := p_\epsilon$ and $\Phi_{A^-}'(0) = \Phi_{A^+}'(0) := q_\epsilon$. Hence,

$$(9.10) \qquad \mathbf{P}(V_A) = p_\epsilon^{\alpha_+} q_\epsilon^{\alpha_-} = p_\epsilon^{\alpha_-} q_\epsilon^{\alpha_+}.$$

As $\epsilon \to 0+$, $p_\epsilon \to 0$ and $q_\epsilon \to q_0 \in (0,1)$. Hence, in order for (9.10) to hold for all ϵ, it must be the case that $\alpha_+ = \alpha_-$. □

The previous propositions do not show whether or not measures satisfying (9.1) and (9.9) exist. If they do, we will denote them by $\mathbf{P}_\alpha^+, \mathbf{P}_\alpha$, and call them the α-right-restriction and α-restriction measure, respectively. If $\mathbf{P}_\alpha$ exists, then $\mathbf{P}_\alpha^+$ can be obtained from $\mathbf{P}^+$ by filling in.

EXAMPLE 9.5. Let $\hat{B}$ be an $\mathbb{H}$-excursion as in §5.3. From (5.17) we know that if $A \in \mathcal{Q}$,

$$\mathbf{P}\{\hat{B}(0,\infty) \cap A = \emptyset\} = \Phi_A'(0).$$

Hence the measures $\mathbf{P}_1, \mathbf{P}_1^+$ are obtained by considering the appropriate hulls generated by $\hat{B}(0,\infty)$.

EXAMPLE 9.6. Suppose $\mathbf{P}_\alpha, \mathbf{P}_\beta$ exists for some $\alpha, \beta > 0$ and let K_α, K_β be independent samples from these distributions. If K is the hull generated by the union of K_α and K_β, then

$$\mathbf{P}\{K \cap A = \emptyset\} = \mathbf{P}\{K_\alpha \cap A = \emptyset, K_\beta \cap A = \emptyset\} = \Phi_A'(0)^{\alpha+\beta}.$$

Hence K has the distribution of $\mathbf{P}_{\alpha+\beta}$. Similarly, $\mathbf{P}^+_{\alpha+\beta}$ can be obtained frm $\mathbf{P}^+_\alpha, \mathbf{P}^+_\beta$ by taking unions. If α is a positive integer, then $\mathbf{P}_\alpha, \mathbf{P}^+_\alpha$ is obtained from the union of α independent Brownian excursions. Hence the parameter α in the restriction measure can be loosely considered as the number of (Brownian) excursions.

EXAMPLE 9.7. Let γ be a chordal $SLE_{8/3}$ path from 0 to infinity in $\mathbb{H}$. By (6.6), if $A \in \mathcal{Q}$,
$$\mathbf{P}\{\gamma(0,\infty) \cap A = \emptyset\} = \Phi'_A(0)^{5/8}.$$
Hence the distribution of $K = \gamma(0, \infty)$ is that of $\mathbf{P}_{5/8}$. In particular, the distribution of the hulls generated by eight independent $SLE_{8/3}$ paths is the same as that generated by five independent Brownian excursions.

9.2. Right-restriction measures

The goal of this section is to prove the following theorem.

THEOREM 9.8. *For every $\alpha > 0$, the measure $\mathbf{P}^+_\alpha$ exists.*

We will prove the theorem by constructing the measure. It suffices to find a construction for all $\alpha \in (0, 1)$, for then we can construct $\mathbf{P}^+_\alpha$ for other α by taking unions. Let $\theta_0 \in (0, \pi)$ and let W_{θ_0} be the wedge
$$W_{\theta_0} = \{e^{i\theta} : 0 < \theta < \theta_0\}.$$
Let $\mathcal{Q}_{+,\theta_0}$ be the set of $A \in \mathcal{Q}_+$ such that $A \subset W_{\theta_0}$. Let $B_t = X_t + iY_t$ be a Brownian motion in W_{θ_0} that is reflected horizontally off the line $\{\theta = \theta_0\}$ and stopped when it reaches $[0, \infty)$; this is an example of oblique reflection (see §C). In this case, the process Y_t has the distribution of a standard one-dimensional Brownian motion. If $A \in \mathcal{Q}_{+,\theta_0}$ let $\rho_A = \inf\{t > 0 : B_t \in A \cup [0, \infty)\}$ and let $\rho = \rho_\emptyset = \inf\{t > 0 : B_t \in [0, \infty)\} = \inf\{t > 0 : Y_t = 0\}$. As before, let $\mathcal{I}_R = \{z : \text{Im}(z) = R\}$ and let $\sigma_R = \inf\{t : \text{Im}(B_t) = R\}$. Note that if $z \in W_{\theta_0} \cap \{\text{Im}(w) < R\}$, then $\mathbf{P}^z\{\sigma_R < \rho\} = \text{Im}(z)/R$. A *(Brownian) excursion (with horizontal reflection) in W_{θ_0} starting at $z \in W_{\theta_0}$ is the process $\hat{B}_t$ such that if $\text{Im}(z) < R$, $\hat{B}_t, 0 \le t \le \sigma_R$, has the same distribution at $B_t, 0 \le t \le \sigma_R$, given $\sigma_R < \rho$.* As in the case of $\mathbb{H}$-excursions, it is not difficult to see that we can also define a Brownian excursion in W_{θ_0} starting at $x \in [0, \infty)$ by appropriate limits.

Let $f_{\theta_0}(z) = z^{\theta_0/\pi}$; note that f_{θ_0} maps $\mathbb{H}$ conformally onto W_{θ_0}. Also $f_{\theta_0}(0) = 0$ and $A \in \mathcal{Q}_+$ if and only if $f_{\theta_0}(A) \in \mathcal{Q}_{+,\theta_0}$.

PROPOSITION 9.9. *If $A \in \mathcal{Q}_+$ and $\hat{B}_t$ is an excursion in W_{θ_0}, then*
$$\mathbf{P}\{\hat{B}(0, \infty) \cap f_{\theta_0}(A) = \emptyset\} = \Phi'_A(0)^{\theta_0/\pi}.$$

PROOF. Let $\tilde{A} = f_{\theta_0}(A)$ and let $\tilde{\Phi}_A = f_{\theta_0} \circ \Phi_A \circ f_{\theta_0}^{-1}$. Note that $\tilde{\Phi}_A$ is the unique conformal transformation of $W_{\theta_0} \setminus A$ onto W_{θ_0} with $\tilde{\Phi}_A(0) = 0$ and such that $\tilde{\Phi}_A(z) \sim z$ as $z \to \infty$. Note that if $\tilde{A} \subset \{\text{Im}(w) < R\}$, then
$$h(z) = h_{R,A}(z) := \mathbf{P}^z\{\tilde{B}[0, \sigma_R] \cap (A \cup \mathbb{R}) = \emptyset\}$$
is the unique harmonic function on $(W_{\theta_0} \cap \{\text{Im}(w) < R\}) \setminus A$ with: $h(z) = 0, z \in A \cup \mathbb{R}$; $h(z) = 1, z \in W_{\theta_0} \cap \mathcal{I}_R$; and $\partial_x h(z) = 0$ if $\arg(z) = \theta_0, 0 < \text{Im}(z) < R$. Hence, as in the case of Brownian excursions, we get
$$\mathbf{P}^z\{\hat{B}(0, \infty) \cap f_{\theta_0}(A) = \emptyset\} = \frac{\text{Im}(\tilde{\Phi}_A(z))}{\text{Im}(z)}.$$

As $z \to 0$, $\tilde{\Phi}_A(z) \sim \Phi'_A(0)^{\theta_0/\pi} z$. □

COROLLARY 9.10. *Suppose $\alpha \in (0,1)$. Then if $\hat{B}$ is the Brownian excursion in $W_{\alpha\pi}$, the distribution of the hull obtained from filling in the left side of $f_{\alpha\pi}^{-1}(\hat{B}(0,\infty))$ has the distribution $\mathbf{P}_\alpha^+$.*

COROLLARY 9.11. *For every $0 < \alpha < 5/8$, the restriction measure $\mathbf{P}_\alpha$ does not exist.*

PROOF. Let $q(\alpha)$ be the probability that i lies in the hull distributed according to $\mathbf{P}_\alpha^+$. The construction above shows that $q(\alpha) \in (0,1)$ for each $\alpha \in (0,1)$. Also, since $\mathbf{P}_{\alpha+\beta}^+$ can be obtained from $\mathbf{P}_\alpha^+$ and $\mathbf{P}_\beta^+$ by taking unions, $q(\alpha+\beta) \geq q(\alpha) + [1 - q(\alpha)] q(\beta)$, and hence $q(\alpha)$ is strictly increasing in a. Since the hull corresponding to $\alpha = 5/8$ is an $SLE_{8/3}$ curve, and the distribution of $SLE_{8/3}$ is symmetric about the imaginary axis, $q(5/8) = 1/2$. Hence $q(\alpha) < 1/2$ for $\alpha < 5/8$. However, if $\mathbf{P}_\alpha$ exists, the measure is symmetric about the imaginary axis, and hence the probability that i is contained in the left-filled hull must be at least $1/2$. □

We will now describe an alternative, simpler method of constructing the measure $\mathbf{P}_\alpha^+$. Let μ denote the excursion measure in $\mathbb{H}$ restricted to excursions that begin and end at $(-\infty, 0]$. We can write

$$\mu = \int_{-\infty}^0 \int_{-\infty}^0 \mu_\mathbb{H}(x_1, x_2) \, dx_1 \, dx_2 = c \int_{-\infty}^0 \int_{-\infty}^0 \mu_\mathbb{H}^\#(x_1, x_2) \, |x_1 - x_2|^{-1} \, dx_1 \, dx_2,$$

where $\mu_\mathbb{H}^\#(x_1, x_2)$ denotes the probability measure corresponding to excursions from x_1 to x_2. If $\alpha \in \mathcal{Q}_+$, let $\mu(A)$ denote μ restricted to curves that intersect A. Similarly, let $\mu_\mathbb{H}^\#(x_1, x_2; A)$ denote $\mu_\mathbb{H}^\#(x_1, x_2)$ restricted to curves that intersect A.

LEMMA 9.12. *If $A \in \mathcal{Q}_+$, then $|\mu(A)| < \infty$.*

PROOF. Using estimates for Brownian motion in $\mathbb{H}$, it is easy to check that there is a $c = c(A) < \infty$ such that if $x < 0$, the probability that a Brownian motion starting at $x + \epsilon i$ hits A before leaving $\mathbb{H}$ is bounded above by $c x^2 \epsilon$. Hence

$$\int_{-\infty}^0 |x_1 - x_2|^{-1} |\mu_\mathbb{H}^\#(x_1, x_2; A)| \, dx_2 \leq c \, x_1^{-2},$$

and the result follows by integrating. □

If $\lambda > 0$, we can take a Poissonian realization from the measure μ. Let K_λ denote the element of $\mathcal{J}_+$ obtained by "left-filling" in the union of all curves in the realization. The previous lemma shows that the probability that this realization avoids $A \in \mathcal{Q}^+$ is $\exp\{-\lambda |\mu(A)|\} > 0$.

PROPOSITION 9.13. *There is a c such that if $A \in \mathcal{J}_+$ and $\lambda > 0$, then $K_\lambda(A)$ has the distribution $\mathbf{P}_{c\lambda}^+$.*

PROOF. Conformal invariance of the excursion measure shows that $K_\lambda(A)$ is a right-restriction measure, i.e., has the distribution of $\mathbf{P}_{\alpha(\lambda)}^+$ for some $\alpha(\lambda) > 0$. Since the probability of avoiding A is $\exp\{-\lambda |\mu(A)|\}$, we see that $\alpha(\lambda) = c\lambda$ for some c. □

9.3. The boundary of restriction hulls

In the previous section we had two constructions of random hulls $K \in \mathcal{J}_+$ with distribution $\mathbf{P}_\alpha^+$. While these constructions are different, they both use paths of Brownian motions "filled in". Here we will give another construction of $\mathbf{P}_\alpha^+$, that focuses on the curve that is the right boundary of K. We will show that this curve is, in some sense, an SLE_κ curve with drift. Since we know that for all α the curve is a boundary of a Brownian curve, Theorem 8.2 tells us that we should expect that the curve would have Hausdorff dimension $4/3$. From §7.4, we see that $\kappa = 8/3$ would be expected.

As before, let us write $a = 2/\kappa$, and we will restrict our consideration to $a \geq 1/2$, i.e., $\kappa \leq 4$. Suppose that γ were the curve that maps out ∂K. Assume, as we would expect, that γ is a simple curve, and that γ is parametrized so that $\text{hcap}(\gamma[0,t]) = at$. Suppose we have observed γ up through time t and let $g_t = g_{\gamma[0,t]}$; $U_t = g_t(\gamma(t))$. Then g_t satisfies

$$(9.11) \qquad \dot{g}_t(z) = \frac{a}{g_t(z) - U_t}, \quad g_0(z) = z.$$

In the case of SLE, the process is started afresh at U_t. For the right boundary, however, it is important that we not lose the information about the part of $\gamma[0,t]$ that points "into" the hull K. For this reason we also wish to keep track of O_t, which is the "left-hand" image of 0 under g_t.[1]

Let B_t denote a standard Brownian motion and let $v \in \mathbb{R}$. Let (O_t, U_t) be a pair of processes satisfying

$$(9.12) \qquad \dot{O}_t = \frac{a}{O_t - U_t}, \quad dU_t = \frac{v}{U_t - O_t} dt + dB_t, \quad O_t = U_t = 0.$$

DEFINITION 9.14. If $a \geq 1/2, u \in \mathbb{R}$, and (O_t, U_t) satisfy (9.12), then the family of maps g_t satisfying the Loewner equation (9.11) is called the $SLE(2/a, 2v/a)$ process (parametrized so that

$$g_t(z) = z + \frac{2a}{z} + O(\frac{1}{|z|^2}), \quad z \to \infty) .$$

REMARK 9.15. These processes were introduced in [59] and called $SLE(\kappa, \rho)$ processes. Here $\kappa = 2/a, \rho = 2v/a$. There the processes were considered with the standard parametrization where the hcap at time t is $2t$. We have chosen the parametrization in (9.12).

REMARK 9.16. To see that one can define processes satisfying (9.12), consider the family of Bessel processes

$$dZ_t^x = \frac{v+a}{Z_t^x} dt + dB_t, \quad Z_0^x = x.$$

For each x, let $T_x = \inf\{t : Z_t^x = 0\}$ and let

$$\bar{Z}_t = \inf\{Z_t^x : T_x > t\},$$

---

[1] Since $\gamma[0,t]$ is a simple curve and $\gamma(0,t] \subset \mathbb{H}$, if $z_n \in \mathbb{H} \setminus \gamma(0,t]$ with $z_n \to 0$, then $g_t(z_n)$ has at most two limit points O_t, O_t^*. The first corresponds to z_n approaching 0 from the left of $\gamma(0,t]$ and the second to z_n approaching 0 from the right of $\gamma(0,t]$.

Then $\bar{Z}_t$ is a Bessel process reflected at the origin with $\bar{Z}_0 = 0$; in particular,
$$d\bar{Z}_t = \frac{v+a}{\bar{Z}_t}\,dt + dB_t, \quad \bar{Z}_t > 0.$$
Given $\bar{Z}_t$, we define
$$O_t = -a \int_0^t \frac{ds}{\bar{Z}_s}, \quad U_t = \bar{Z}_t + O_t.$$
From this construction we see that
$$\mathbf{P}\{O_t < U_t \text{ for all } t > 0\} = 1$$
if and only if $v + a \geq 1/2$.

The following theorem can be considered as a generalization of Theorem 6.17.

THEOREM 9.17. *Suppose g_t satisfies (9.12) with $a = 3/4, v > -3/2$ and let J_t denote the hulls generated by g_t. If $A \in \mathcal{Q}_+$, then*
$$\mathbf{P}\{J_\infty \cap A \neq \emptyset\} = \Phi'_A(0)^\alpha,$$
where
$$\alpha = \frac{(4v+5)(4v+3)}{24}.$$
In particular, the distribution of the right boundary of a hull distributed according to $\mathbf{P}_\alpha^+$ is given by $SLE(8/3, 8v/3)$ where
$$v = \frac{1}{4}\sqrt{1+24\alpha} - 1.$$

PROOF. Let
$$b = \frac{v(1+2v)}{3},$$
and let Φ_t be as in §4.6 with $\Phi_0 = \Phi_A$. Let
$$M_t = \Phi'_t(U_t)^{5/8}\,\Phi'_t(O_t)^b\left[\frac{\Phi_t(U_t) - \Phi_t(O_t)}{U_t - O_t}\right]^v.$$
Recall from (4.34), (4.35), (4.36), and (4.37) that
$$\dot{\Phi}_t(z) = \frac{3}{4}\left[\Phi'_t(U_t)\frac{\Phi'_t(U_t)}{\Phi_t(z) - \Phi_t(U_t)} - \Phi'_t(z)\frac{1}{z - U_t}\right],$$
$$\dot{\Phi}_t(U_t) = -\frac{9}{8}\Phi''_t(U_t),$$
$$\dot{\Phi}'_t(z) = \frac{3}{4}\left[-\frac{\Phi'_t(U_t)^2\,\Phi'_t(z)}{(\Phi_t(z) - \Phi_t(U_t))^2} + \frac{\Phi'_t(z)}{(z-U_t)^2} - \frac{\Phi''_t(z)}{z-U_t}\right],$$
$$\dot{\Phi}'_t(U_t) = \frac{3\,\Phi''_t(U_t)}{16\,\Phi'_t(U_t)} - \frac{\Phi'''_t(U_t)}{4}.$$
Given this, a long but straightforward computation using Itô's formula shows that M_t is a local martingale. (The "2" in the formulas has become an $a = 3/4$ because of our choice of parametrization.)

We will now show that $M_t \leq 1$. By (5.21) and the mean-value theorem,

$$\Phi'_t(U_t) \leq \frac{\Phi_t(U_t) - \Phi_t(O_t)}{U_t - O_t} \leq \Phi'_t(O_t) \leq 1. \tag{9.13}$$

We consider three different cases: (1) $\alpha \geq 5/8$ ($b, v \geq 0$); (2) $1/8 \leq \alpha < 5/8$ ($-1/2 \leq v \leq 0$, $b \leq 0$); (3) $0 < \alpha < 1/8$ $-1 < v < -1/2$, $0 < b < 1/3$). In case (1), $M_t \leq \Phi'_t(U_t)^{(5/8)} \leq 1$. In case (2), (9.13) implies

$$M_t \leq \Phi'_t(U_t)^{(5/8)+b+v} = \Phi'_t(U_t)^\alpha \leq 1.$$

In case (3), note that

$$\epsilon := \frac{5}{8} - b + v + 2\sqrt{b} = \frac{5}{8} - \frac{v(1+2v)}{3} + v + 2\sqrt{\frac{v(1+2v)}{3}} > 0.$$

Then (5.22) gives us $M_t \leq \Phi'_t(O_t)^\epsilon \leq 1$. In particular, M_t is a bounded martingale.

By the martingale convergence theorem, $M_\infty = \lim_{t \to \infty} M_t$ exists, and as in the proof of Theorem 6.17, we can see that M_∞ is the indicator function of the event $\{J_\infty \cap A \neq \emptyset\}$. Therefore,

$$\mathbf{P}\{J_\infty \cap A \neq \emptyset\} = \mathbf{E}(M_\infty) = \mathbf{E}(M_0) = \Phi'_A(0)^\alpha.$$

□

COROLLARY 9.18. *Consider the hull created by filling in the union of k independent $\mathbb{H}$-excursions. Then the distribution of the right boundary of the hull is the same as that of $SLE(8/3, 8v/3)$ where*

$$v = \frac{1}{4}\sqrt{1 + 24k} - 1.$$

9.4. Constructing restriction measures

We have already seen that the restriction measures $\mathbf{P}_\alpha$ do not exist if $\alpha < 5/8$. In this section we will show how to construct the (unique) restriction measure $\mathbf{P}_\alpha$ for $\alpha \geq 5/8$. Recall that $\mathbf{P}_{5/8}$ is given by $SLE_{8/3}$; the construction we describe in this section will also make use of SLE_κ, but with different values of κ. If α is a nonnegative integer, we have already commented that we can obtain $\mathbf{P}_\alpha$ by considering the union of α $\mathbb{H}$-excursions. The construction in this section for these α will be different, but by uniqueness will give the same measure $\mathbf{P}_\alpha$.

For $\alpha \geq 5/8$, define $\kappa = \kappa(\alpha)$ and $\lambda = \lambda(\alpha)$ as in (6.9):

$$\kappa = \frac{6}{2\alpha + 1}, \quad \lambda = (8 - 3\kappa)\alpha.$$

Note that as α increases from $5/8$ to ∞, κ decreases from $8/3$ to 0 and λ increases from 0 to ∞. Let g_t denote the conformal maps of an SLE_κ with driving function $U_t = \kappa B_t$, and let $\mathcal{O}_\lambda$ be an independent realization of the Brownian bubble soup as described in §5.7. The elements of $\mathcal{O}_\lambda$ can be written as (s, η^s) where $\eta^s : [0, s] \to \mathbb{C}$ is a curve with $\eta^s(0) = \eta^s(s) = 0$ and $\eta(0, s) \subset \mathbb{H}$. Let K be the hull generated by

$$\gamma[0, \infty) \cup \left[\bigcup_{(s, \eta^s) \in \mathcal{O}_\lambda} g_s^{-1}(\eta^s) \right].$$

We claim that K is distributed according to $\mathbf{P}_\alpha$. To show this, we need to show that if $A \in \mathcal{Q}$, then
$$\mathbf{P}\{K \cap A = \emptyset\} = \Phi'_A(0)^\alpha.$$

The strategy to prove this is very similar to that used in §6.4 to establish the restriction property for $SLE_{8/3}$. In fact, the important computation was done in that section. Let $\mathcal{F}_t = \sigma\{B_s : 0 \le s \le t\}$ denote the filtration of the Brownian motion, which is the same as the filtration for γ. Let $V = V_A$ denote the event $\{K \cap A = \emptyset\}$ and let M_t denote the bounded martingale
$$\tilde{M}_t = \mathbf{E}[1_V \mid \mathcal{F}_t].$$
The martingale convergence theorem implies that w.p.1 $\tilde{M}_t \to \tilde{M}_\infty = \mathbf{E}[1_V \mid \mathcal{F}_\infty] = \mathbf{E}[1_V \mid \gamma[0, \infty)]$. If we find another bounded martingale M_t with $M_\infty = \mathbf{E}[1_V \mid \mathcal{F}_\infty]$, then it must hold that $M_t = \tilde{M}_t$ w.p.1. Let $t_A = \inf\{t : \gamma(t) \in A\}$. On the event $\{t < t_A\}$, define $\Phi_t = \Phi_{t,A} = g_{g_t(A)}$.
$$M_t = 1\{\gamma(0,t] \cap A = \emptyset\} \, \Phi'_t(U_t)^\alpha \, \exp\left\{\frac{\lambda}{6} \int_0^t S\Phi_s(U_s) \, ds\right\}.$$
(Recall that $\Phi'_t(U_t) \le 1$ and $S\Phi_s(U_s) \le 0$, so $M_t \le 1$.) In Proposition 6.19, it was shown that M_t is a martingale. Using the argument in Theorem 6.17, we can see that
$$M_\infty = 1\{\gamma(0,t] \cap A = \emptyset\} \, \exp\left\{\frac{\lambda}{6} \int_0^\infty S\Phi_s(U_s) \, ds\right\}.$$

To finish the proof, we use the fact that if $\gamma : [0, \infty) \to \overline{\mathbb{H}}$ is any curve with: $\gamma(0) = 0$; $\gamma(0, \infty) \to \mathbb{H}$; $\gamma(t) \to \infty$ as $t \to \infty$, and $\mathrm{hcap}[\gamma(0,t]] = 2t$; and $A \in \mathcal{Q}$ with $\gamma(0,\infty) \cap A = \emptyset$, then the probability that there is no s such the image under $g_s^{-1} = g_{\gamma(0,s]}^{-1}$ of the boundary bubble at time s in $\mathcal{O}_\lambda$ intersects A is given by
$$\exp\left\{\frac{\lambda}{6} \int_0^\infty S\Phi_s(U_s) \, ds\right\}.$$
Here U_t is the driving function for γ and $\Phi_s = g_{g_s(A)}$. Let $\bar{\Phi}_s(z) = \Phi_s(z) - \Phi_s(U_s)$, and note that $S\Phi_s(U_s) = S\bar{\Phi}_s(0)$. Let $\mathcal{O}_{\lambda,t}$ denote the bubbles in the realization up to time t, and let us write
$$g^{-1} \circ \mathcal{O}_{\lambda,t} = \{g_{s(\eta)}^{-1} : \eta \in \mathcal{O}_{\lambda,t}, s(\eta) \le t\}.$$
Let Y_t denote the the probability that $\mathcal{O}_{\lambda,t} \cap A = \emptyset$. Then the Markov property tells us that
$$\mathbf{P}\{g^{-1} \circ \mathcal{O}_{\lambda,\infty} = \emptyset\} = Y_t \, \mathbf{P}\{\gamma^{(t)}(0,\infty) \cap \bar{\Phi}_s(A) = \emptyset\}.$$
Here $\gamma(s) = g_t(\gamma(s+t)) - g_t(\gamma(t)) = g_t(\gamma(s+t)) - U_t$. Hence, it suffices to show that
$$\dot{Y}_0 = \frac{\lambda}{6} S\Phi_A(0).$$
But this follows from Proposition 5.22.

Proposition 5.34 gives an equivalent way of constructing $\mathbf{P}_\alpha$. Let κ, λ be as above and let g_t, γ denote a chordal SLE_κ path. Let $\mathcal{L}_\lambda$ denote an independent realization of the Brownian loop soup with parameter λ. Let K be the hull generated by $\gamma(0, \infty)$ and the union of all the loops $\eta \in \mathcal{L}_\lambda$ such that $\eta[0, t_\eta] \cap \gamma(0, \infty) \ne \emptyset$. Then K is distributed according to $\mathbf{P}_\alpha$.

More generally, if D is a Jordan domain[2] and z, w are distinct points on D, we can define a measure $\nu_D^\#(z, w)$ on "hulls in D connecting z and w" by conformal transformation of $\mathbf{P}_\alpha$. We can also define $\nu_D^\#(z,w)$ directly by taking chordal SLE_κ connecting z and w in D (which is defined as the conformal image of chordal SLE_κ in $\mathbb{H}$) and attaching to the curve all the loops in a Brownian loop soup of parameter λ that lie in D and intersect the path. If ∂D is locally analytic at z, w we define

$$\nu_D(z,w) = |f'(1)|^{-\alpha} |f'(-1)|^{-\alpha} \nu_D^\#(z,w),$$

where $f : \mathbb{D} \to D$ is a conformal transformation with $f(1) = z, f(-1) = w$. (Although, the transformation f is not unique, it is an easy exercise to show that $f'(1) f'(-1)$ is independent of the choice of transformation.) We call $\{\nu_D(z,w)\}$ the *(chordal) restriction family with exponent* α. This family is unique if we normalize so that $|\nu_\mathbb{D}(1, -1)| = 1$.

PROPOSITION 9.19. *Let $\nu_D(z,w)$ denote the restriction family with exponent α, where D ranges over Jordan domains D and distinct boundary points z, w at which ∂D is locally analytic. Then the family satisfies:*
- **Restriction property.** *If $D \subset D'$ and $\partial D, \partial D'$ agree in neighborhoods of z, w, then $\mu_D(z,w)$ is $\mu_{D'}(z,w)$ restricted to hulls K that lie in D.*
- **Conformal covariance.** *If $f : D \to D'$ is a conformal transformation, then*

$$f \circ \mu_D(z, w) = |f'(z)|^\alpha |f'(w)|^\alpha \mu_{D'}(f(z), f(w)).$$

PROOF. The conformal covariance property holds by definition. To show the restriction property, let $h : D' \to \mathbb{H}$ be a conformal transformation with $h(z) = 0, h(w) = \infty$. Let $A = \mathbb{H} \setminus h(D)$. Then $f := h^{-1} \circ \Phi_A \circ h$ is a conformal transformation of D onto D' with $f(z) = z, f(w) = w$. Note that $f'(z) = \Phi'_A(0), f'(w) = 1$ (the second equality uses the fact that "$\Phi'_A(\infty) = 1$" in the sense that $\Phi_A(\tilde{w}) = \tilde{w} + O(1)$ as $\tilde{w} \to \infty$). We now use the restriction property of $\mathbf{P}_\alpha$ on $\mathbb{H}$. □

---

[2] The assumption that the domain is Jordan is more than is needed.

APPENDIX A

Hausdorff dimension

A.1. Definition

If $V \subset \mathbb{R}^d$ and $\alpha, \epsilon > 0$, let

$$\mathcal{H}^\alpha_\epsilon(V) = \inf \sum_{n=1}^\infty [\mathrm{diam}(U_n)]^\alpha,$$

where the infimum is over all countable collections of sets $U_1, U_2, \ldots$ with $V \subset \cup U_n$ and $\mathrm{diam}(U_n) < \epsilon$. It is easy to see that $\mathcal{H}^\alpha_\epsilon$ is an outer measure, i.e., a monotone function from subsets of $\mathbb{R}^d$ into $[0, \infty]$ with $\mathcal{H}^\alpha_\epsilon(\emptyset) = 0$ and

$$\mathcal{H}^\alpha_\epsilon(\bigcup_{j=1}^\infty V_j) \leq \sum_{j=1}^\infty \mathcal{H}^\alpha_\epsilon(V_j).$$

The *Hausdorff α-measure* is defined by

$$\mathcal{H}^\alpha(V) = \lim_{\epsilon \to 0+} \mathcal{H}^\alpha_\epsilon(V).$$

Since $\mathcal{H}^\alpha_\epsilon(V)$ is increasing in ϵ, the limit on the right exists with infinity being a possible value. Note that $\mathcal{H}^\alpha$ is an outer measure. (It is also true that $\mathcal{H}^\alpha$ restricted to Borel subsets of $\mathbb{R}^d$ is a Borel measure, see [14, Section 19]; we will not use this fact.) It is easy to check that if $\mathcal{H}^\alpha(V) < \infty$, then $\mathcal{H}^\beta(V) = 0$ for $\beta > \alpha$, and if $\mathcal{H}^\alpha(V) > 0$, then $\mathcal{H}^\beta(V) = \infty$ for $\beta < \alpha$. The *Hausdorff dimension* of V is defined by

$$\dim_h(V) = \inf\{\alpha : \mathcal{H}^\alpha(V) = 0\} = \sup\{\alpha : \mathcal{H}^\alpha(V) = \infty\}.$$

In this section we will discuss methods to compute or estimate $\dim_h(V)$. Note that monotonicity and subadditivity of $\mathcal{H}^\alpha$ imply that

$$\dim_h[\bigcup_{n=1}^\infty V_n] = \sup_n \dim_h(V_n).$$

A *dyadic ball* will be a closed ball $\mathcal{B}$ of radius 2^k for some $k \in \mathbb{Z}$. If U is any set, then U is contained in a ball of diameter at most $2\,\mathrm{diam}(U)$ and hence in a dyadic ball of diameter at most $4\,\mathrm{diam}(U)$. Hence,

$$4^{-\alpha} \lim_{k \to \infty} [\,\inf \sum_{n=1}^\infty \mathrm{diam}(U_n)^\alpha] \leq \mathcal{H}^\alpha_\epsilon(V) \leq \lim_{k \to \infty} [\,\inf \sum_{n=1}^\infty \mathrm{diam}(U_n)^\alpha],$$

where the infimums are over dyadic balls $U_1, U_2, \ldots$ with $V \subset \cup U_n$ and $\mathrm{diam}(U_n) \leq 2^{-k}$. Upper bounds for Hausdorff dimension tend to be easier to give since they only require finding some nice cover of the set.

LEMMA A.1. $\mathcal{H}^\alpha(V) = 0$ if and only if there exist sequences $\epsilon_n, \delta_n \to 0+$ and sets $U_{n,1}, U_{n,2}, \ldots$ such that $\text{diam}[U_{n,j}] < \epsilon_n$, $V \subset \cup_{j=1}^\infty U_{n,j}$ and

(A.1) $$\sum_{j=1}^\infty \text{diam}(U_{n,j})^\alpha < \delta_n$$

In particular, if for every $\epsilon > 0$, V can be covered by $N(\epsilon)$ balls of diameter ϵ, then

(A.2) $$\dim_h(V) \leq b := \liminf_{\epsilon \to 0+} \frac{\log N(\epsilon)}{\log(1/\epsilon)}.$$

PROOF. This is immediate from the definition since (A.1) implies $\mathcal{H}^\alpha_{\epsilon_n}(V) < \delta_n$ and (A.2) implies that $\mathcal{H}^\alpha(V) = 0$ for $\alpha > b$. □

LEMMA A.2. Suppose $V \subset \mathbb{R}^m$ and $f : V \to \mathbb{R}^d$ is a Hölder b-continuous function, i.e., there exists a c such that for all $z, w \in V$, $|f(z) - f(w)| \leq c|z - w|^b$. Then $\dim_h[f(V)] \leq b^{-1} \dim_h(V)$.

PROOF. Let $\alpha > \dim_h(V)$, and let ϵ_n, δ_n, and $U_{n,j}$ be as in the previous lemma for α. Let $\tilde{U}_{n,j} = f(U_{n,j} \cap V)$, Then $\tilde{U}_{n,1}, \tilde{U}_{n,2}, \cdots$ covers $f(V)$, $\text{diam}[\tilde{U}_{n,j}] \leq c \, \text{diam}[U_{n,j}]^b < c\epsilon_n^b$, and

$$\sum_{j=1}^\infty \text{diam}[\tilde{U}_{n,j}]^{\alpha/b} \leq \sum_{j=1}^\infty c^{\alpha/b} \, \text{diam}[U_{n,j}]^\alpha < c^{\alpha/b} \, \delta_n.$$

□

Lower bounds on dimension are harder to give. We will give two lemmas. The first can be considered a converse to the last lemma; it gives a way to give a lower bound on $\dim_h[f(V)]$ in terms of $\dim_h(V)$. The second, which goes back to Frostman [36], is particularly useful for giving lower bounds for dimensions of random sets. Roughly, it says that if one can put a measure supported on set V that is "at least s-dimensional", then V must have dimension as least s.

LEMMA A.3. Suppose $V \subset \mathbb{R}^m$ and $f : V \to \mathbb{R}^d$ is a function satisfying the following: there is a decreasing function $\delta \mapsto N_\delta$ such that

(A.3) $$\lim_{\delta \to 0+} \frac{\log N_\delta}{\log(1/\delta)} = 0,$$

and such that for each ball $\mathcal{B} \subset \mathbb{R}^d$ of diameter δ, $f^{-1}(\mathcal{B})$ is contained in the union of at most N_δ balls of diameter δ^a. Then $\dim_h[f(V)] \geq a \dim_h(V)$.

PROOF. Let $\alpha > \dim_h[f(V)]$. Find $\epsilon_n, \delta_n, U_{n,j}$ as in Lemma A.1 with

$$\text{diam}[U_{n,j}] < \epsilon_n, \quad f(V) \subset \bigcup_{j=1}^\infty U_{n,j},$$

$$\sum_{j=1}^\infty [\text{diam}(U_{n,j})]^\alpha < \delta_n.$$

Without loss of generality we may assume that the $U_{n,j}$ are dyadic balls (perhaps replacing ϵ_n with $4\epsilon_n$ and δ_n with $4^\alpha \delta_n$). Let $K_n(k)$ denote the number of balls

$U_{n,1}, U_{n,2}, \ldots$ with diameter 2^k; then

$$\sum_{k=-\infty}^{\infty} K_n(k) \, 2^{k\alpha} < \delta_n.$$

Since $K_n(k) = 0$ if $2^k \geq \epsilon_n$, (A.3) implies that for every $\beta > \alpha$,

$$\lim_{n \to \infty} \sum_{k=-\infty}^{\infty} N_{2^k} \, K_n(k) \, 2^{k\beta} = 0.$$

Since $f^{-1}(U_{n,1}), f^{-1}(U_{n,2}), \ldots$ covers V and $f^{-1}(U_{n,j})$ is contained in the union of $N_{\text{diam}(U_{n,j})}$ balls of diameter $[\text{diam}(U_{n,j})]^a$, these balls give a cover $\mathcal{U}_{n,1}, \mathcal{U}_{n,2}, \ldots$ of V with

$$\sum_{l=1}^{\infty} \text{diam}[\mathcal{U}_{n,l}]^{\beta/a} \leq \sum_{k=-\infty}^{\infty} N_{2^k} \, K_n(k) \, 2^{k\beta} \longrightarrow 0.$$

Therefore $\dim_h(V) \leq \beta/a$, and since this holds for all $\beta > \alpha > \dim_h[f(V)]$, we conclude that $\dim_h(V) \leq \dim_h[f(V)]/a$. $\square$

LEMMA A.4. *Suppose $s > 0$, $V \subset \mathbb{R}^d$ is a Borel set, and μ is a positive Borel measure with $0 < \mu(V) < \infty$, $\mu(\mathbb{R}^d \setminus V) = 0$, and*

$$(A.4) \qquad \int_V \int_V \frac{\mu(dz) \, \mu(dw)}{|z - w|^s} = I < \infty.$$

Then $\mathcal{H}^s(V) = \infty$. In particular, $\dim_h(V) \geq s$.

PROOF. Without loss of generality assume $\mu(V) = 1$. Note that (A.4) implies that μ gives zero measure to points. For any z, let

$$\phi(z) = \int_V \frac{\mu(dw)}{|z - w|^s}.$$

Let $c_1 > 0$. We first claim that if

$$(A.5) \qquad \limsup_{\epsilon \to 0+} \epsilon^{-s} \mu[\mathcal{B}(z, \epsilon)] \geq c_1 > 0,$$

then $\phi(z) = \infty$. To see this note that (A.5) and $\mu(\{z\}) = 0$ imply that we can find $r_1 > t_1 > r_2 > t_2 > r_3 > t_3 > \cdots$ such that $\mu(A_j) \geq (c_1/2) r_j^s$, where $A_j = \mathcal{B}(z, r_j) \setminus \mathcal{B}(z, t_j)$. This implies

$$\int_{A_j} \frac{\mu(dw)}{|z - w|^s} \geq c_1/2.$$

Since the A_j are disjoint, this gives $\phi(z) = \infty$. Using (A.4), we see that

$$\mu\{z \in V : \limsup_{\epsilon \to 0+} \epsilon^{-s} \mu[\mathcal{B}(z, \epsilon)] > 0\} \leq \mu\{z \in V : \phi(z) = \infty\} = 0,$$

and hence there is an $\epsilon_0 > 0$ such that $\mu(\hat{V}) \geq 1/2$ where $\hat{V} = \{z \in V : \mu[\mathcal{B}(z, \epsilon)] \leq c_1 \epsilon^s \text{ for all } 0 < \epsilon < \epsilon_0\}$. If V_j is a ball with $\text{diam}[V_j] < \epsilon_0$ and $V_j \cap \hat{V} \neq \emptyset$, then $\mu(V_j) \leq c_1 [\text{diam}(V_j)]^s$. Therefore if $V_1, V_2, \ldots$ is any sequence of sets with $\hat{V} \subset \cup V_j$, $\text{diam}[V_j] < \epsilon_0$, $V_j \cap \hat{V} \neq \emptyset$,

$$\sum_{j=1}^{\infty} [\text{diam}(V_j)]^s \geq c_1^{-1} \sum_{j=1}^{\infty} \mu(V_j) \geq c_1^{-1} \mu(\tilde{V}) \geq c_1^{-1}/2.$$

Therefore, $\mathcal{H}^s(\hat{V}) \geq \mathcal{H}^s_{\epsilon_0}(\hat{V}) \geq c_1^{-1}/2$. Since this holds for every $c_1 > 0$, $\mathcal{H}^s(V) = \infty$. □

REMARK A.5. This lemma is very useful for giving lower bounds of Hausdorff dimensions of random sets V. For such sets one can often define a random measure μ supported on V such that

$$\mathbf{E}[\int_V \int_V \frac{\mu(dz)\,\mu(dw)}{|z-w|^s}] < \infty,$$

which implies that w.p.1

$$I = \int_V \int_V \frac{\mu(dz)\,\mu(dw)}{|z-w|^s} < \infty.$$

On the event that $\mu(V) > 0$ and $I < \infty$, we conclude that $\dim_h(V) \geq s$.

A.2. Dimension of Brownian paths

In this section we show that the Hausdorff dimension of the paths of complex Brownian motion is 2. We will prove a stronger statement, originally due to Kaufman [40], that with probability one, $\dim_h[B(V)] = 2 \dim_h(V)$ for every $V \subset [0,1]$. The event of probability zero for which this does not hold will be independent of V. For the upper bound, we will consider the event that the Brownian paths are Hölder-α continuous for all $\alpha < 1/2$. For the lower bound, the event is discussed in the next lemma.

LEMMA A.6. *There exists a $c < \infty$ such that if B_t is a complex Brownian motion, then w.p.1 there is an $\epsilon_0 = \epsilon_0(\omega) > 0$ such that the following holds. If $\mathcal{B}$ is a disk of radius $\epsilon < \epsilon_0$ in $\mathbb{C}$, then $\{s \in [0,1] : B_s \in \mathcal{B}\}$ is contained in the union of at most $c\,[\log(1/\epsilon)]^2$ intervals of length ϵ^2.*

PROOF. Let $\mathcal{T}_n$ denote the set of closed disks of radius 2^{-n+1} centered at points $2^{-n}(j_1 + ij_2)$ with $j_1, j_2 \in \mathbb{Z}$. Then every closed disk of radius less than 2^{-n} is contained in the union of at most four disks in $\mathcal{T}_n$. Hence it suffices to find a c such that w.p.1 there is a $k_0 = k_0(\omega)$ such that for all $n \geq k_0$ and every $\mathcal{B} \in \mathcal{T}_n$, $B^{-1}(\mathcal{B}) \cap [0,1]$ is contained in the union of at most cn^2 intervals of length 2^{-2n}. Let $V_{\mathcal{B}} = V_{\mathcal{B}}(c)$ denote the event that $B^{-1}(\mathcal{B}) \cap [0,1]$ is not contained in the union of cn^2 such intervals, and let $V^n = \cup_{\mathcal{B} \in \mathcal{T}_n} V_{\mathcal{B}}$.

Using Exercise 2.13, we see that there is a $c_1 > 0$ such that for any n, $\mathcal{B} \in \mathcal{T}_n$, and $z \in \mathbb{C}$,

(A.6) $$\mathbf{P}^z\{B[2^{-2n}, 1] \cap \mathcal{B} = \emptyset\} \geq c_1/n.$$

Let $\tau_0 = 0$, and for $j > 0$, let τ_j be the first time t after $\tau_{j-1} + 2^{-2n}$ that $B_t \in \mathcal{B}$. Then by iterating the estimate above we can see that for c sufficiently large and all z,

$$\mathbf{P}^z[V_{\mathcal{B}}] \leq \mathbf{P}^z\{\tau_{cn^2} \leq 1\} \leq [1 - \frac{c_1}{n}]^{cn^2} \leq 8^{-n}.$$

By using the strong Markov property, we see that for any $\mathcal{B} \in \mathcal{T}_n$,

$$\mathbf{P}^0[V_{\mathcal{B}}] \leq \mathbf{P}^0\{B[0,1] \cap \mathcal{B} \neq \emptyset\}\, 8^{-n}.$$

For $k \geq 0$, $\mathcal{T}_{n,k}$ be the set of $\mathcal{B} \in \mathcal{T}_n$ such that $\mathcal{B} \cap \mathcal{B}(0,k) \neq \emptyset$; for convenience let $\mathcal{T}_{n,-1} = \emptyset$. Note that the cardinality of $\mathcal{T}_{n,k}$ is $(k+1)^2\, O(4^n)$. Hence,

$$\begin{aligned}
\mathbf{P}^0[V^n] &\leq \sum_{\mathcal{B} \in \mathcal{T}_n} \mathbf{P}^0[V_\mathcal{B}] \\
&= \sum_{k=0}^{\infty} \sum_{\mathcal{B} \in \mathcal{T}_{n,k} \setminus \mathcal{T}_{n,k-1}} \mathbf{P}^0[V_\mathcal{B}] \\
&\leq c'\, 2^{-n} \sum_{k=0}^{\infty} (k+1)^2\, \mathbf{P}^0\{B[0,1] \cap \{|z| \geq k-1\} \neq \emptyset\} < c''\, 2^{-n}.
\end{aligned}$$

The lemma then follows from the Borel-Cantelli Lemma. $\square$

THEOREM A.7. *Suppose B_t is a complex Brownian motion. Then w.p.1 for every $V \subset [0,1]$,*

$$\dim_h[B(V)] = 2\dim_h[V].$$

PROOF. Corollary 1.40 states that there is an event of probability one such that Brownian paths are Hölder α-continuous for all $\alpha < 1/2$. Lemma A.2 tells us that on this event $\dim_h[B(V)] \leq \alpha^{-1} \dim_h[V]$ for all $\alpha < 1/2$ and hence $\dim_h[B(V)] \leq 2\dim_h[V]$. For the other direction consider the event of probability one in Lemma A.6. Lemma A.3 tells us that on this event $\dim_h[B(V)] \geq 2\dim_h[V]$. Note that the event of probability one does not depend on the set V. $\square$

REMARK A.8. For Brownian motion in $\mathbb{R}^d$, $d \geq 3$, Lemma A.6 holds with virtually the same proof with $[\log(1/\epsilon)]^2$ replaced with $\log(1/\epsilon)$. The difference comes in (A.6) where the right hand side becomes c_1 rather than c_1/n. Hence, Theorem A.7 holds for $d \geq 3$ (one could also conclude it by considering projections onto $\mathbb{R}^2$). For $d = 1$, a version of Lemma A.6 holds, but in this case $[\log(1/\epsilon)]^2$ must be replaced with $\epsilon^{-1} \log(1/\epsilon)$; in this case, the right hand side of (A.6) is $c_1\, 2^{-n}$. It is easy to see that Theorem A.7 does not hold for $d = 1$ since $\dim_h[B[0,1]] = 1$; in fact, it is possible for $\dim_h[B(V)] < \min\{1, 2\dim_h(V)\}$. For example, if $V = \{s \in [0,1] : B_s = 0\}$, then $\dim_h(V) = 1/2$ and $\dim_h[B(V)] = 0$.

REMARK A.9. In Chapter 8 we use the fact that the event of probability one in Theorem A.7 does not depend on V. In particular, to compute the Hausdorff dimension of a random subset V of $B[0,1]$, it suffices to compute the dimension of $\{s \in [0,1] : B_s \in V\}$.

A.3. Dimension of random "Cantor sets" in $[0,1]$

We first consider deterministic Cantor sets. Suppose $0 < K < M$ are positive integers. Suppose $A_0 \supset A_1 \supset A_2 \supset \cdots$ such that

- $A_0 = [0,1]$.
- Given A_{n-1}, which is the union of K^{n-1} distinct closed intervals of the form $[(j-1)M^{-(n-1)}, jM^{-(n-1)}]$, A_n is obtained by dividing each of these small intervals into M equal pieces of the form $[(k-1)M^{-n}, kM^{-n}]$, and selecting K of them to be in A_n.

We may use any rule to choose the K intervals and may use different rules for different intervals, but we always choose exactly K intervals of length M^{-n} from each of the intervals of length $M^{-(n-1)}$ in A_{n-1}. This guarantees that if r is a positive integer, then every interval of length M^{-r} contains at most $2K^{n-r}$ of the intervals at level n. Note that the "Cantor set"

$$A = \bigcap_{n=0}^{\infty} A_n$$

is a nonempty compact set. We will now prove that $\dim_h(A) = \log K/\log M$. The upper bound follows immediately from Lemma A.1 since $A \subset A_n$ and A_n is the union of K^n intervals of length M^{-n}. One way to get the lower bound is to consider the probability measure μ supported on A that gives measure K^{-n} to each of the K^n intervals of length M^{-n} at level n. Then $\mu = \lim_{n \to \infty} \mu_n$, where μ_n is the measure whose Radon-Nikodym derivative with respect to Lebesgue measure is $(M/K)^n 1_{A_n}$, and the limit is the usual weak convergence of finite measures. We will show that

$$\int_0^1 \int_0^1 \frac{\mu(dx)\,\mu(dy)}{|x-y|^s} < \infty$$

for all $s < \log K/\log M$. It suffices to prove that

$$\sup_n \int_0^1 \int_0^1 \frac{\mu_n(dx)\,\mu_n(dy)}{|x-y|^s} \leq I_s < \infty,$$

since this implies

$$\int_0^1 \int_0^1 \frac{\mu(dx)\,\mu(dy)}{|x-y|^s} = \lim_{\epsilon \to 0+} \int_0^1 \int_0^1 \frac{\mu(dx)\,\mu(dy)}{\epsilon \vee |x-y|^s}$$

(A.7)
$$= \lim_{\epsilon \to 0+} \lim_{n \to \infty} \int_0^1 \int_0^1 \frac{\mu_n(dx)\,\mu_n(dy)}{\epsilon \vee |x-y|^s} \leq I_s.$$

We will show the stronger fact

$$\sup_n \sup_{y \in [0,1]} \int_0^1 \frac{\mu_n(dx)}{|x-y|^s} < \infty.$$

We need only the simple estimate for $s \in (0,1)$,

$$\frac{C_1(s)\,\epsilon}{(|z-y|+\epsilon)^s} \leq \int_z^{z+\epsilon} \frac{dx}{|x-y|^s} \leq \frac{C_2(s)\,\epsilon}{(|z-y|+\epsilon)^s}.$$

Then,

$$\int_0^1 \frac{\mu_n(dx)}{|x-y|^s} = \left(\frac{M}{K}\right)^n \sum_{j=1}^{K^n} \int_{I_j} \frac{dx}{|x-y|^s} \leq c(s)\,K^{-n} \sum_{j=1}^{K^n} \rho(y, I_j)^{-s},$$

where the sum is over all the intervals $I_1, \ldots, I_{K^n}$ in A_n and $\rho(y, I_j) = \rho_n(y, I_j) = \text{dist}(y, I_j) + M^{-n}$. The number of intervals with $\rho(y, I_j) \leq M^{-r}$ is bounded above by $c(M)\,K^{n-r}$. Therefore,

$$K^{-n} \sum_{j=1}^{K^n} \rho(y, I_j)^{-s} \leq K^{-n} \sum_{r=0}^{n} \sum_{M^{-(r+1)} \leq \rho(y, I_j) \leq M^{-r}} \rho(y, I_j)^{-s}$$

$$\leq C(M, s) \sum_{r=0}^{n} M^{rs} K^{-r} \leq C(M, s, K),$$

provided $M^s < K$, i.e., $s < \log K/\log M$. Therefore, by Lemma A.4, $\dim_h(A) \geq \log K/\log M$.

REMARK A.10. The argument we give for $\dim_h(A) \geq \log K/\log M$ is not the easiest argument, but it is the one we adapt for random A.

We now consider random Cantor sets. Fix an integer $M \geq 2$. For each $n = 0, 1, 2, \ldots$ and $k = 1, 2, \ldots, M^n$, let $I(n, k) = I(n, k; M)$, denote the M-adic interval $[(k-1)M^{-n}, kM^{-n}]$. Suppose $J(n, k)$ are $0-1$ random variables that are nested in the sense that if $n_1 \geq n$, $I(n_1, k_1) \subset I(n, k)$, and $J(n, k) = 0$, then $J(n_1, k_1) = 0$. Let A_n denote the random set

$$A_n = \bigcup_{J(n,k)=1} I(n, k).$$

Then $A_0 \supset A_1 \supset A_2 \supset \cdots$ and $A = \cap_{n=0}^{\infty} A_n$ is a compact subset that is nonempty if and only if each $A_n \neq \emptyset$. Let Y_n be the number of intervals of length M^{-n} comprising A_n, i.e.,

$$Y_n = \sum_{k=1}^{M^n} J(n, k).$$

If $\epsilon \in (0, 1/2)$, let

$$J_\epsilon(n, k) = J(n, k)\, 1\{I(n, k) \cap [\epsilon, 1-\epsilon] \neq \emptyset\},$$

where 1 denotes indicator function. Then

$$A_\epsilon := \bigcap_{n=1}^{\infty} \bigcup_{J_\epsilon(n,k)=1} I(n, k) = A \cap [\epsilon, 1-\epsilon].$$

For the remainder of this section we assume this setup.

EXAMPLE A.11. Let B_t be a one-dimensional Brownian motion and $J(n, k)$ the indicator function of the event $0 \in B[I(n, k)]$. Then $A = \{t \in [0, 1] : B_t = 0\}$, which is called the zero set of the Brownian motion.

EXAMPLE A.12. Suppose $X(n, k)$ are i.i.d. $0-1$ random variables with

$$\mathbf{P}\{X(n, k) = 1\} = p \in (0, 1).$$

Let

$$J(n, k) = \prod_{j=1}^{n} X(j, l(j; n, k)),$$

where $l(j; n, k)$ is defined so that $I(n, k) \subset I(j, l(j; n, k))$, i.e., $I(j, l(j; n, k))$ is the "ancestor" of $I(k, n)$ at the jth level. Then this random Cantor set can be considered as a branching (Galton-Watson) process. The set A is nonempty if and only if the branching process does not die out.

PROPOSITION A.13. *Suppose there is a $\zeta \in (0, 1)$ and a $c < \infty$ such that for all n, k, $\mathbf{E}[Y_n] \leq c\, M^{(1-\zeta)n}$. Then*

(A.8) $$\mathbf{P}\{\dim_h(A) \leq 1 - \zeta\} = 1.$$

REMARK A.14. One standard way to show that $\mathbf{E}[Y_n] \leq c\, M^{(1-\zeta)n}$ is to show that $\mathbf{P}\{J(n, j) = 1\} \leq c'\, M^{-\zeta n}$.

PROOF. It suffices to show for every $\epsilon > 0$, $\mathbf{P}\{\dim_h(A) \geq 1-\zeta+\epsilon\} = 0$. By the Markov inequality, $\mathbf{P}\{Y_n \geq M^{n(1-\zeta+\epsilon)}\} \leq M^{-n(1-\zeta+\epsilon)} \mathbf{E}[Y_n] \leq c'M^{-n\epsilon}$. By the Borel-Cantelli Lemma, w.p.1 for all n sufficiently large $Y_n \leq M^{n(1-\zeta+\epsilon)}$. Lemma A.1 shows that on this event $\dim_h(A) \leq 1-\zeta+\epsilon$. □

The lower bound will use a standard technique known as a second moment method.

LEMMA A.15. *If X is a nonnegative random variable with $\mathbf{E}[X^2] < \infty$, then*

$$\mathbf{P}\{X \geq \frac{1}{2}\mathbf{E}[X]\} \geq \frac{(\mathbf{E}[X])^2}{4\mathbf{E}[X^2]}.$$

PROOF. Without loss of generality assume $\mathbf{E}[X] = 1$. Since $\mathbf{E}[X; X < 1/2] \leq 1/2$, $\mathbf{E}[X; X \geq 1/2] \geq 1/2$. Then

$$\begin{aligned}
\mathbf{E}[X^2] &\geq \mathbf{E}[X^2; X \geq 1/2] \\
&= \mathbf{P}\{X \geq 1/2\} \mathbf{E}[X^2 \mid X \geq 1/2] \\
&\geq \mathbf{P}\{X \geq 1/2\} \mathbf{E}[X \mid X \geq 1/2]^2 \\
&= \mathbf{E}[X; X \geq 1/2]^2 / \mathbf{P}\{X \geq 1/2\} \\
&\geq (1/4)/\mathbf{P}\{X \geq 1/2\}.
\end{aligned}$$

□

PROPOSITION A.16. *Suppose there is a $\zeta \in (0,1)$ and $\beta_1, \beta_2 < \infty$ such that for all n,*

(A.9) $$\mathbf{E}(Y_n) \geq \beta_1 M^{n(1-\zeta)},$$

and for all n, j, k,

(A.10) $$\mathbf{P}\{J(n,j) J(n,k) = 1\} \leq \beta_2 M^{-\zeta n} [|j-k|+1]^{-\zeta}.$$

Then there exists a $\rho = \rho(\beta_1, \beta_2, \zeta, M) > 0$ such that

$$\mathbf{P}\{\dim_h(A) = 1-\zeta\} \geq \rho.$$

PROOF. Note that (A.10) with $j = k$ and Proposition A.13 imply $\mathbf{P}\{\dim_h(A) \leq 1-\zeta\} = 1$. Hence we only need to show that that there is a ρ such that for every $s < 1-\zeta$,

$$\mathbf{P}\{\dim_h(A) \geq s\} \geq \rho.$$

All constants in this proof are allowed to depend on $\beta_1, \beta_2, \zeta, M$. Some constants may also depend on s in which case this dependence will be explicitly noted.

Let μ_n be the random measure $M^{\zeta n}$ times Lebesgue measure restricted to A_n so that $\mu_n[I(n,k)] = M^{(\zeta-1)n} J(n,k)$. Then (A.9) implies that

$$\mathbf{E}(\mu_n[0,1]) = \mathbf{E}[M^{(\zeta-1)n} Y_n] \geq \beta_1.$$

Also,
$$\begin{aligned}
\mathbf{E}[(\mu_n[0,1])^2] &= M^{2(\zeta-1)n}\,\mathbf{E}[Y_n^2] \\
&= M^{2(\zeta-1)n} \sum_{j=1}^{M^n}\sum_{k=1}^{M^n} \mathbf{P}\{J(j,n)\,J(k,n)=1\} \\
&\leq \beta_2\, M^{2(\zeta-1)n} \sum_{j=1}^{M^n}\sum_{k=1}^{M^n} M^{-\zeta n}\,[|j-k|+1]^{-\zeta} \\
&\leq 2\beta_2\, M^{(\zeta-1)n} \sum_{j=0}^{M^n} (j+1)^{-\zeta} \leq c.
\end{aligned}$$

Therefore, by Lemma A.15, there is a $\tilde{c} > 0$ such that

(A.11) $$\mathbf{P}\{\mu_n[0,1] \geq \beta_1/2\} \geq \tilde{c}.$$

Let
$$\mathcal{E}(n,s) = \int_0^1 \int_0^1 \frac{\mu_n(dx)\,\mu_n(dy)}{|x-y|^s}.$$

We claim that for every $s < 1-\zeta$, there is an $R_s < \infty$ such that for each n,

(A.12) $$\mathbf{E}[\mathcal{E}(n,s)] \leq R_s.$$

To see this, note that

$$\begin{aligned}
\mathbf{E}&\Big[\int_0^1\int_0^1 \frac{\mu_n(dx)\,\mu_n(dy)}{|x-y|^s}\Big] \\
&= \sum_{j=1}^{M^n}\sum_{k=1}^{M^n} \mathbf{P}\{J(n,j)J(n,k)=1\}\, M^{2\zeta n} \int_{(j-1)M^{-n}}^{jM^{-n}} \int_{(k-1)M^{-n}}^{kM^{-n}} \frac{dx\,dy}{|x-y|^s} \\
&\leq c_s \sum_{j=1}^{M^n}\sum_{k=1}^{M^n} M^{-\zeta n}[|j-k|+1]^{-\zeta}\, M^{2\zeta n}\, M^{-2n}\, M^{sn}[|j-k|+1]^{-s} \\
&\leq c_s M^n \sum_{j=0}^{M^n} M^{n(\zeta+s-2)}\,(j+1)^{-(\zeta+s)} \leq R_s.
\end{aligned}$$

(A.12) implies that
$$\mathbf{P}\{\mathcal{E}(n,s) \geq 2R_s/\tilde{c}\} \leq \tilde{c}/2,$$

where $\tilde{c}$ is as in (A.11). If V_n is the event $V_n = \{\mu_n[0,1] \geq \beta_1/2;\ \mathcal{E}(n,s) \leq 2R_s/\tilde{c}\}$, then $\mathbf{P}(V_n) \geq \tilde{c}/2$ and hence
$$\mathbf{P}\{V_n \text{ i.o.}\} \geq \rho,$$

where $\rho = \tilde{c}/2$. Note that ρ does not depend on s. On the event $\{V_n \text{ i.o.}\}$, we can find a subsequence μ_{n_j} (the subsequence can depend on ω, the realization) with μ_{n_j} supported on A_{n_j}, $\mu_{n_j}[0,1] \geq \beta_1/2$ and
$$\int_0^1 \int_0^1 \frac{\mu_{n_j}(dx)\,\mu_{n_j}(dy)}{|x-y|^s} \leq 2R_s/\tilde{c}.$$

By taking a further subsequence if necessary, which we denote by just μ_{n_j}, we can find a measure μ with $\mu_{n_j} \to \mu$. Note that $\mu[0,1] \geq \beta_1/2$, μ is supported on A, and (see (A.7)),

$$\int_0^1 \int_0^1 \frac{\mu(dx)\,\mu(dy)}{|x-y|^s} \leq 2\,R_s/\tilde{c}.$$

Hence, by Lemma A.4, on this event $\dim_h(A) \geq s$. □

EXAMPLE A.17. Let B_t be a one-dimensional Brownian motion starting at the origin, $\epsilon \in (0, 1/2)$, and $A_\epsilon = \{t \in [\epsilon, 1-\epsilon] : B_t = 0\}$. We claim that there exist constants c_1, c_2 such that for all n, j,

(A.13) $\quad c_1\,j^{-1/2} \leq \mathbf{P}\{0 \in B[(j-1)\,M^{-n}, j\,M^{-n}]\} \leq c_2\,j^{-1/2}$.

It is easy to see that the probability is independent of n; hence we can assume $n = 0$. In fact, the probability can be computed exactly using the Markov property and the reflection principle giving

$$1 - (2/\pi)\arctan\sqrt{j-1}.$$

However, (A.13) can be established by a cruder argument — if $n = 0$, we would expect that the probability would be comparable to $\mathbf{P}\{|B_{j-1}| \leq 1\}$ which is of order $j^{-1/2}$. If $j < k$, then the strong Markov property can be used with (A.13) to show that

$$\mathbf{P}\{0 \in B[(k-1)\,M^{-n}, k\,M^{-n}] \mid 0 \in B[(j-1)\,M^{-n}, j\,M^{-n}]\}$$

$$\leq \mathbf{P}\{0 \in B[(k-j-1)\,M^{-n}, (k-j)\,M^{-n}]\} \leq c_2\,(k-j)^{-1/2}.$$

In other words,

$$\mathbf{P}\{J_\epsilon(n,j) = 1\} \geq c_1(\epsilon)\,M^{-n/2},$$

$$\mathbf{P}\{J_\epsilon(n,j)\,J_\epsilon(n,k) = 1\} \leq c_2(\epsilon)\,M^{-n/2}\,(|k-j|+1)^{-1/2}$$

provided that

$$[(j-1)M^{-n}, jM^{-n}] \cap [\epsilon, 1-\epsilon] \neq \emptyset, \quad [(k-1)M^{-n}, kM^{-n}] \cap [\epsilon, 1-\epsilon] \neq \emptyset.$$

Proposition A.16 gives that $\mathbf{P}\{\dim_h(A_\epsilon) = 1/2\} = q_\epsilon > 0$. Note that $q_\epsilon < 1$ since there is a positive probability that $A_\epsilon = \emptyset$. However, it is not very difficult to extend this argument to show that w.p.1

$$\dim_h\{t \in [0,1] : B_t = 0\} = 1/2.$$

EXAMPLE A.18. Let $X(n,j)$, p, $J(n,j)$ be as in the second example above and define ζ by $p = M^{-\zeta}$. The expected number of intervals at level n is $M^{(1-\zeta)n}$. If $\zeta \geq 1$, then the corresponding branching process dies out, i.e., $\mathbf{P}\{A = \emptyset\} = 1$. Let us assume $0 < \zeta < 1$. Then

$$\mathbf{P}\{J(n,j) = 1\} = p^n = M^{-\zeta n},$$

$$\mathbf{P}\{J(n,j)\,J(n,k) = 1\} = p^n p^{s(j,k,n)} = M^{-\zeta n}\,M^{-\zeta\,s(j,k,n)},$$

where $s(j,k,n)$ is defined by saying that j, k have the same "ancestor interval" at level $n - s(j,k,n)$ but have different ancestors after that. Note that if $s(j,k,n) = m$, then $|k-j| + 1 \leq M^m$. Hence,

$$\mathbf{P}\{J(n,j)\,J(n,k) = 1\} \leq M^{-\zeta n}\,(|k-j|+1)^{-\zeta}.$$

Proposition A.16 then implies that $\mathbf{P}\{\dim_h(V) = 1 - \zeta\} \geq \rho > 0$. In fact, we can improve this. Let q be the survival probability for the corresponding branching process, i.e., $\mathbf{P}\{A \neq \emptyset\} = q$. Then it is easy to see that, except for an event of probability zero, if $A \neq \emptyset$, then $Y_n \to \infty$. But for every $N < \infty$,

$$\mathbf{P}\{\dim_h(A) \geq 1 - \zeta\} \geq \mathbf{P}\{Y_n \geq N \text{ for some } n\} \left[1 - (1 - \rho)^N\right],$$

and hence, $\mathbf{P}\{\dim_h(A) = 1 - \zeta\} = q$.

APPENDIX B

Hypergeometric functions

Hypergeometric functions arise as solutions to differential equations arising in the study of Bessel processes and SLE. Here we list some of the basic definitions and properties of the functions we will need. We follow mainly the treatment in [**64**], and the reader can check that book (or other books on special functions) for details. The *hypergeometric series* is

$$F(\alpha, \beta, \gamma; z) = 1 + \sum_{k=1}^{\infty} \frac{(\alpha)_k \, (\beta)_k}{(\gamma)_k \, k!} \, z^k,$$

where $(c)_k = \Gamma(c+k)/\Gamma(c) = c_k \, (c_k + 1) \cdots (c_k + (k-1))$. Here we assume γ is not a nonpositive integer. The definition is symmetric in α and β. If α or β is a nonpositive integer, then the series is a polynomial. Otherwise, it is easy to check that the radius of convergence is 1 so that $F(\alpha, \beta, \gamma; z)$ is an analytic function on $\mathbb{D}$ called the *hypergeometric function*. In fact, there is an analytic extension of $F(\alpha, \beta, \gamma; z)$ to $D := \mathbb{C} \setminus [1, \infty)$. We will only need to consider the hypergeometric function for parameters with nonnegative real parts; in fact, we will assume from now on that

(B.1) $\qquad\qquad \mathrm{Re}[\alpha], \mathrm{Re}[\beta] \geq 0 \quad \mathrm{Re}[\gamma] > \mathrm{Re}[\alpha + \beta] > 0.$

In this case [**64**, (9.1.6)], we can define $F(\alpha, \beta, \gamma; z)$ for all $z \in \mathbb{D}$ by

(B.2) $\qquad F(\alpha, \beta, \gamma; z) = \frac{\Gamma(\gamma)}{\Gamma(\beta)\, \Gamma(\gamma - \beta)} \int_0^1 t^{\beta-1} \, (1-t)^{\gamma-\beta-1} \, (1 - tz)^{-\alpha} \, dt.$

This formula assumes $\mathrm{Re}[\beta] > 0$; if $\mathrm{Re}[\alpha] > 0$, we can interchange α and β.

The hypergeometric function arises as a solution to the hypergeometric equation

(B.3) $\quad x\,(1-x)\,\phi''(x) + [\gamma - (\alpha + \beta + 1)\,x]\,\phi'(x) - \alpha\,\beta\,\phi(x) = 0, \quad 0 < x < 1.$

If γ is not an integer, then two linearly independent solutions to this equation are [**64**, (7.2.6)]

(B.4) $\qquad F(\alpha, \beta, \gamma; x), \quad x^{1-\gamma}\, F(1 - \gamma + \alpha, 1 - \gamma + \beta, 2 - \gamma; z).$

(These are also solutions for integer γ but are not linearly independent.) If $\alpha = 0$, the first solution is the constant function. Using the identity

(B.5) $\qquad\qquad \int_0^1 t^{a-1}\,(1-t)^{b-1}\,dx = \frac{\Gamma(a)\,\Gamma(b)}{\Gamma(a+b)},$

we see that under the assumption (B.1),

(B.6) $\qquad\qquad F(\alpha, \beta, \gamma; 1-) = \frac{\Gamma(\gamma)\,\Gamma(\gamma - \alpha - \beta)}{\Gamma(\gamma - \alpha)\,\Gamma(\gamma - \beta)}.$

B.1. The case $\alpha = 2/3, \beta = 1/3, \gamma = 4/3$

Let $F^*(z) = F(2/3, 1/3, 4/3; z)$, which comes up in studying SLE_6. In this case, substituting $w = tz$ in (B.2) gives

$$\text{(B.7)} \qquad 3\,F^*(z) = z^{-1/3} \int_0^z w^{-2/3}(1-w)^{-2/3}\,dw.$$

If $z = -x$ with $x > 0$,

$$3\,(-x)^{1/3}\,F^*(-x) = e^{-\pi i/3} \int_0^x y^{-2/3}(1+y)^{-2/3}\,dy.$$

Also, if $y > 0$,

$$\text{(B.8)} \qquad F^*(-y) = (1+y)^{-1/3}\,F^*\!\left(\frac{y}{y+1}\right).$$

Using the Schwarz-Christoffel transformation (see, e.g., [1, §6.2.2]), one can see that the map

$$z \longmapsto \frac{\Gamma(2/3)}{\Gamma(1/3)^2} \int_0^z w^{-2/3}(1-w)^{-2/3}\,dw = \frac{\Gamma(2/3)}{\Gamma(1/3)\,\Gamma(4/3)}\,z^{1/3}\,F^*(z)$$

is the conformal transformation of $\mathbb{H}$ onto the equilateral triangle with vertices $0, 1$, and $(1 + i\sqrt{3})/2$. Hence, we get the following.

PROPOSITION B.1. *Let*

$$\phi(z) = \frac{\Gamma(2/3)}{\Gamma(1/3)\,\Gamma(4/3)}\,z^{1/3}\,F^*(z).$$

Then ϕ is the conformal transformation of $\mathbb{H}$ onto the equilateral triangle with vertices $0, 1$, and $(1 + i\sqrt{3})/2$ satisfying $\phi(0) = 0, \phi(1) = 1, \phi(\infty) = (1 + i\sqrt{3})/2$. If $x > 0$, then

$$\phi(-x) = \frac{\Gamma(2/3)}{\Gamma(1/3)\,\Gamma(4/3)}\,F^*\!\left(\frac{x}{1+x}\right) e^{i\pi/3}.$$

REMARK B.2. See [19, Part Seven, II] for a detailed treatment of the relationship between hypergeometric functions and conformal maps of $\mathbb{H}$ onto triangles.

B.2. Confluent hypergeometric functions

In this section we will discuss the solution of

$$\text{(B.9)} \qquad \psi''(x) + \left[x + \frac{2a}{x}\right]\psi'(x) - \frac{2b}{x^2}\psi(x) = 0, \quad 0 < x < \infty,$$

with boundary conditions $\phi(0) = 0, \phi(\infty) = 1$. Here $a \in \mathbb{R}, b \geq 0$ or $a < 1/2, b > 0$. If we write $\psi(x) = \phi(x^2/2)$, then ϕ satisfies

$$y^2\,\phi''(y) + [y^2 + (a + \tfrac{1}{2})\,y]\phi'(y) - \frac{b}{2}\,\phi(y) = 0.$$

Let

$$r = \frac{(1/2) - a + \sqrt{((1/2) - a)^2 + 2b}}{2} > \frac{1}{2} - a.$$

If we write $\phi(y) = e^{-y}\,y^r\,v(y)$, then v satisfies

$$\text{(B.10)} \qquad y\,v''(y) + [(2r + a + \tfrac{1}{2}) - y]\,v'(y) - (r + a + \tfrac{1}{2})\,v(y) = 0.$$

The boundary conditions become $v(y) = o(y^{-r})$ as $y \to 0+$ and $v(y) \sim e^y y^{-r}$ as $y \to \infty$. The solution to this is $v(y) = [\Gamma(\alpha)/\Gamma(\alpha + r)] \Phi(\alpha, \alpha + r; y)$, where $\alpha = r + a + (1/2) > 1$, and $\Phi(\alpha, \gamma; z)$ denotes the confluent hypergeometric function (of the first kind)

$$\Phi(\alpha, \gamma; z) = 1 + \sum_{k=1}^{\infty} \frac{(\alpha)_k}{(\gamma)_k k!} z^k$$

(see [**64**, Section 9.10 and (9.12.8)]). There is a second solution to (B.10) but it blows up like $y^{1-\alpha-r}$ as $y \to 0+$, and hence does not satisfy the boundary condition. Hence the solution to (B.9) with the boundary conditions is

$$\psi(x) = e^{-x^2/2} \frac{x^q \Gamma(\alpha)}{2^{q/2} \Gamma(\alpha + \frac{q}{2})} \Phi(\alpha, \alpha + \frac{q}{2}; \frac{x^2}{2}),$$

where

$$q = 2r = \frac{1 - 2a + \sqrt{(1-2a)^2 + 8b}}{2}$$

and

$$\alpha = \frac{2a + 1 + q}{2} = 1 + \frac{\sqrt{(1-2a)^2 + 8b}}{4}.$$

B.3. Another equation

Consider the equation

(B.11) $\qquad \phi''(x) + a \cot(x/2) \phi'(x) - \dfrac{ab}{2\sin^2(x/2)} \phi(x) = 0, 0 < x < 2\pi.$

In Lemma 1.29, we need to find a solution with $\phi(0+) = 0$. Note that the solution of (B.11) with $\phi(\pi) = 1, \phi'(\pi) = 1$ has the property that $\phi''(x) > 0, \phi'(x) < 0$ for $0 < x < \pi$; in particular, this solution does not vanish as $x \to 0+$. Therefore there is at most one solution (up to multiplicative constant) of (B.11) with $\phi(0+) = 0$. If we restrict to $0 < x < \pi$, the substitution $u = \sin(x/2)$, turns (B.11) into

$$u^2(1-u^2) \phi''(u) + [2au - (2a+1)u^3] \phi'(u) - 2ab \phi(u) = 0, \quad 0 < u < 1$$

With the aid of Maple, we can find two linearly independent solutions,

$$u^{(1/2)-a} P_\nu^\mu(\sqrt{1-u^2}), \quad u^{(1/2)-a} Q_\nu^\mu(\sqrt{1-u^2}).$$

where $\nu = a - (1/2), \mu = \sqrt{(1-2a)^2 + 8ab}/2$, and P_ν^μ, Q_ν^μ denote associated Legendre functions (see [**64**, Section 7.12]). These functions can be written in terms of hypergeometric functions; in fact, we can also write two linearly independent solutions in the form (see [**6**, 3.4 (6), 3.4 (10)])

$$\phi_1(u) := u^{(1/2)-a} \left(\frac{1 + \sqrt{1-u^2}}{1 - \sqrt{1-u^2}} \right)^{\mu/2} F(-\nu, \nu+1, 1-\mu; [1-\sqrt{1-u^2}]/2),$$

$$\phi_2(u) := u^{(1/2)-a} \left(\frac{1 - \sqrt{1-u^2}}{1 + \sqrt{1-u^2}} \right)^{\mu/2} F(-\nu, \nu+1, 1-\mu; [1-\sqrt{1-u^2}]/2),$$

As $u \to 0+$, $\sqrt{1-u^2} = 1 - u^2 + O(u^4)$, so $\phi_2(u) = u^q + O(u^{q+1})$ where $q = (1/2) - a + \sqrt{(1-2a)^2 + 8ab}$. Simplifying and substituting we get

$$\phi_2(x) = \sin^q(x/2)\left[1 + \cos(x/2)\right]^{-\mu} F(-\nu, \nu+1, 1-\mu; [1-\cos(x/2)]/2).$$

Although the exact solution requires special functions, there is another way to see why there should exist a solution that behaves like x^q as $x \to 0+$. For x small, (B.11) looks like

$$\phi''(x) + \frac{2a}{x}\phi'(x) - \frac{2ab}{x^2}\phi(x) = 0.$$

The function $\phi(x) = x^q$ satisfies this (see (1.10)-(1.11)).

APPENDIX C

Reflecting Brownian motion

In §6.8, Brownian motion in a wedge with oblique (non-perpendicular) reflection was considered. In this appendix, we will explain how one derives the conditions (6.24) - (6.27) for the transition functions. Since these are local conditions, we will consider only the case of Brownian motion in $\mathbb{H}$ reflected at angle $\theta \in (0, \pi)$ off of $\mathbb{R}$. We will start by discussing one-dimensional reflecting Brownian motion.

Let Y_t be a standard one-dimensional Brownian motion. The process $|Y_t|$ is called (one-dimensional) reflecting Brownian motion. The *local time (at 0)* of Y_t is the unique, continuous increasing process l_t with the property that $\tilde{Y}_t := |Y_t| - l_t$ is a standard Brownian motion. This process increases only on the set $\{t : Y_t = 0\}$. For every $\epsilon > 0$, define a sequence of stopping times η_j, σ_j by $\eta_0 = 0$, $\sigma_j = \inf\{t > \eta_j : |Y_t| = \epsilon\}$ and $\eta_{j+1} = \inf\{t > \sigma_j : Y_t = 0\}$. Let $U(\epsilon, t)$ be the largest j such that $\sigma_j \leq t$. Then

$$l_t = \lim_{n \to \infty} 2^{-n} U(2^{-n}, t).$$

There is another way to construct a reflecting Brownian motion. Fix an $\epsilon > 0$. Suppose $\tilde{Y}_t$ is a Brownian motion with $\tilde{Y}_0 \geq 0$. Let $\sigma_0 = 0$ and for $k > 0$, $\sigma_k = \inf\{t \geq 0 : \tilde{Y}_t = -(k-1)\epsilon\}$. Define Z_t^ϵ to be the right continuous process

$$Z_t^\epsilon = \tilde{Y}_t + k\epsilon, \qquad \sigma_k \leq t < \sigma_{k+1}.$$

Note that Z_t^ϵ acts like an ordinary Brownian motion except that when it reaches the origin it moves instantaneously to ϵ. Note that

$$\lim_{\epsilon \to 0+} Z_t^\epsilon = \tilde{Y}_t + |m_t|,$$

where $m_t = \inf\{\tilde{Y}_s \wedge 0 : 0 \leq s \leq t\}$. The distribution of $(Z_t, |m_t|)$ is exactly the same as that of $(|Y_t|, l_t)$ above.

We will now consider Brownian motion in $\mathbb{H}$ reflected at angle $\theta \in (0, \pi)$ off of $\mathbb{R}$. We start with independent one-dimensional Brownian motions $X_t, \tilde{Y}_t$ and let Z_t, m_t be as in the previous paragraph. Let

$$W_t = (X_t + i\tilde{Y}_t) + |m_t|(\cot\theta + i).$$

Note that W_t is a semimartingale since $|m_t|$ is increasing and hence has paths of bounded variation on each interval. If we write $l_t = |m_t|$, then l_t is the local time of W_t at $\mathbb{R}$. we can write

$$dW_t = dB_t + (\cot\theta + i)\, dl_t,$$

where $B_t = X_t + i\tilde{Y}_t$ is a standard complex Brownian motion. Note that $\text{Im}(W_t)$ is a reflecting Brownian motion, but $\text{Re}(W_t), \text{Im}(W_t)$ are not independent.

Let $p(t, z, \cdot)$ denote the transition probability density for W_t, i.e., the density of W_t given $W_0 = z$. For fixed $T < \infty$, $w \in \mathbb{H}$, the process $R_t := p(T - t, W_t, w)$ must be a martingale. Itô's formula gives

$$R_t = \int_0^t [-\dot{p}(T - s, W_s, w) + \frac{1}{2}\Delta_z p(T - s, W_s, w)]\, dt +$$

$$\int_0^t [(\cot\theta)\partial_x p(T - s, W_s, w) + \partial_y p(T - s, W_s, w)]\, dl_s + [\text{martingale}].$$

From this we can see that

$$\dot{p}(t, z, w) = \frac{1}{2}\Delta_z p(t, z, w), \quad z \in \mathbb{H},$$

$$\partial_{\mathbf{v},z} p(t, z, w) = 0, \quad z \in \mathbb{R},$$

where $\mathbf{v} = (\cot\theta + i)/|\cot\theta + i|$ and $\partial_{\mathbf{v},z}$ denotes the directional derivative in the variable z. Since W_t acts like usual Brownian motion in $\mathbb{H}$, we can see that

$$\dot{p}(t, z, w) = \frac{1}{2}\Delta_w p(t, z, w), \quad z \in \mathbb{H}.$$

We claim that

(C.1) $$\partial_{\mathbf{v}',w} p(t, z, w) = 0, \quad w \in \mathbb{R},$$

where $\mathbf{v}' = (-\cot\theta + i)/|\cot\theta + i|$. If ϕ, ψ are C^2 functions decaying rapidly at infinity, define

$$P_t^* \phi(z) = \int_{\mathbb{H}} p(t, w, z)\, \phi(w)\, dA(w), \quad P_t \psi(w) = \int_{\mathbb{H}} p(t, w, z)\, \psi(z)\, dA(z).$$

To prove (C.1) it suffices to show that $\partial_{\mathbf{v}'} P_t^* \phi(x)$ for $x \in \mathbb{R}$. Note that

$$\int_{\mathbb{H}} [P_{s+t}^* \phi(z)]\, \psi(z)\, dA(z) = \int_{\mathbb{H}} [P_s^* \phi(z)]\, [P_t \psi(z)]\, dA(z)$$

$$= \int_{\mathbb{H}} \phi(z)\, [P_{s+t} \psi(z)]\, dA(z).$$

In particular,

$$\frac{d}{dt} \int_{\mathbb{H}} [P_t^* \phi(z)]\, \psi(z)\, dA(z) = \int_{\mathbb{H}} [P_t^* \phi(z)]\, [\frac{d}{ds} P_s \psi\,|_{s=0}\, (z)]\, dA(z)$$

$$= \frac{1}{2} \int_{\mathbb{H}} [P_t^* \phi(z)]\, \Delta \psi(z)\, dA(z).$$

Using Green's identity we can write the last quantity as

$$\frac{1}{2} \int_{\mathbb{H}} [\Delta P_t^* \phi(z)]\, \psi(z)\, dA(z) + \frac{1}{2} [\int_{\mathbb{R}} \{\, [P_t^* \phi(x)]\partial_y \psi(x) - [\partial_y P_t^* \phi(x)]\, \psi(x)\, \}\, dx\,].$$

The first term is just

$$\frac{d}{dt} \int_{\mathbb{H}} [P_t^* \phi(z)]\, \psi(z)\, dA(z).$$

Now assume that ψ has been chosen with compact support with $\partial_{\mathbf{v}}\psi(z) = 0$ for $z \in \mathbb{R}$. Then $\partial_y \psi(x) = -[\cot \theta]\, \partial_x \psi(x)$. Hence by integration by parts,

$$\int_{\mathbb{R}} [\,[\cot \theta]\, \partial_x P_t^* \phi(x) - \partial_y P_t^* \phi(x)\,]\psi(x)\, dx = 0.$$

By appropriate choice of ψ we see that $\partial_{\mathbf{v}'} P_t^* \phi(x) = 0$.

Bibliography

[1] L. Ahlfors (1979), *Complex Analysis*, 3rd ed., McGraw-Hill.
[2] L. Ahlfors (1973), *Conformal Invariants, Topics in Geometric Function Theory*, McGraw-Hill.
[3] M. Aizenman, A. Burchard (1999), Hölder continuity and dimension bounds for random curves, Duke Math. J. **99**, 419–453.
[4] M. Aizenman, A. Burchard, C. Newman, D. Wilson (1999), Scaling limits for minimal and random spanning trees in two dimensions, Random Structures and Algorithms **15**, 319–367.
[5] R. Bass (1995), *Probabilistic Techniques in Analysis*, Springer-Verlag.
[6] H. Bateman (1953), *Higher Transcendental Functions, Vol. I*, McGraw-Hill.
[7] M. Bauer, D. Bernard (2002), SLE_κ growth processses and conformal field theories, Phys. Lett. **B543**, 135–138.
[8] P. Berg, J. McGregor (1966), *Elementary Partial Differential Equations*, Holden-Day.
[9] V. Beffara (2004), Hausdorff dimenson for SLE_6, Ann. Probab **32**, 2606–2629.
[10] V. Beffara, The dimension of the SLE curves, preprint.
[11] A. Belavin, A. Polyakov, A. Zamolodchikov (1984), Infinite conformal symmetry of critical fluctuations in two dimensions, J. Stat. Phys. **34**, 763–774.
[12] A. Belavin, A. Polyakov, A. Zamolodchikov (1984), Infinite conformal symmetry in two-dimensonal quantum field theory. Nuclear Phys. B **241**, 333–380.
[13] P. Billingsley (1968), *Convergence of Probability Measures*, Wiley.
[14] P. Billingsley (1986), *Probability and Measure*, 2nd ed., Wiley.
[15] L. de Branges (1985), A proof of the Bieberbch conjecture, Acta Math. **154**, 137–152.
[16] K. Burdzy (1987), *Multidimensional Brownian Excursions and Potential Theory*, Pitman Research Notes in Mathematics 164, John Wiley & Sons.
[17] K. Burdzy, G. Lawler (1990), Non-intersection exponents for random walk and Brownian motion I: Existence and an invariance principle, Probab. Theor. Related Fields **84**, 393–410.
[18] K. Burdzy, G. Lawler (1990), Non-intersection exponents for random walk and Brownian motion II: Estimates and applications to a random fractal, Ann. Probab. **18**, 981–1009.
[19] C. Carathéodory (1954), *Theory of Functions of a Complex Variable, Vol. II*, Chelsea.
[20] J. Cardy (1984), Conformal invariance and surface critical behavior, Nucl. Phys. B **240** (FS12), 514–532.
[21] J. Cardy (1992), Critical percolation in finite geometries, J. Phys. A **25**, L201–L206.
[22] J. Cardy (1996), *Scaling and Renormalization in Statistical Physics*, Cambridge Univ. Press.
[23] J. B. Conway (1995), *Functions of One Complex Variable II*, Springer-Verlag.
[24] J. Dubédat (2003), SLE and triangles, Electron. Comm. Probab. **8**, 28–42.
[25] J. Dubédat (2003), $SLE(\kappa, \rho)$ martingales and duality, preprint.
[26] J. Dubédat (2004), Critical percolation in annuli and SLE_6, Comm.Math. Phys. **245**, 627–637.
[27] B. Duplantier (1998), Random walks and quantum gravity in two dimensions, Phys. Rev. Lett. **81**, 5489–5492.
[28] B. Dupantier (2003), Conformal fractal geometry and boundary quantum gravity, preprint.
[29] B. Duplantier, K.-H. Kwon (1988), Conformal invariance and intersection of random walks, Phys. Rev. Lett. **61**, 2514–2517.
[30] P. Duren (1983), *Univalent functions*, Springer-Verlag.
[31] R. Durrett (1996), *Probability: Theory and Examples*, 2nd ed., Duxbury.
[32] R. Durrett (1996), *Stochastic Calculus: A Practical Introduction*, CRC Press.
[33] K. Falconer (1990), *Fractal Geometery: Mathematical Foundations and Applications*, Wiley.

[34] R. Friedrich, W. Werner (2002), Conformal fields, restriction properties, degenerate representations and SLE, C. R. Ac. Sci. Paris Ser. I Math **335**, 947–952.

[35] R. Friedrich, W. Werner (2003), Conformal restriction, highest-weight representations and SLE, Comm. Math. Phys. **243**, 105–122.

[36] O. Frostman (1935), Potential d'équilbre et capacité des ensembles avec quelques applications à la théorie des fonctions, Meddl. Lunds Univ. Math. Sem. **3**, 1–118.

[37] I. Gruzberg, and L. Kadanoff (2004), The Loewner equation: maps and shapes, J. Stat. Phys, **114**, 1183-1198.

[38] W. Kager, B. Nienhuis (2004), A guide to stochastic Loewner evolution and its applicatons, J. Stat. Phys. **115**, 1149–1229.

[39] I. Karatzas and S. Shreve (1991), *Brownian Motion and Stochastic Calculus*, 2nd ed., Springer-Verlag.

[40] R. Kaufman (1969). Une propriété métrique du mouvement brownien. C. R. Acad. Sci. Paris **268**, 727–728.

[41] T. Kennedy (2003), Monte Carlo tests of SLE predictions for $2D$ self-avoiding walks, Phys. Rev. Lett. **88**, 130601.

[42] H. Kober (1957), *Dictionary of Conformal Representations*, Dover.

[43] M. Kozdron (2004), Simple random walk excursion measure in the plane, Ph.D. dissertation, Duke University.

[44] R. Langlands, Y. Pouillot, Y. Saint-Aubin (1994), Conformal invariance in two-dimensional percolation, Bull. AMS **30**, 1–61.

[45] G. Lawler (1991), *Intersections of Random Walks*, Birkhäuser-Boston.

[46] G. Lawler (1996), Hausdorff dimension of cut points for Brownian motion, Electron. J. Probab. **1** (1996), #2.

[47] G. Lawler (1996), The dimension of the frontier of planar Brownian motion, Electron. Comm. Probab. **1**, 29–47.

[48] G. Lawler (1999), Geometric and fractal properties of Brownian motion and random walk paths in two and three dimensions, Bolyai Mathematical Society Studies **9**, 219–258.

[49] G. Lawler (2004), An introduction to the stochastic Loewner evolution, in *Random Walks and Geometry*, V. Kaimanovich, ed., de Gruyter.

[50] G. Lawler (2004), Conformally invariant processes in the plane, in *School and Conference on Probability Theory*, ICTP Lecture Notes **17**, G. Lawler. ed., 305–351.

[51] G. Lawler, O. Schramm, W. Werner (2001), Values of Brownian intersection exponents I: Half-plane exponents, Acta Math. **187**, 237–273.

[52] G. Lawler, O. Schramm, W. Werner (2001), Values of Brownian intersection exponents II: Plane exponents, Acta Math. **187**, 275–308.

[53] G. Lawler, O. Schramm, W. Werner (2002), Values of Brownian intersection exponents III: Two-sided exponents, Ann. Inst. Henri Poincaré PR **38**, 109-123.

[54] G. Lawler, O. Schramm, W. Werner (2002), Analyticity of intersection exponents for planar Brownian motion, Acta Math. **189**, 179–201.

[55] G. Lawler, O. Schramm, W. Werner (2001), The dimension of the planar Brownian frontier is 4/3, Math. Res. Lett. **8**, 401-411

[56] G. Lawler, O. Schramm, W. Werner (2002), Sharp estimates for Brownian non-intersection probabilities, in *In and out of equilibrium. Probability with a Physics flavour*, V. Sidoravicius, ed., Birkhäuser.

[57] G. Lawler, O. Schramm, W. Werner (2004), Conformal invariance of planar loop-erased random walks and uniform spanning trees, Annals of Probab. **32**, 939–995.

[58] G. Lawler, O. Schramm, W. Werner (2004), On the scaling limit of planar self-avoiding walk, in *Fractal Geometry and Applications: A Jubilee of Benoit Mandelbrot*, Vol II., M. Lapidus, M. van Frankenhuijsen, ed., Amer. Math. Soc., 339–364.

[59] G. Lawler, O. Schramm, W. Werner (2003), Conformal restriction: the chordal case, J. Amer. Math. Soc. **16**, 917–955.

[60] G. Lawler, J. A. Trujillo Ferreras, Random walk loop soup, preprint.

[61] G. Lawler, W. Werner (1999), Intersection exponents for planar Brownian motion, Annals of Probability **27**, 1601–1642.

[62] G. Lawler, W. Werner (2000), Universality for conformally invariant intersection exponents, J. Europ. Math. Soc. **2**, 291-328.

[63] G. Lawler, W. Werner (2004), The Brownian loop soup, Probab. Theory Related Fields **128**, 565–588.
[64] N. N. Lebedev (1972), *Special Functions & their Applications*, Dover.
[65] P. Lévy (1946), *Processus Stochastiques et Mouvement Brownien*, Gauthier-Villars, Paris.
[66] K. Löwner (1923), Untersuchungen über schlichte konforme Abbildungen des Einheitskreises I., Math. Ann. **89**, 103–121.
[67] N. Madras, G. Slade (1993), *The Self-Avoiding Walk*, Birkhäuser.
[68] B. Mandelbrot (1982), *The Fractal Geometry of Nature*, Freeman.
[69] D. Marshall, S. Rohde, The Loewner differential equation and slit mappings, preprint.
[70] M. H. A. Newman (1951), *Elements of the Topology of Plane Sets of Points*, 2nd ed., Cambridge U. Press (reprinted by Dover, 1992)
[71] Ch. Pommerenke (1992) *Boundary Behaviour of Conformal Maps*, Springer-Verlag.
[72] D. Revuz, M. Yor (1994), *Continuous Martingales and Brownian Motion*, 2nd ed., Springer-Verlag
[73] L. Rogers, D. Williams (2000), *Diffusions, Markov Properties and Martingales*, Vol I and II, Cambridge Univ. Press.
[74] S. Rohde, O. Schramm, Basic properties of SLE, preprint.
[75] W. Rudin (1987), *Real and Complex Analysis*, 3rd ed.
[76] O. Schramm (2000), Scaling limits of loop-erased random walks and uniform spanning trees, Israel J. Math. **118**, 221–288.
[77] O. Schramm (2001), A percolation formula, Electron. Comm. Probab. **6**, 115–120.
[78] O. Schramm, S. Sheffield, The harmonic explorer and its convergence to $SLE(4)$, preprint.
[79] S. Smirnov (2001), Critical percolation in the plane: Conformal invariance, Cardy's formula, scaling limits, C. R. Acad. Sci. Paris S. I Math. **333** no. 3, 239–244.
[80] B. Virág (2003), Brownian beads, Probab. Theory Related Fields **127**, 267–387.
[81] W. Werner (1997), Asymptotic behaviour of disconnection and non-intersection exponents, Probab. Theory Related Fields **108**, 131–152.
[82] W. Werner (2001), Critical exponents, conformal invariance and planar Brownian motion, in *Proceedings of the 4th ECm Barcelona 2000*, Prog. Math, **202**, Birkhäuser.
[83] W. Werner (2004), Random planar curves and Schramm-Loewner evolutions, Ecole d'Eté de Probabilités de Saint-Flour XXXII - 2002, Lecture Notes in Mathematics **1840**, Springer-Verlag, 113–195.
[84] W. Werner (2004), Girsanov's transformation for $SLE(\kappa, \rho)$ processes, intersection exponents and hiding exponents. Ann. Fac. Sci. Toulouse Math. (6) **13**, 121–147.

Index

adapted, 11
analytic function, 43
Area Theorem, 61

Bessel process, 23, 25, 130
Beurling estimate, 84
Bieberbach Conjecture (de Branges's Theorem), 66, 103
boundary bubble measure, 137
Brownian bridge, 123
Brownian loop measure, 141
Brownian motion, 11
    complex, 11
    cut point, 187
    dimension of paths, 221
    frontier, 187
    Hölder continuity, 41
    intersection exponent, 192, 201
    modulus of continuity, 39, 41
    pioneer point, 187
    reflected, 209
    reflecting, 168, 233
bubble soup, 144

capacity, 61
    half-plane, 69
Cauchy integral formula, 43
Cauchy-Riemann equations, 43
compact $\mathbb{H}$-hull, 69
compact hull, 61
concatenation, 121
conformal annulus, 88
conformal rectangle, 81
conformal transformation, 44
continuously increasing hulls, 96
convergence
    in the Carathéodory sense, 78
    uniformly on compact sets, 78
curve, 43
    closed, 43
    smooth, 43

Distortion Theorem, 64
domain, 11, 43
driving function, 95

excursion measure, 127

    one-dimensional, 135
extremal distance (extremal length), 81

Feynman-Kac formula, 39

gambler's ruin, 1, 51
Girsanov's transformation, 23
Green's function, 53
    disk, 55
    for excursions, 133
    half plane, 55
Green's Theorem, 44
Growth Theorem, 65

$\mathbb{H}$-excursion, 130, 208
h-process, 25
harmonic function, 22, 46
harmonic measure, 48
Hausdorff dimension, 217
Hausdorff measure, 217
Hölder continuity, 41
    of SLE, 182
holomorphic, 43
hyperbolic metric, 66
hypergeometric function, 33, 229
    confluent, 30, 230

inradius, 60
Itô's Formula, 17, 21

Jordan curve, 43, 60
Jordan domain, 60

Koebe 1/4 Theorem, 63
Koebe function, 61

Laplacian, 21
Laplacian random walk, 3
linear fractional transformation, 44
locality property, 8
    for SLE, 152
locally analytic, 48
locally connected, 59
locally real, 109
Loewner chain
    chordal, 95

radial, 99
Loewner differential equation
    chordal, 92
    radial, 98
    whole-plane, 101
loop soup, 144
loop-erased random walk, 3, 147
loops
    rooted, 122
    unrooted, 122

martingale
    continuous, 11
    exponential, 23
    local, 12
    square integrable, 12
maximum principle, 44
Möbius transformation, 44
module, 81

optional sampling theorem, 22
oscillation, 39

percolation exploration process, 8
piecewise analytic, 48
Poisson kernel, 48
    boundary, 126
    disk, 48
    half-infinite strip, 49
    half plane, 49
Poisson point process, 144
Prohorov metric, 120

quadratic variation, 12

r-adjacent, 51
random walk excursions, 2
regular, 46
restriction measure, 206, 208
    right-restriction, 206, 208, 209
restriction property, 5, 141
    for excursion measure, 128
    for SLE, 153, 159
Riemann mapping theorem, 58
Riemann sphere, 44
right-domain, 205

Schramm-Loewner evolution (SLE)
    $SLE(\kappa,\rho)$, 211
    Cardy's formula, 163

chordal, 147
crossing exponents, 166, 171, 175
double points, 150
Hausdorff dimension, 183
Hölder continuity, 182
locality property, 152
path, 148, 181
radial, 156
restriction property, 153, 159
scaling, 148, 162
whole-plane, 162
Schwarz lemma, 45
Schwarz reflection principle, 48
Schwarzian derivative, 138
    and SLE, 156
self-avoiding walk, 5
semimartingale, 19
separation lemma, 199
simple process, 12
simple random walk, 1
simply connected, 57
smooth Jordan hull, 133
subadditivity, 190

univalent, 60

w.p.1, 11

Index of symbols

$\oplus$, 121
$\langle \cdot \rangle$, 12
A^*, 69
$\mathcal{A}$, 60
$\mathcal{A}_1$, 60
a, 26, 150, 177
$\mathcal{B}(z,\epsilon)$, 43
$\overline{\mathcal{B}}(z,\epsilon)$, 43
$b\mathcal{I}$, 12
$b\mathcal{M}$, 12
$\hat{C}$, 44
$\tilde{\mathcal{C}}$, 122
$\tilde{\mathcal{C}}_U$, 122
$\xrightarrow{Cara}$, 78
cap, 61
$\text{cap}_\mathbb{H}$, 74
$\mathbb{D}$, 44
$d_\mathcal{K}$, 119
$d_\mathcal{X}$, 119
diam, 59
$\dim_h$, 21
$\text{dist}(z,g)$, 108
$F(\alpha,\beta,\gamma;z)$, 229
F_K, 61
$\mathcal{F}_t$, 11
f_{Koebe}, 61
g_A, 69
g_t, 147
$\tilde{g}_A$, 76
γ^R, 121
H_D, 48
H_t, 93, 147
$\mathcal{H}$, 61
$\mathcal{H}^*$, 61
$\mathcal{H}_0^*$, 61
$\mathcal{H}_0$, 61
$\mathbb{H}$, 44
hcap, 69
hm, 48
$\mathcal{I}$, 12
inrad, 60
$\mathcal{J}$, 205
$\mathcal{J}_+$, 205
K_t, 147
$\mathcal{K}$, 119

κ, 147
$L(A_1, A_2; D)$, 81
$\mathcal{L}$, 144
$\mathcal{L}M$, 12
$\mathcal{M}$, 12
$\mathcal{M}^2$, 12
mod, 81
$\mu^\#$, 120
$\mu_D(z,w;t)$, 123
μ^{loop}, 141
$\mu_\mathbb{H}^{\text{bub}}$, 138
$\mathcal{N}$, 152
$\hat{\mathcal{N}}$, 152
$\mathcal{O}$, 144
osc, 13
$\mathbf{P}_\alpha$, 208
$\mathbf{P}_\alpha^+$, 208
$\Phi(\alpha,\gamma;z)$, 231
Φ_A, 133
$\mathcal{Q}$, 69
$\mathcal{Q}_+$, 133
$\mathcal{Q}_-$, 133
$\mathcal{Q}_\pm$, 133
rad, 67
$Sf(z)$, 138
SLE_κ, 147
$SLE(\kappa,\rho)$, 211
$\mathcal{S}$, 60
$\mathcal{S}^*$, 60
$\mathcal{S}M$, 19
T_x, 25
T_z, 93
τ_D, 11
$\Theta(\mathcal{R}_L; \partial_1, \partial_2)$, 81
$\xrightarrow{u.c.}$, 78
$\mathcal{X}$, 119
ξ, 201
$\tilde{\xi}$, 192

Titles in This Series

116 **Alexander Koldobsky,** Fourier analysis in convex geometry, 2005
115 **Carlos Julio Moreno,** Advanced analytic number theory: L-functions, 2005
114 **Gregory F. Lawler,** Conformally invariant processes in the plane, 2005
113 **William G. Dwyer, Philip S. Hirschhorn, Daniel M. Kan, and Jeffrey H. Smith,** Homotopy limit functors on model categories and homotopical categories, 2004
112 **Michael Aschbacher and Stephen D. Smith,** The classification of quasithin groups II. Main theorems: The classification of simple QTKE-groups, 2004
111 **Michael Aschbacher and Stephen D. Smith,** The classification of quasithin groups I. Structure of strongly quasithin K-groups, 2004
110 **Bennett Chow and Dan Knopf,** The Ricci flow: An introduction, 2004
109 **Goro Shimura,** Arithmetic and analytic theories of quadratic forms and Clifford groups, 2004
108 **Michael Farber,** Topology of closed one-forms, 2004
107 **Jens Carsten Jantzen,** Representations of algebraic groups, 2003
106 **Hiroyuki Yoshida,** Absolute CM-periods, 2003
105 **Charalambos D. Aliprantis and Owen Burkinshaw,** Locally solid Riesz spaces with applications to economics, second edition, 2003
104 **Graham Everest, Alf van der Poorten, Igor Shparlinski, and Thomas Ward,** Recurrence sequences, 2003
103 **Octav Cornea, Gregory Lupton, John Oprea, and Daniel Tanré,** Lusternik-Schnirelmann category, 2003
102 **Linda Rass and John Radcliffe,** Spatial deterministic epidemics, 2003
101 **Eli Glasner,** Ergodic theory via joinings, 2003
100 **Peter Duren and Alexander Schuster,** Bergman spaces, 2004
99 **Philip S. Hirschhorn,** Model categories and their localizations, 2003
98 **Victor Guillemin, Viktor Ginzburg, and Yael Karshon,** Moment maps, cobordisms, and Hamiltonian group actions, 2002
97 **V. A. Vassiliev,** Applied Picard-Lefschetz theory, 2002
96 **Martin Markl, Steve Shnider, and Jim Stasheff,** Operads in algebra, topology and physics, 2002
95 **Seiichi Kamada,** Braid and knot theory in dimension four, 2002
94 **Mara D. Neusel and Larry Smith,** Invariant theory of finite groups, 2002
93 **Nikolai K. Nikolski,** Operators, functions, and systems: An easy reading. Volume 2: Model operators and systems, 2002
92 **Nikolai K. Nikolski,** Operators, functions, and systems: An easy reading. Volume 1: Hardy, Hankel, and Toeplitz, 2002
91 **Richard Montgomery,** A tour of subriemannian geometries, their geodesics and applications, 2002
90 **Christian Gérard and Izabella Łaba,** Multiparticle quantum scattering in constant magnetic fields, 2002
89 **Michel Ledoux,** The concentration of measure phenomenon, 2001
88 **Edward Frenkel and David Ben-Zvi,** Vertex algebras and algebraic curves, second edition, 2004
87 **Bruno Poizat,** Stable groups, 2001
86 **Stanley N. Burris,** Number theoretic density and logical limit laws, 2001
85 **V. A. Kozlov, V. G. Maz'ya, and J. Rossmann,** Spectral problems associated with corner singularities of solutions to elliptic equations, 2001
84 **László Fuchs and Luigi Salce,** Modules over non-Noetherian domains, 2001

TITLES IN THIS SERIES

83 **Sigurdur Helgason,** Groups and geometric analysis: Integral geometry, invariant differential operators, and spherical functions, 2000

82 **Goro Shimura,** Arithmeticity in the theory of automorphic forms, 2000

81 **Michael E. Taylor,** Tools for PDE: Pseudodifferential operators, paradifferential operators, and layer potentials, 2000

80 **Lindsay N. Childs,** Taming wild extensions: Hopf algebras and local Galois module theory, 2000

79 **Joseph A. Cima and William T. Ross,** The backward shift on the Hardy space, 2000

78 **Boris A. Kupershmidt,** KP or mKP: Noncommutative mathematics of Lagrangian, Hamiltonian, and integrable systems, 2000

77 **Fumio Hiai and Dénes Petz,** The semicircle law, free random variables and entropy, 2000

76 **Frederick P. Gardiner and Nikola Lakic,** Quasiconformal Teichmüller theory, 2000

75 **Greg Hjorth,** Classification and orbit equivalence relations, 2000

74 **Daniel W. Stroock,** An introduction to the analysis of paths on a Riemannian manifold, 2000

73 **John Locker,** Spectral theory of non-self-adjoint two-point differential operators, 2000

72 **Gerald Teschl,** Jacobi operators and completely integrable nonlinear lattices, 1999

71 **Lajos Pukánszky,** Characters of connected Lie groups, 1999

70 **Carmen Chicone and Yuri Latushkin,** Evolution semigroups in dynamical systems and differential equations, 1999

69 **C. T. C. Wall (A. A. Ranicki, Editor),** Surgery on compact manifolds, second edition, 1999

68 **David A. Cox and Sheldon Katz,** Mirror symmetry and algebraic geometry, 1999

67 **A. Borel and N. Wallach,** Continuous cohomology, discrete subgroups, and representations of reductive groups, second edition, 2000

66 **Yu. Ilyashenko and Weigu Li,** Nonlocal bifurcations, 1999

65 **Carl Faith,** Rings and things and a fine array of twentieth century associative algebra, 1999

64 **Rene A. Carmona and Boris Rozovskii, Editors,** Stochastic partial differential equations: Six perspectives, 1999

63 **Mark Hovey,** Model categories, 1999

62 **Vladimir I. Bogachev,** Gaussian measures, 1998

61 **W. Norrie Everitt and Lawrence Markus,** Boundary value problems and symplectic algebra for ordinary differential and quasi-differential operators, 1999

60 **Iain Raeburn and Dana P. Williams,** Morita equivalence and continuous-trace C^*-algebras, 1998

59 **Paul Howard and Jean E. Rubin,** Consequences of the axiom of choice, 1998

58 **Pavel I. Etingof, Igor B. Frenkel, and Alexander A. Kirillov, Jr.,** Lectures on representation theory and Knizhnik-Zamolodchikov equations, 1998

57 **Marc Levine,** Mixed motives, 1998

56 **Leonid I. Korogodski and Yan S. Soibelman,** Algebras of functions on quantum groups: Part I, 1998

55 **J. Scott Carter and Masahico Saito,** Knotted surfaces and their diagrams, 1998

For a complete list of titles in this series, visit the
AMS Bookstore at **www.ams.org/bookstore/**.